Generative Complexity in Algebra

MEMOIRS
of the
American Mathematical Society

Number 828

Generative Complexity in Algebra

Joel Berman
Paweł M. Idziak

May 2005 • Volume 175 • Number 828 (end of volume) • ISSN 0065-9266

American Mathematical Society
Providence, Rhode Island

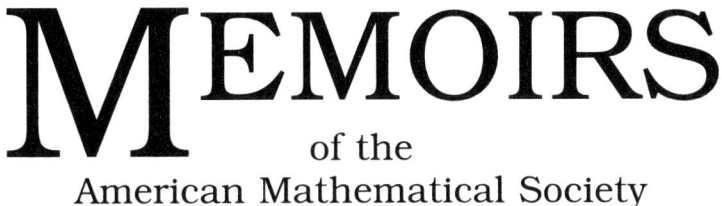

2000 *Mathematics Subject Classification.*
Primary 08A05; Secondary 03C13, 03C45, 05A16, 08B20.

Library of Congress Cataloging-in-Publication Data

Berman, Joel, 1943–
Generative complexity in algebra / Joel Berman, Pawel M. Idziak.
p. cm. — (Memoirs of the American Mathematical Society, ISSN 0065-9266 ; no. 828)
"Volume 175, number 828 (end of volume)."
Includes bibliographical references.
ISBN 0-8218-3707-9 (alk. paper)
1. Ordered algebraic structures. 2. Model theory. 3. Algebraic varieties—Classification theory. I. Idziak, Pawel. II. Title. III. Series.

QA3.A57 no. 828
[QA172.4]
510 s—dc22
[511.3′3] 2005041979

Memoirs of the American Mathematical Society is published bimonthly (each volume consisting usually of more than one number) by the American Mathematical Society at 201 Charles Street, Providence, RI 02904-2294, USA. Periodicals postage paid at Providence, RI. Postmaster: Send address changes to Memoirs, American Mathematical Society, 201 Charles Street, Providence, RI 02904-2294, USA.

10 9 8 7 6 5 4 3 2 1 10 09 08 07 06 05

Contents

Abstract

The G-*spectrum* or *generative complexity* of a class $\mathcal{C}$ of algebraic structures is the function $\mathrm{G}_{\mathcal{C}}(k)$ that counts the number of non-isomorphic models in $\mathcal{C}$ that are generated by at most k elements. We consider the behavior of $\mathrm{G}_{\mathcal{C}}(k)$ when $\mathcal{C}$ is a locally finite equational class (variety) of algebras and k is finite. We are interested in ways that algebraic properties of $\mathcal{C}$ lead to upper or lower bounds on generative complexity. Some of our results give sharp upper and lower bounds so as to place a particular variety or class of varieties at a precise level in an exponential hierarchy. We say $\mathcal{C}$ has *many models* if there exists $c > 0$ such that $\mathrm{G}_{\mathcal{C}}(k) \geqslant 2^{2^{ck}}$ for all but finitely many k, $\mathcal{C}$ has *few models* if there is a polynomial $p(k)$ with $\mathrm{G}_{\mathcal{C}}(k) \leqslant 2^{p(k)}$, and $\mathcal{C}$ has *very few models* if $\mathrm{G}_{\mathcal{C}}(k)$ is bounded above by a polynomial in k. Much of our work is motivated by a desire to know which locally finite varieties have few or very few models, and to discover conditions that force a variety to have many models. We present characterization theorems for a very broad class of varieties including most known and well-studied types of algebras, such as groups, rings, modules, lattices. Two main results of our work are: a full characterization of locally finite varieties omitting the tame congruence theory type **1** with very few models as the affine varieties over a ring of finite representation type, and a full characterization of finitely generated varieties omitting type **1** with few models. In particular, we show that a finitely generated variety of groups has few models if and only if it is nilpotent and has very few models if and only if it is Abelian.

Received by the editor March 6, 2002.

2000 *Mathematics Subject Classification.* Primary 08A05; Secondary 03C13, 03C45, 05A16, 08B20.

Key words and phrases. enumeration of finite algebras, classification of finite algebras, locally finite varieties, free spectra.

CHAPTER 1

Introduction

Given a class $\mathcal{C}$ of structures two basic questions about the class $\mathcal{C}$ are:

- what are the cardinalities of structures in $\mathcal{C}$,
- how many nonisomorphic structures of a given cardinality are in $\mathcal{C}$.

The first question is a problem of determining the *spectrum* Spec($\mathcal{C}$) of the class $\mathcal{C}$, that is, the class of cardinalities that occur as sizes of structures in $\mathcal{C}$. This problem appears in all sorts of contexts in mathematics. For example, if $\mathcal{C}$ is the class of models of a first order theory in a countable language then, by the Löwenheim–Skolem theorems, every infinite cardinality is in Spec($\mathcal{C}$). Thus, Spec($\mathcal{C}$) is usually defined to be the set of sizes of finite models in $\mathcal{C}$. Let us here mention a result of R. Fagin [**21**] that characterizes those sets of integers that can expressed as Spec($\mathcal{C}$), where $\mathcal{C}$ is a class of all models of a first order sentence in a first order language. They are exactly the sets that can be recognized in Nondeterministic Exponential time. He also proved a similar connection between Nondeterministic Polynomial time and what are called generalized spectra.

The second basic question is about the *fine spectrum* of the class $\mathcal{C}$, i.e., about the function that assigns to each cardinal k the number of nonisomorphic k-element structures in $\mathcal{C}$. The problem of determining the fine spectrum of a class, either exactly or asymptotically, is a natural one, and is a problem that has been considered for various classes $\mathcal{C}$ and in various contexts over the years. Combinatorial enumeration problems such as finding the number of k-vertex graphs in some specific class of graphs are familiar examples of such fine spectra problems.

The terminology "fine spectrum" was introduced by W. Taylor in [**66**] in the case that $\mathcal{C}$ is a variety, that is, a class of algebraic structures or algebras closed under the formation of homomorphisms, subalgebras and direct products. Among the results of Taylor's paper is a characterization of those varieties that have exactly one model of size 2^k for each finite k and no other finite models. All such varieties must be generated by a 2-element algebra. Subsequent papers by R. Quackenbush [**60**] and D. Clark and P. Krauss [**18**] extended Taylor's work to n-element algebras that generate varieties having minimal fine spectra.

Despite this work on minimal fine spectra there has been relatively little success on general problems involving fine spectra of varieties. In [**67**] Taylor writes of fine spectra functions that "Characterizations of such functions seems hopeless" and Quackenbush states in his survey on enumeration problems for ordered structures [**61**] that the fine spectrum problem here "is usually hopeless".

There are many reasons for the difficulties in determining the fine spectra of varieties. One reason that interests us is that algebras are often described by means of a set of generators. Once we know the generators of an algebra $\mathbf{A}$ in a variety $\mathcal{V}$ and some of the conditions that these generators satisfy, our freedom in building the rest of the model is heavily restricted. This effect is widely used in group theory where a group is usually presented by a (finite) set of generators and a set of relations the generators must obey. The constraints put on the behavior of the generators place restrictions on the structure of the entire algebra **A**. However, there is in general no obvious or transparent way to determine the cardinality of **A**. This makes the counting of all n-element or at most n-element algebras difficult, if at all possible, even if we content ourselves with an asymptotic estimate.

Another area of research in which fine spectra appear is in finite model theory. When investigating asymptotic probabilities in finite model theory, some of the results rely on counting all finite models, up to isomorphism, for a given theory T. This usually is not hard when T has no axioms. However there are only a few results on zero–one (or more generally on limit) laws for specific theories T. One reason is that for such counting a deep insight into the structure of finite models of T is required. The counting is even more difficult if the language of T contains function symbols. Except for unary functions [**49**] (in which case the models behave much more like relational structures than algebras) and Abelian groups [**19**] (where we completely understand the structure) there are only a few other results on limit laws for algebras to report. The reader may wish to consult [**12, 13, 14, 15, 16, 38**] for related work.

Again, there are various reasons for the difficulties here with fine spectra for classes arising from theories in a language containing function symbols. Some important techniques for asymptotic probabilities rely on extension axioms. These techniques work perfectly well for purely relational structures when often the resulting random structure is model complete. However (randomly) adding a new element a to the universe of an algebra **A** in a variety $\mathcal{V}$ and keeping the resulting extended algebra in $\mathcal{V}$ requires a much bigger extension $\mathbf{A}^*$ of **A**. The number and the behavior of all those new elements in $\mathbf{A}^*$ is fully determined by the old algebra **A** and the interaction of the new element a with the elements of A. Thus, in vector spaces by adding a new element we actually increase the dimension.

One possible way to overcome these difficulties is to count k-generated models instead of k-element ones. This is a more tractable problem when the classes are varieties of algebras, and we believe is the proper setting for asymptotic probabilities in algebra. Note however that these numbers are the same for purely relational languages.

Thus, we introduce the G-*spectrum*, or *generative complexity*, of a class $\mathcal{C}$, which is the function $\mathrm{G}_{\mathcal{C}}(k)$ that counts the number of nonisomorphic (at most) k-generated models in $\mathcal{C}$.

We concentrate on the case that $\mathcal{C}$ is a variety of algebras and restrict ourselves to finite k. Even in this new setting the counting remains hard. It requires an understanding of the 'generating power' in a given variety $\mathcal{V}$. This is related to the old problem of determining the *free spectrum* of $\mathcal{V}$, that is, the sizes of free algebras $\mathbf{F}_{\mathcal{V}}(k)$ in $\mathcal{V}$ with $k = 1, 2, \ldots$ free generators. This is because every k-generated algebra in $\mathcal{V}$ is isomorphic to a quotient of $\mathbf{F}_{\mathcal{V}}(k)$ by a congruence relation. However, two different congruences may give rise to isomorphic quotients. Thus, the second problem we meet here is to measure the amount of homogeneity in $\mathbf{F}_{\mathcal{V}}(k)$. One cannot hope to solve these two problems without understanding the structure of algebras from $\mathcal{V}$.

Let us also mention that the infinite counterpart of the problem of counting nonisomorphic models is widely studied in Model Theory, and is one of the fundamental topics for Shelah's classification theory and for stability theory. Note that in the infinite realm (and for a countable language) being κ-generated and having κ elements are the same.

For example the celebrated Vaught conjecture says that the number of nonisomorphic countable models of any first order theory in a countable language is either countable or 2^ω. In [**28**] and [**29**] B. Hart, S. Starchenko and M. Valeriote have been able to prove this conjecture for varieties of algebras. They actually determined the possible infinite fine spectra (which, in this case, are the same as infinite G-spectra) of varieties and correlated them with algebraic properties of varieties. A characterization of locally finite varieties with strictly less than 2^ω nonisomorphic countable models can be easily inferred from a deep work on decidability done by R. McKenzie and M. Valeriote [**56**].

However, one cannot easily (if at all) transfer infinite methods and results into the finite world. Thus, we focus on the G-spectrum of a variety rather than its fine spectrum. We further restrict ourselves to locally finite varieties, i.e., varieties in which all finitely generated algebras are finite. This gives that the G-spectrum of the variety $\mathcal{V}$ is integer-valued. Even with this finiteness restriction G-spectra can be arbitrarily large: for any sequence $(p_k)_{k\in\omega}$ of integers there is a locally finite variety $\mathcal{V}$ of groupoids such that $\mathrm{G}_{\mathcal{V}}(k) \geqslant p_k$ for all k (see Example 5.9). On the other hand, if $\mathcal{V}$ is a finitely generated variety of finite type then easily established upper bounds for the G-spectrum, free spectrum and fine spectrum of $\mathcal{V}$ are $2^{2^{2^{ck}}}$, $2^{2^{ck}}$ and 2^{ck^s}, respectively, for some constants c and s.

The main problems that stimulate our research in this area are:

- *How does the growth of the* G*-spectrum of a locally finite variety* $\mathcal{V}$ *affect the structure of algebras in* $\mathcal{V}$*?*
- *In what way do algebraic properties of* $\mathcal{V}$ *influence the behavior of* $\mathrm{G}_{\mathcal{V}}(k)$*?*
- *For a given variety* $\mathcal{V}$*, can an explicit formula for* $\mathrm{G}_{\mathcal{V}}(k)$ *be found?*
- *Can the asymptotic behavior of* $\mathrm{G}_{\mathcal{V}}(k)$ *be determined?*

For some $\mathcal{V}$, the value of $\mathrm{G}_{\mathcal{V}}(k)$ is easily determined as the following examples show.

EXAMPLE 1.1. For the variety $\mathcal{V}$ of sets, that is, algebras with no fundamental operations, $\mathrm{G}_{\mathcal{V}}(k) = k$. In Example 3.9 we describe the G-spectra of varieties generated by arbitrary multi-unary algebras. Let $\mathcal{V}$ be the variety of vector spaces over a fixed field. A k-generated algebra here is a vector space of dimension at most k. So $\mathrm{G}_{\mathcal{V}}(k) = k + 1$. If $\mathcal{A}_p$ is the variety generated by the p-element group for a prime p, then a k-generated group in $\mathcal{A}_p$ has size p^m, for $0 \leqslant m \leqslant k$. So $\mathrm{G}_{\mathcal{A}_p}(k) = k + 1$. In Example 6.7 we determine $\mathrm{G}_{\mathcal{V}}(k)$ for $\mathcal{V}$ an arbitrary, finitely generated variety of Abelian groups. For the variety $\mathcal{B}$ of Boolean algebras we have $|\mathrm{F}_{\mathcal{B}}(k)| = 2^{2^k}$. Every k-generated member of $\mathcal{B}$ is a Boolean algebra with m atoms, $0 \leqslant m \leqslant 2^k$. Thus $\mathrm{G}_{\mathcal{B}}(k) = 1 + 2^k$. In contrast to these varieties that have small G-spectra, we show in Chapter 4 that semilattices and distributive lattices have $\mathrm{G}_{\mathcal{V}}(k)$ that are doubly exponential functions of k while in Chapter 5 we exhibit locally finite varieties with arbitrarily large G-spectra.

This work reports our results on G-spectra. We introduce some classes of functions in order to describe the possible behavior of the $\mathrm{G}_{\mathcal{C}}(k)$.

DEFINITION 1.2. Let f be a real-valued function of the positive integers.

- We say f is *at most 0-fold exponential* if there exists a polynomial p such that $f(k) \leqslant p(k)$ for all k.
- For $m > 0$ the function f is *at most m-fold exponential* if there is an at most $(m-1)$-fold exponential function g such that $f(k) \leqslant 2^{g(k)}$ for all k.
- The function f is called *at least 0-fold exponential* if there exists a constant $c > 0$ such that $f(k) \geqslant ck$ for all but finitely many k.
- For $m > 0$ the function f is *at least m-fold exponential* if there is an at least $(m-1)$-fold exponential function g such that $f(k) \geqslant 2^{g(k)}$ for all but finitely many k.
- The function f is of *m-fold exponential complexity* if f is both at most and at least m-fold exponential.
- A class $\mathcal{C}$ of structures has m-fold exponential generative complexity if the function $\mathrm{G}_{\mathcal{C}}(k)$ is of m-fold exponential complexity.

We usually write polynomial, exponential, doubly exponential, and triply exponential in place of 0-fold exponential, 1-fold exponential, 2-fold exponential, and 3-fold exponential.

DEFINITION 1.3. *Let $\mathcal{V}$ be a locally finite variety.*

- *$\mathcal{V}$ has* many models *if $\mathrm{G}_{\mathcal{V}}(k)$ is at least doubly exponential.*
- *$\mathcal{V}$ has* few models *if $\mathrm{G}_{\mathcal{V}}(k)$ is at most exponential.*
- *$\mathcal{V}$ has* very few models *if $\mathrm{G}_{\mathcal{V}}(k)$ is at most polynomial.*

The research reported in this work was motivated by a desire to know which locally finite varieties have few and very few models, respectively. Although we have not managed to solve these problems in the most general setting we have obtained such a characterization for a very broad class of varieties including most known and well-studied types of algebras, such as groups, rings, modules, lattices.

The proofs of our results give a deep insight into the structure of locally finite varieties with few and very few models. Our analysis relies heavily on two major developments of the late 70's and early 80's. One of them is *modular commutator theory*. The theory had been introduced by J.D.H. Smith [**65**] for congruence permutable varieties. It was further developed by J. Hagemann and Ch. Herrmann [**27**], H.P. Gumm [**26**] and R. Freese and R. McKenzie [**22**]. The Freese and McKenzie book contains several important results and techniques that are extremely useful when studying congruence modular varieties. A binary operation on congruences that simultaneously generalizes the concept of a commutator $[H, K]$ of two normal subgroups H, K of a group as well as the ideal multiplication in rings is defined. The theory shows how some information about algebras or varieties can be recovered from congruence lattices endowed with this binary operation. Moreover the concept of the commutator allows us to speak about a solvable, nilpotent or Abelian congruence (or algebra) as well as about the center of an algebra or the centralizer of a congruence relation.

The second big development in universal algebra that we use is *tame congruence theory*, which was created and described in D. Hobby and R. McKenzie [**32**]. Tame congruence theory is a tool for studying the local structure of finite algebras. Instead of considering the whole algebra and all of its operations at once, tame congruence theory allows us to localize to small subsets on which the structure is much simpler to understand and to handle. According to this theory there are only five possible ways a finite algebra can behave locally. The local behavior must be one of the following:

1. a finite set with a group action on it,
2. a finite vector space over a finite field,
3. a two element Boolean algebra,
4. a two element lattice,
5. a two element semilattice.

Now, if from our point of view a local behavior of an algebra is 'bad' then we can often show that the algebra itself behaves 'badly'. For example, since the varieties of distributive lattices or semilattices have many models (see Theorems 4.1, 4.2) then one can argue that in any locally finite variety with few models structures of type **4** or **5** cannot occur (see Theorem 8.1).

On the other hand it is not true that if the local behavior is 'good' then the global one is as well. Several kinds of interactions between these small sets can produce a fairly messy global behavior. Such interactions often contribute to produce many models. Also the relative 'geographical layout' of those small sets can result in unpredictable phenomena.

The two main results of our work are:

- a full characterization of locally finite varieties omitting type **1** with very few models as the affine varieties over a ring of finite representation type, (Theorem 13.4),
- a full characterization of finitely generated varieties omitting type **1** with few models (Theorem 17.2).

The necessity of the conditions in the above theorems was shown by detecting more than a dozen different ways in which a 'bad' local behavior can occur in an algebra **A**. In each such situation we are able to produce many nonisomorphic k–generated algebras in the variety generated by **A**. After detecting all those instances of 'bad' local behavior we formulate global algebraic conditions that forbid such undesirable behavior. More surprisingly we are able to show the list of conditions we obtain is actually complete, at least for finitely generated varieties, i.e., these conditions taken together are sufficient for a variety to have few (or very few) models.

The characterization in Theorem 13.4 was obtained by P. Idziak and R. McKenzie in [**36**]. The conditions in this situation are very simple and easily stated.

The conditions involved in the second characterization (Theorem 17.2) are more complicated, although they also have a natural algebraic meaning. In both cases we know that the bound for the number of algebras implies a very transparent and manageable structure. For example, when specializing our results to groups we get the following:

- every finitely generated variety of groups has at most doubly exponentially many models,
- a finitely generated variety of groups has few models if and only if it is nilpotent,
- a finitely generated variety of groups has very few models if and only if it is Abelian.

while for commutative rings with unit our characterization reduces to:

- every finitely generated variety of rings has at most doubly exponentially many models,
- a finitely generated variety of commutative rings with unit has few models if and only if the Jacobson radical in the generating ring squares to 0,
- no nontrivial variety of rings with unit has very few models.

In the first part of the book we consider the general behavior of the G-spectrum for arbitrary varieties of algebras. We are especially interested in ways that algebraic properties of a variety lead to upper or lower bounds on generative complexity.

The second part is entirely devoted to the proof of our characterization of locally finite varieties with few and very few models. Most of the results from the first part of the book are used in the second part either in our proofs or in providing a context for our main results.

Acknowledgments: The work on this project started in the fall of 1996 when both authors were visiting the Fields Institute in Toronto. The support and the research environment provided by the Fields Institute is acknowledged with gratitude. The research was also supported in part by the Polish KBN.

The authors thank the referee for a very careful reading of the manuscript and for providing suggestions that substantially improved the presentation of the final version of the book.

CHAPTER 2

Background Material

An *algebra* $\mathbf{A} = \langle A, \mathbf{f}_i (i \in I) \rangle$ is a nonvoid set A together with a collection of finitary operations $\mathbf{f}_i$ on A indexed by a set I. The set A is called the *universe* of the algebra and the $\mathbf{f}_i$ are the *fundamental operations* of $\mathbf{A}$. For $i \in I$ the operation $\mathbf{f}_i$ maps A^{n_i} to A, that is, $\mathbf{f}_i$ is n_i-ary. The function from I to the integers given by $i \mapsto n_i$ is the *similarity type* of the algebra $\mathbf{A}$. If I is finite, then the algebra is said to be of *finite similarity type*. For algebras of finite similarity type we often just list the operations, e.g., a Boolean algebra might be given as $\mathbf{B} = \langle B, \wedge, \vee, ', 0, 1 \rangle$. An algebra is *finite* if its universe is finite and is *trivial* if its universe has only one element. If $n_i \leqslant 1$ for all $i \in I$ the algebra is called *multiunary*. A multiunary algebra $\mathbf{A}$ in which the set of fundamental operations forms a group of permutations on A is called a G-*set*.

An algebra $\mathbf{A} = \langle A, \mathbf{f}_i (i \in I) \rangle$ may also be viewed as a model in the language L where L consists of all the function symbols $\mathbf{f}_i$ for $i \in I$. When necessary, we distinguish the function symbol $\mathbf{f}_i$ in L from the fundamental operation $\mathbf{f}_i$ on A by writing $\mathbf{f}_i^{\mathbf{A}}$ to denote the n_i-ary operation on the algebra $\mathbf{A}$. A *term* for L over a set of variables $X = \{x_1, x_2, \ldots\}$ is defined inductively by letting every $x_j \in X$ be a term and if $i \in I$ and $\mathbf{t}_1, \ldots, \mathbf{t}_{n_i}$ are terms, then $\mathbf{f}_i(\mathbf{t}_1, \ldots, \mathbf{t}_{n_i})$ is also a term. If the variables that appear in a term $\mathbf{t}$ are in the set $\{x_1 \ldots, x_n\}$, then we say $\mathbf{t}$ is n-ary and denote this by writing $\mathbf{t}(x_1, \ldots, x_n)$. If $\mathbf{t}(x_1, \ldots, x_n)$ is an n-ary term for L over X and $\mathbf{A}$ is an algebra in the language L, then the *term operation* $\mathbf{t}^{\mathbf{A}}$ on $\mathbf{A}$ corresponding to $\mathbf{t}$ is defined by letting $x_i^{\mathbf{A}}$ be the projection on the i-th coordinate and if

$$\mathbf{t}(x_1, \ldots, x_n) = \mathbf{f}_i(\mathbf{t}_1(x_1, \ldots, x_n), \ldots, \mathbf{t}_{n_i}(x_1, \ldots, x_n)),$$

then

$$\mathbf{t}^{\mathbf{A}}(a_1, \ldots, a_n) = \mathbf{f}_i^{\mathbf{A}}(\mathbf{t}_1^{\mathbf{A}}(a_1, \ldots, a_n), \ldots, \mathbf{t}_{n_i}^{\mathbf{A}}(a_1, \ldots, a_n))$$

for all $(a_1, \ldots, a_n) \in A^n$. To simplify notation we often suppress the subscript on fundamental operations and just write $\mathbf{f}$ or $\mathbf{f}(x_1, \ldots, x_r)$. Likewise, we often omit the algebra superscript on term operations. We also use the bar convention by writing $\overline{a}$ for $(a_1, \ldots, a_n)$. However, if $a \in A$, then $\overline{a}$ denotes the n-tuple $(a, \ldots, a)$.

The collection of all term operations on an algebra forms a *clone*, that is, a family of operations on a set that contains all the projection operations and is closed under composition. Thus the set of term operations of $\mathbf{A}$ is called the *clone of* $\mathbf{A}$ and is denoted $\operatorname{Clo} \mathbf{A}$. The clone of all operations on A that can be obtained from the term operations of $\mathbf{A}$ and all the constant operations is called the *clone of polynomial operations* of $\mathbf{A}$ and is denoted $\operatorname{Pol} \mathbf{A}$. The set of n-ary polynomial operations of $\mathbf{A}$ is written $\operatorname{Pol}_n \mathbf{A}$. Two algebras $\mathbf{A}$ and $\mathbf{B}$ are said to be *polynomially equivalent* if they have the same universe and $\operatorname{Pol} \mathbf{A} = \operatorname{Pol} \mathbf{B}$.

A *subuniverse* of an algebra $\mathbf{A}$ is a set $S \subseteq A$ that is closed under the fundamental operations of $\mathbf{A}$, that is, $\mathbf{f}(\overline{a}) \in S$ for every fundamental operation $\mathbf{f}$ of $\mathbf{A}$ and every $\overline{a} \in S^r$. An algebra $\mathbf{B}$ is a *subalgebra* of $\mathbf{A}$ if $\mathbf{B}$ and $\mathbf{A}$ have the same similarity type, the universe of $\mathbf{B}$ is a subuniverse of $\mathbf{A}$, and for every operation symbol $\mathbf{f}_i$, the operation $\mathbf{f}_i^{\mathbf{B}}$ is the restriction to B of the operation $\mathbf{f}_i^{\mathbf{A}}$. Since the intersection of an arbitrary family of subuniverses of an algebra $\mathbf{A}$ is a subuniverse it follows that the set of all subuniverses of $\mathbf{A}$, denoted Sub $\mathbf{A}$, forms a complete lattice when ordered by inclusion. For a subset X of A, the *subuniverse of* $\mathbf{A}$ *generated by* X, denoted $\mathrm{Sg}^{\mathbf{A}}(X)$, is the intersection of all subuniverses of $\mathbf{A}$ that contain X. Another way to describe $\mathrm{Sg}^{\mathbf{A}}(X)$ is to observe that the subuniverse generated by X consists of all elements of the form $\mathbf{t}^{\mathbf{A}}(\overline{x})$ where $\mathbf{t}$ ranges over all terms for the language of $\mathbf{A}$ and the $\overline{x}$ are tuples from X. An algebra $\mathbf{A}$ is said to be *k-generated* if there exists a set X of cardinality at most k such that the universe of $\mathbf{A}$ is $\mathrm{Sg}^{\mathbf{A}}(X)$.

Given two algebras $\mathbf{A}$ and $\mathbf{B}$ of the same similarity type, a function $h : A \to B$ is called a *homomorphism* if $h(\mathbf{f}(a_1, \ldots, a_r)) = \mathbf{f}(h(a_1), \ldots, h(a_r))$ for every fundamental operation $\mathbf{f}$ of $\mathbf{A}$ and all $a_i \in A$. A homomorphism is an *isomorphism* if it is both one-to-one and onto. If h is a homomorphism from $\mathbf{A}$ to $\mathbf{B}$, then $h(A)$ is a subuniverse of $\mathbf{B}$ and if $A = \mathrm{Sg}^{\mathbf{A}}(X)$, then the subuniverse $h(A)$ is generated by $h(X)$.

A congruence relation on an algebra $\mathbf{A}$ is an equivalence relation θ on A that is preserved by the fundamental operations of $\mathbf{A}$, that is, if f is an r-ary fundamental operation and $(a_1, b_1), \ldots, (a_r, b_r) \in \theta$, then $(\mathbf{f}(\overline{a}), \mathbf{f}(\overline{b})) \in \theta$. Notation that is often used to express that (a, b) is in the congruence relation θ includes $a\theta b$ and $a \overset{\theta}{\equiv} b$. For a congruence relation θ on $\mathbf{A}$ the congruence class containing an element a is denoted a/θ and A/θ is the set of all congruence classes of θ. The intersection of a family of congruence relations of an algebra is again a congruence relation so the set of all congruence relations of $\mathbf{A}$, when ordered by inclusion, forms a complete lattice. The lattice of congruence relations of $\mathbf{A}$ is denoted $\mathsf{Con}\,\mathbf{A}$. The top element of this lattice is $A \times A$ and is written as 1_A; the bottom element is the diagonal 0_A, which consists of all pairs (a, a) for $a \in A$. We frequently omit the subscripts in 0_A and 1_A.

Some terminology from lattice theory is used in describing $\mathsf{Con}\,\mathbf{A}$. For $a \leqslant b$ in a lattice $\mathbf{L}$ the ordered pair (a, b) is called a *quotient* in $\mathbf{L}$ and the *interval* from a to b, written $I\,[a, b]$, is the subuniverse of $\mathbf{L}$ consisting of $\{c \in L : a \leqslant c \leqslant b\}$. The element a is *covered* by b if $a < b$ and $I[a, b] = \{a, b\}$. If a is covered by b, then we write $a \prec b$ and call $I[a, b]$ a *prime interval* or a *prime quotient.* A *subcover* of an element b is any element covered by b. An *atom* in a lattice with least element 0 is any element that covers 0 and a *coatom* or *dual atom* in a lattice with largest element 1 is any element covered by 1. If $I[a, b]$ and $I[c, d]$ are intervals such that $b \wedge c = a$ and $b \vee c = d$, then $I[a, b]$ is said to *transpose up* to $I[c, d]$, written $I[a, b] \nearrow I[c, d]$; and $I[c, d]$ is said to *transpose down* to $I[a, b]$, written $I[c, d] \searrow I[a, b]$; and the two intervals are called *transposes* of one another. Two intervals are said to be *projective* if one can be obtained from the other by a finite sequence of transposes. A fundamental fact in lattice theory is that a lattice is modular if and only if its projective intervals are isomorphic. Another equivalent condition for modularity is that the lattice has no elements a, b, c satisfying $a < b$, $a \vee c = b \vee c$ and $a \wedge c = b \wedge c$. Such a 5-element sublattice generated by a, b, c will be called an *$[a, b, c]$-pentagon.*

For a set $X \subseteq A \times A$ the *congruence relation on* $\mathbf{A}$ *generated by* X is the intersection of all $\theta \in \mathsf{Con}\,\mathbf{A}$ for which $X \subseteq \theta$. We write $\mathrm{Cg}^{\mathbf{A}}(X)$ for this congruence relation but in the case that $X = \{(a, b)\}$ we write $\mathrm{Cg}^{\mathbf{A}}(a, b)$. A congruence relation is said to be *finitely generated* if it is of the form $\mathrm{Cg}^{\mathbf{A}}(X)$ for some finite X and is *principal* if it is of the form $\mathrm{Cg}^{\mathbf{A}}(a, b)$. Note that every congruence relation on $\mathbf{A}$ is the least upper bound in $\mathsf{Con}\,\mathbf{A}$ of all principal congruence relations contained in it.

For arbitrary elements a, b, c, d in an algebra $\mathbf{A}$ the following useful characterization of when $(c, d) \in \mathrm{Cg}^{\mathbf{A}}(a, b)$ is due to Maltsev:

$$(c, d) \in \mathrm{Cg}^{\mathbf{A}}(a, b) \text{ iff } \exists z_0, \dots, z_n \in A, \exists \mathbf{p}_1, \dots, \mathbf{p}_n \in \mathrm{Pol}_1\mathbf{A},$$

$$c = z_0, d = z_n, \text{ and } \{z_{i-1}, z_i\} = \{\mathbf{p}_i(a), \mathbf{p}_i(b)\} \text{ for all } 1 \leqslant i \leqslant n.$$

The elements $z_0, z_1, \dots, z_n$ form what is called a *Maltsev chain* from c to d.

An algebra $\mathbf{A}$ is *simple* if it is nontrivial and $\mathsf{Con}\,\mathbf{A}$ consists solely of 1_A and 0_A. An algebra is called *congruence distributive* or *congruence modular* if its congruence lattice satisfies the distributive identity or the modular identity. Two congruence relations $\theta, \tau \in \mathsf{Con}\,\mathbf{A}$ *permute* if $\theta \circ \tau = \tau \circ \theta$. If θ and τ permute, then $\theta \vee \tau = \theta \circ \tau$ in $\mathsf{Con}\,\mathbf{A}$. An algebra is *congruence permutable* if every pair of its congruence relations permute.

Homomorphisms and congruence relations are naturally linked: If h is a homomorphism on $\mathbf{A}$, then the *kernel of* h, denoted $\ker(h)$, is the set of all $(a_1, a_2) \in A^2$ for which $h(a_1) = h(a_2)$. For every homomorphism h the relation $\ker(h)$ is a congruence on $\mathbf{A}$. On the other hand, if $\theta \in \mathsf{Con}\,\mathbf{A}$, then the congruence classes of θ form the elements of an algebra $\mathbf{A}/\theta$ and the map $a \mapsto a/\theta$ is a homomorphism from $\mathbf{A}$ onto $\mathbf{A}/\theta$ with kernel θ.

We next consider direct products of algebras. Suppose $\mathbf{A}_j$, for $j \in J$, are algebras of the same similarity type indexed by a set J. The *direct product* of these algebras, denoted $\prod_{j\in J} \mathbf{A}_j$, is an algebra of the same similarity type as the $\mathbf{A}_j$ with universe $\prod_{j\in J} A_j$ and fundamental operations defined coordinatewise: $\mathbf{f}(\overline{a}, \overline{b}, \overline{c}, \dots)_j = \mathbf{f}(a_j, b_j, c_j, \dots)$ for all $j \in J$. Often the index set J is finite, say $J = \{1, \dots, n\}$, and we write $\mathbf{A}_1 \times \cdots \times \mathbf{A}_n$ for the direct product in this situation. If J is the empty set, then $\prod_{j\in J} \mathbf{A}_j$ is a trivial algebra. A direct product of copies of a single algebra $\mathbf{A}$ is called a *direct power* of $\mathbf{A}$. We write $\mathbf{A}^J$ for a direct power of $\mathbf{A}$ indexed by a set J and we often view the elements of this algebra as functions from J to A.

If U is an ultrafilter on a set J and if $\mathbf{A} = \prod_{j\in J} \mathbf{A}_j$, then the equivalence relation θ_U on A defined by

$$(\overline{a}, \overline{b}) \in \theta_U \text{ iff } \{j \in J : a_j = b_j\} \in U$$

is, in fact, a congruence relation on $\mathbf{A}$ and the quotient $\mathbf{A}/\theta_U$ is called an *ultraproduct* of the $\mathbf{A}_j$.

If $\mathbf{A} = \prod_{j\in J} \mathbf{A}_j$, then the *$j$-th projection* map π_j is a homomorphism of $\mathbf{A}$ onto $\mathbf{A}_j$. The kernel of π_j is usually written as η_j and thus for $a, b \in A$ we have $(a, b) \in \eta_j$ if and only if $a(j) = b(j)$. The η_j are called *projection kernels*. It is easily checked that if J_1 and J_2 are nonvoid complementary subsets of J and $\alpha_i = \bigwedge_{j\in J_i} \eta_j$ for $i = 1, 2$, then in $\mathsf{Con}\,\mathbf{A}$ we have:

(1) The congruences α_1 and α_2 permute,
(2) $\alpha_1 \vee \alpha_2 = 1_A$,
(3) $\alpha_1 \wedge \alpha_2 = 0_A$.

Conversely, if $\mathbf{A}$ is any algebra and α_1 and α_2 are any two congruences for which these three conditions hold, then α_1 and α_2 are each called *factor congruences* of $\mathbf{A}$ and $\mathbf{A} \simeq \mathbf{A}/\alpha_1 \times \mathbf{A}/\alpha_2$. An algebra is called *directly indecomposable* if it is nontrivial and is not isomorphic to the direct product of two nontrivial algebras. Every finite algebra is isomorphic to the direct product of directly indecomposable algebras but this is not necessarily the case for infinite algebras.

We use the following notation for dealing with subalgebras of a direct power of an algebra $\mathbf{A}$. Let T be a nonvoid set. For $f \in A^T$ and $t \in T$ we sometimes write f_t for $f(t)$. If $a \in A$, then as previously mentioned, we let $\overline{a} \in A^T$ be such that $\overline{a}_t = a$ for all $t \in T$. The *diagonal* of A^T is $\Delta = \{\overline{a} : a \in A\}$. If $\mathbf{B}$ is a subalgebra of $\mathbf{A}^T$ and C is a subset of A, we write $\mathbf{B}(C)$ for $B \cap C^T$. Often, if $C = \{a_1, \ldots, a_n\}$ we write $\mathbf{B}(a_1, \ldots, a_n)$ for $\mathbf{B}(C)$. A subalgebra of $\mathbf{A}^T$ containing Δ is called a *diagonal subalgebra* of $\mathbf{A}^T$ or a *diagonal subpower* of $\mathbf{A}$. An important property of any diagonal subpower $\mathbf{D}$ of $\mathbf{A}$ is that every polynomial operation of $\mathbf{A}$ extends coordinatewise to a polynomial operation of $\mathbf{D}$. Also, for every projection kernel η_t we have $\mathbf{D}/\eta_t \simeq \mathbf{A}$.

For a congruence relation θ of $\mathbf{A}$ and a subalgebra $\mathbf{B}$ of $\mathbf{A}^T$, let $\overline{\theta} \in \mathsf{Con}\,\mathbf{B}$ be defined by $\overline{\theta} = \big\{(f,g) \in B^2 : (f_t, g_t) \in \theta \text{ for all } t \in T\big\}$. An element $f \in A^T$ is called *θ-constant* if $(f_s, f_t) \in \theta$ for all $s, t \in T$. We use the following easily proved facts repeatedly in our work.

PROPOSITION 2.1. *Let θ be a congruence relation on an algebra $\mathbf{A}$ and let T be a nonvoid set. For a diagonal subalgebra $\mathbf{D}$ of $\mathbf{A}^T$ generated by a set of θ-constant elements we have:*

(1) *Every $f \in D$ is θ-constant;*
(2) $\mathbf{A}/\theta \simeq \mathbf{D}/\overline{\theta}$.

Certain subalgebras of a direct product called subdirect products play an important role in our work. An algebra $\mathbf{A}$ is a *subdirect product* of the algebras $\mathbf{A}_j$, for $j \in J$, if $\mathbf{A}$ is a subalgebra of $\prod_{j\in J} \mathbf{A}_j$ and for each $j \in J$ the projection map from $\mathbf{A}$ to $\mathbf{A}_j$ is onto. Thus, if $\mathbf{A}$ is a subdirect product of the $\mathbf{A}_j$ for $j \in J$ and γ_j is the kernel of the j-th projection homomorphism from $\mathbf{A}$ to $\mathbf{A}_j$, then $\bigcap_{j\in J} \gamma_j = 0_A$ and each $\mathbf{A}_j$ is isomorphic to $\mathbf{A}/\gamma_j$. Conversely, if a family of congruence relations, γ_j for $j \in J$, on an algebra $\mathbf{A}$ has the property that $\bigcap_{j\in J} \gamma_j = 0_A$, then $\mathbf{A}$ is isomorphic to an algebra that is the subdirect product of $\mathbf{A}/\gamma_j$ for $j \in J$. A *subdirect representation* of $\mathbf{A}$ with subdirect factors $\mathbf{A}_j$ is a homomorphic embedding h of $\mathbf{A}$ into $\prod_{j\in J} \mathbf{A}_j$ for which $h(A)$ is the universe of an algebra that is a subdirect product of the $\mathbf{A}_j$.

An algebra $\mathbf{A}$ is *subdirectly irreducible* if it is nontrivial and in any subdirect representation of $\mathbf{A}$ at least one of the projection maps is an isomorphism. We use the following internal characterization of a subdirectly irreducible algebra: An algebra $\mathbf{A}$ is subdirectly irreducible if and only if there is a $\mu \in \mathsf{Con}\,\mathbf{A}$ such that $0_A < \mu$ and $\mu \leqslant \theta$ for all $0_A < \theta \in \mathsf{Con}\,\mathbf{A}$. The congruence relation μ is called the *monolith* of the subdirectly irreducible algebra $\mathbf{A}$. Thus, $\mathbf{A}$ is subdirectly irreducible if and only if in the lattice $\mathsf{Con}\,\mathbf{A}$ the element 0_A is strictly meet irreducible. A theorem of Birkhoff states that every algebra is a subdirect product of subdirectly irreducible algebras. This theorem is equivalent to the statement that in any algebra $\mathbf{A}$ the congruence relation 0_A is the intersection of all strictly meet irreducible members of $\mathsf{Con}\,\mathbf{A}$. If $\mathbf{A}$ is a subdirectly irreducible algebra with monolith μ, then $\mu = \mathrm{Cg}^{\mathbf{A}}(a,b)$ for every $(a,b) \in \mu$ with $a \neq b$ and for such an ordered pair (a,b) we say $\mathbf{A}$ is (a,b)-*irreducible.*

An *equation* or *identity* in a language L is an expression $\mathbf{p} \approx \mathbf{q}$ where $\mathbf{p}$ and $\mathbf{q}$ are terms for L. If $\mathbf{A}$ is an algebra in the language L we say that an identity $\mathbf{p} \approx \mathbf{q}$ *holds for* $\mathbf{A}$, denoted $\mathbf{A} \models \mathbf{p} \approx \mathbf{q}$, if $\mathbf{p}^{\mathbf{A}}(\overline{a}) = \mathbf{q}^{\mathbf{A}}(\overline{a})$ for all $\overline{a} \in A^n$. Alternate phrasings for this include $\mathbf{A}$ *satisfies* $\mathbf{p} \approx \mathbf{q}$ or $\mathbf{p} \approx \mathbf{q}$ *is an identity for* $\mathbf{A}$. The identity $\mathbf{p} \approx \mathbf{q}$ holds for a class $\mathcal{C}$ of algebras if $\mathbf{p} \approx \mathbf{q}$ holds for every algebra in $\mathcal{C}$ and in which case we write $\mathcal{C} \models \mathbf{p} \approx \mathbf{q}$. For a set Σ of identities in a language L and $\mathcal{C}$ a class of algebras in this language, we write $\mathcal{C} \models \Sigma$ if every identity in Σ holds for every algebra in $\mathcal{C}$. A class $\mathcal{C}$ of algebras in a language L is called an *equational class* if there is a set Σ of identities in L such that $\mathcal{C}$ consists of all algebras in this language that satisfy every identity in Σ.

Let $\mathcal{C}$ be an arbitrary class of algebras of the same similarity type. The class of all homomorphic images of algebras in $\mathcal{C}$ is denoted $\mathsf{H}(\mathcal{C})$, the class of all algebras isomorphic to a subalgebra of an algebra in $\mathcal{C}$ is $\mathsf{S}(\mathcal{C})$, the class of all algebras isomorphic to a direct product of algebras in $\mathcal{C}$ is $\mathsf{P}(\mathcal{C})$, and the class of all ultraproducts of algebras in $\mathcal{C}$ is $\mathsf{P_U}(\mathcal{C})$. A class $\mathcal{V}$ of algebras of the same similarity type is called a *variety* if $\mathcal{V}$ is closed under the formation of homomorphic images, subalgebras, and direct products. It is not hard to show that $\mathsf{HSP}(\mathcal{C})$ is always variety. We say that $\mathcal{C}$ *generates* the variety $\mathsf{HSP}(\mathcal{C})$. We usually write $\mathcal{V}(\mathcal{C})$ for the variety generated by $\mathcal{C}$. A variety is *finitely generated* if it is generated by a finite set of finite algebras, or equivalently by a single finite algebra. Every variety contains the trivial algebra; a variety is called *nontrivial* if it contains nontrivial algebras.

A class of algebras that contains the trivial algebra and is closed under the operations of S, P, and $\mathsf{P_U}$ is called a *quasivariety.* Clearly, every variety is a quasivariety. In our work we focus mainly on varieties of algebras but we sometimes consider quasivarieties as well.

A classic preservation theorem of Birkhoff states that a class of algebras is a variety if and only if it is an equational class. Since identities are preserved by the formation of homomorphism, subalgebra, and direct product, it follows that a variety $\mathcal{V}$ is generated by a class $\mathcal{C}$ of algebras if and only if $\mathcal{V}$ consists of all algebras that satisfy every identity that holds for every member of $\mathcal{C}$. Note that since every algebra $\mathbf{A}$ is a subdirect product of subdirectly irreducible algebras and these subdirectly irreducible algebras are homomorphic images of $\mathbf{A}$, it follows that every algebra in a variety $\mathcal{V}$ is the subdirect product of subdirectly irreducible algebras in $\mathcal{V}$ and the collection of subdirectly irreducible algebras in $\mathcal{V}$ generates the variety $\mathcal{V}$.

A variety is called *residually small* if there is a bound on the cardinalities of its subdirectly irreducible algebras; it is *residually large* if no such bound exists. A variety $\mathcal{V}$ is *locally finite* if every finitely generated algebra in $\mathcal{V}$ is finite. If every subdirectly irreducible algebra in $\mathcal{V}$ is a simple algebra, then $\mathcal{V}$ is called *semisimple.*

A variety is *congruence distributive*, *congruence modular*, or *congruence permutable* if every algebra in the variety is congruence distributive, congruence modular, or has permuting congruence relations, respectively. An *arithmetical* variety is one that is congruence distributive and congruence permutable. Congruence permutable varieties are known to be congruence modular. A classic result of Maltsev states that a variety $\mathcal{V}$ is congruence permutable if and only if $\mathcal{V}$ has a ternary term $\mathbf{p}$ for which $\mathcal{V} \models \mathbf{p}(x, x, y) \approx \mathbf{p}(y, x, x) \approx y$. Such a term is called a *Maltsev term* for $\mathcal{V}$.

Principal congruence relations in congruence permutable varieties have an especially simple form. A proof of the following proposition may be found in [**55**, Theorem 4.70]

PROPOSITION 2.2. *If an algebra* $\mathbf{A}$ *is in congruence permutable variety, then*

- *Maltsev chains for principal congruences can always be chosen with length 1, in fact, for all* $a, b, c, d \in A$, *we have* $(c, d) \in \mathrm{Cg}^{\mathbf{A}}(a, b)$ *if and only if there exists* $\mathbf{p} \in \mathrm{Pol}_1\mathbf{A}$ *such that* $c = \mathbf{p}(a)$ *and* $d = \mathbf{p}(b)$.
- *More generally, for every* $X \subseteq A^2$ *and* $c, d \in A$, *we have* $(c, d) \in \mathrm{Cg}^{\mathbf{A}}(X)$ *if and only if there exist* $(a_1, b_1), \ldots, (a_n, b_n) \in X$ *and* $\mathbf{p} \in \mathrm{Pol}_n\mathbf{A}$ *such that* $c = \mathbf{p}(a_1, \ldots, a_n)$ *and* $d = \mathbf{p}(b_1, \ldots, b_n)$.

Varieties that are congruence distributive possess a number of special properties. One such property, discovered by B. Jónsson, is a restriction on the incidence of subdirectly irreducible algebras in congruence distributive varieties:

Jónsson's Lemma: *If* $\mathcal{V}(\mathcal{C})$ *is a congruence distributive variety, then every subdirectly irreducible algebra in* $\mathcal{V}(\mathcal{C})$ *is in* $\mathsf{HSP_U}(\mathcal{C})$.

In the special case that $\mathcal{C}$ is a finite set of finite algebras it follows from Jónsson's Lemma that all subdirectly irreducible algebras in the variety generated by $\mathcal{C}$ are in $\mathsf{HS}(\mathcal{C})$ and therefore are finite.

Let $\mathcal{K}$ be an arbitrary class of algebras. A *transversal* of $\mathcal{K}$ is a collection $\mathcal{T}$ of pairwise nonisomorphic members of $\mathcal{K}$ for which every $\mathbf{A} \in \mathcal{K}$ is isomorphic to a member of $\mathcal{T}$. For a variety $\mathcal{V}$ we will have occasion to consider transversals of the k-generated subdirectly irreducible and the k-generated directly indecomposable algebras in $\mathcal{V}$.

An algebra $\mathbf{F}$ is freely generated by a set X in a class $\mathcal{C}$ of algebras if $\mathbf{F}$ is a member of $\mathcal{C}$, the set X generates $\mathbf{F}$, and for every $\mathbf{A} \in \mathcal{C}$ and every function $g : X \to A$ there exists a homomorphism $h : \mathbf{F} \to \mathbf{A}$ such that $h(x) = g(x)$ for all $x \in X$. We write $\mathbf{F}_{\mathcal{C}}(X)$ to denote this free algebra. For a cardinal k we let $\mathbf{F}_{\mathcal{C}}(k)$ denote the algebra, up to isomorphism, freely generated for $\mathcal{C}$ by a set of cardinality k. If $\mathcal{V}$ is a variety, then every k-generated algebra in $\mathcal{V}$ is a homomorphic image of $\mathbf{F}_{\mathcal{V}}(k)$. In fact, for any variety the free algebra on k free generators for the variety is determined, up to isomorphism, by this property. The *free spectrum* of $\mathcal{V}$ is the sequence of cardinalities $|\mathrm{F}_{\mathcal{V}}(k)|$ for $k = 1, 2, \ldots$. We sometimes write $f_{\mathcal{V}}(k)$ for the free spectrum of $\mathcal{V}$. A variety $\mathcal{V}$ is locally finite if and only if $|\mathrm{F}_{\mathcal{V}}(k)|$ is finite for every finite k.

Let $\mathcal{V}$ be a variety, $X = \{x_1, \ldots, x_n\}$, and suppose that $\mathbf{p}(x_1, \ldots, x_n)$ and $\mathbf{q}(x_1, \ldots, x_n)$ are two terms in the language of $\mathcal{V}$. An important property of free algebras is that $\mathbf{F}_{\mathcal{V}}(X) \models \mathbf{p} \approx \mathbf{q}$ if and only if $\mathcal{V} \models \mathbf{p} \approx \mathbf{q}$. So elements of $\mathrm{F}_{\mathcal{V}}(X)$

may be viewed as terms in the language of $\mathcal{V}$, and distinct terms correspond to distinct elements in this free algebra. That is, if $\mathbf{p}$ and $\mathbf{q}$ are terms in X and $\mathbf{F}$ denotes $\mathbf{F}_{\mathcal{V}}(X)$, then the term functions $\mathbf{p}^{\mathbf{F}}$ and $\mathbf{q}^{\mathbf{F}}$ evaluated at the n-tuple $(x_1, \dots, x_n) \in F^n$ give distinct elements of F if and only if $\mathbf{p} \approx \mathbf{q}$ is not an identity for $\mathcal{V}$.

In the event that a variety $\mathcal{V}$ is generated by a finite algebra $\mathbf{A}$, the free algebra $\mathbf{F}_{\mathcal{V}}(n)$ for finite n has a very explicit, concrete representation. Namely, let T denote all n-tuples of elements of A. Then $\mathbf{F}_{\mathcal{V}}(n)$ is the subalgebra of the direct power $\mathbf{A}^T$ that is generated by $g_1, \dots, g_n \in A^T$, where for $(a_1, \dots, a_n) \in T$ and $1 \leqslant i \leqslant n$ the value of g_i at $(a_1, \dots, a_n)$ is a_i.

The material on universal algebra presented so far in this Chapter is "classical" and was all known by the mid 1960s. In our work we will require some deep results that have come out of two more recent developments: generalized commutator theory and tame congruence theory. We now present the basics of these two topics.

Fuller discussions of the generalized commutator may be found in [**22**], [**55**, Section 4.13], [**32**, Chapter 3], and [**42**]. The main reference for tame congruence theory is [**32**].

We begin with the theory of the commutator. Let $\mathbf{A}$ be an algebra, $\gamma \in \mathsf{Con}\,\mathbf{A}$, and $R, S \subseteq A^2$. We say R *centralizes* S *modulo* γ, denoted $C(R, S; \gamma)$, if for every $n \geqslant 1$, every $(n+1)$-ary term $\mathbf{t}$, every $(a, b) \in R$, and every $(c_1, d_1), \dots, (c_n, d_n) \in S$ we have

$$(\mathbf{t}(a, \bar{c}), \mathbf{t}(a, \bar{d})) \in \gamma \text{ iff } (\mathbf{t}(b, \bar{c}), \mathbf{t}(b, \bar{d})) \in \gamma.$$

The following facts are easily verified.

PROPOSITION 2.3. *For binary relations that are congruence relations on* $\mathbf{A}$*:*

(1) *If* $\alpha' \subseteq \alpha$ *and* $\beta' \subseteq \beta$, *then* $C(\alpha, \beta; \gamma)$ *implies* $C(\alpha', \beta'; \gamma)$.
(2) *If* $C(\alpha, \beta, \gamma_i)$ *for all* $i \in I$, *then* $C(\alpha, \beta; \bigcap_{i \in I} \gamma_i)$.
(3) $C(\alpha, \beta; \alpha)$ *and* $C(\alpha, \beta; \beta)$.
(4) *If* $C(\alpha_i, \beta, \gamma)$ *for all* $i \in I$, *then* $C(\bigvee_{i \in I} \alpha_i, \beta; \gamma)$.
(5) *If* $\theta \subseteq \alpha, \beta, \gamma$ *then* $C(\alpha, \beta; \gamma)$ *holds in* $\mathbf{A}$ *iff* $C(\alpha/\theta, \beta/\theta; \gamma/\theta)$ *holds in the quotient* $\mathbf{A}/\theta$.

Moreover, (1) *and* (2) *hold for arbitrary binary relations* $\alpha, \alpha', \beta, \beta'$, *and* (3) *holds if* α *and* β *are binary relations that are preserved by the fundamental operations of* $\mathbf{A}$.

An algebra $\mathbf{A}$ is *Abelian*, or is said to satisfy the *term condition*, if $C(1_A, 1_A; 0_A)$ holds. Note that if $C(1_A, 1_A; \gamma)$, then $\mathbf{A}/\gamma$ is Abelian.

If α and β are congruence relations on an algebra $\mathbf{A}$, then the *commutator* of α and β, denoted $[\alpha, \beta]$, is the least congruence γ for which $C(\alpha, \beta; \gamma)$. The *centralizer of* β *modulo* α, denoted $(\alpha : \beta)$, is the largest congruence δ for which $C(\delta, \beta; \alpha)$. The centralizer of 1_A modulo 0_A is called the *center* of $\mathbf{A}$ and the algebra $\mathbf{A}$ is *centerless* if its center is 0_A.

We will appeal, often without reference, to the following facts about the centralizer and the commutator:

PROPOSITION 2.4. *For congruence relations in an arbitrary algebra* $\mathbf{A}$

(1) $C(\alpha, \beta; \gamma)$ *if and only if* $\alpha \leqslant (\gamma : \beta)$,
(2) $(\alpha : 0_A) = 1_A$,
(3) $\alpha \leqslant (\alpha : \beta)$.

If $\mathbf{A}$ *belongs to a congruence modular variety then we additionally have* (see [**22**])

(4) $[\alpha, \beta] = [\beta, \alpha]$,
(5) $C(\alpha, \beta; \gamma)$ *if and only if* $[\alpha, \beta] \leqslant \gamma$,
(6) $[\alpha, \bigvee_i \beta_i] = \bigvee_i [\alpha, \beta_i]$,
(7) $(\alpha : \bigvee_i \beta_i) = \bigwedge_i (\alpha : \beta_i)$,
(8) $(\bigwedge_i \alpha_i : \beta) = \bigwedge_i (\alpha_i : \beta)$,
(9) *if the intervals* $I[\alpha_1, \beta_1]$ *and* $I[\alpha_2, \beta_2]$ *are projective in the lattice* $\mathsf{Con}\,\mathbf{A}$, *then* $(\alpha_1 : \beta_1) = (\alpha_2 : \beta_2)$.

A consequence of items (2) and (3) in Proposition 2.3 is that $[\alpha, \beta] \leqslant \alpha \cap \beta$ for all congruence relations in an arbitrary algebra, however, for algebras in congruence distributive varieties it is known that $[\alpha, \beta] = \alpha \cap \beta$, see e.g., [**55**, p. 258].

By means of the commutator it is possible define notions of Abelian, solvable and nilpotence for arbitrary algebras. Let $\alpha \leqslant \beta$ be congruence relations of an algebra $\mathbf{A}$. The congruence relation β is *Abelian over* α if $C(\beta, \beta; \alpha)$ and β is *Abelian* if $C(\beta, \beta; 0_A)$. We say β is *solvable over* α if there exists a finite chain of congruence relations $\beta = \gamma_0 \geqslant \gamma_1 \geqslant \ldots \geqslant \gamma_m = \alpha$ such that γ_i is Abelian over γ_{i+1} for all $i < m$. A congruence relation β is *solvable* if it is solvable over 0_A. An algebra $\mathbf{A}$ is *solvable* if 1_A, and hence every congruence relation of $\mathbf{A}$, is solvable. An algebra $\mathbf{A}$ is *locally solvable* if every finitely generated subalgebra of $\mathbf{A}$ is solvable. It can be argued that in the congruence lattice of a finite algebra $\mathbf{A}$ the join of all the solvable congruence relations is itself solvable. This largest solvable congruence relation is called the *solvable radical* of $\mathbf{A}$.

For a congruence θ and $k = 1, 2, \ldots$ we write

$$\begin{aligned} \theta^{(1)} &= \theta & \theta^{[1]} &= \theta \\ \theta^{(k+1)} &= [\theta, \theta^{(k)}] & \theta^{[k+1]} &= [\theta^{[k]}, \theta^{[k]}]. \end{aligned}$$

A congruence relation θ on $\mathbf{A}$ is called *m-step left nilpotent* if $\theta^{(m+1)} = 0_A$ and the algebra $\mathbf{A}$ is *left nilpotent* if 1_A is m-step left nilpotent for some finite m. In the congruence modular varieties we use the word nilpotent rather than left nilpotent. Note that θ is solvable if $\theta^{[m]} = 0_A$ for some m.

Let α and β be congruence relations of an algebra $\mathbf{A}$ with $\alpha < \beta$. The congruence β is said to be *strongly Abelian* over α if for every $n > 1$, for every $\mathbf{p} \in \mathrm{Pol}_n\mathbf{A}$, for every $(a_1, b_1) \in \beta, \ldots, (a_n, b_n) \in \beta$ and for all $c_2, \ldots, c_n \in A$, the following implication holds:

$$\begin{aligned} &(\mathbf{p}(a_1, a_2, \ldots, a_n), \mathbf{p}(b_1, b_2, \ldots, b_n)) \in \alpha \Rightarrow \\ &\qquad (\mathbf{p}(a_1, c_2, \ldots, c_n), \mathbf{p}(b_1, c_2, \ldots, c_n)) \in \alpha. \end{aligned}$$

A congruence relation β of an algebra $\mathbf{A}$ is *strongly Abelian* if it is strongly Abelian over 0_A; an algebra is strongly Abelian if every one of its congruence relations is strongly Abelian; and a variety is strongly Abelian if every algebra in it is strongly Abelian. The definitions of *strongly solvable* and *locally strongly solvable* are exactly like their solvable counterparts, except with strongly Abelian replacing Abelian.

We next sketch the material on tame congruence theory that we will need.

For a nonvoid subset U of an algebra $\mathbf{A}$ the *algebra induced by* $\mathbf{A}$ *on* U is the algebra $\mathbf{A}|_U$ whose universe is U and whose fundamental operations are all polynomials $\mathbf{p} \in \mathrm{Pol}\,\mathbf{A}$ for which $\mathbf{p}|_U$ maps U into U. The algebra $\mathbf{A}|_U$ is nonindexed, that is, there is no index set specified for the set of fundamental operations. Note that every polynomial operation of $\mathbf{A}|_U$ is a fundamental operation. Two nonvoid subsets U and V of $\mathbf{A}$ are called *polynomially isomorphic* if there exist $\mathbf{f}, \mathbf{g} \in \mathrm{Pol}_1\mathbf{A}$ such that $\mathbf{f}(U) = V$, $\mathbf{g}(V) = U$, $\mathbf{fg}$ is the identity on V, and $\mathbf{gf}$ is the identity on U. If U and V are polynomially isomorphic, then the algebras $\mathbf{A}|_U$ and $\mathbf{A}|_V$ are isomorphic as nonindexed algebras, that is, it is possible to index the fundamental operations of each with one index set so that the resulting algebras are isomorphic in the usual sense.

An idempotent polynomial for an algebra $\mathbf{A}$ is any $\mathbf{e} \in \mathrm{Pol}_1\mathbf{A}$ such that $\mathbf{e}^2(x) = \mathbf{e}(x)$ for all $x \in A$. For an idempotent polynomial $\mathbf{e}$ the restriction $\mathbf{e}|_{\mathbf{e}(A)}$ is the identity map on $\mathbf{e}(A)$. Algebras induced by $\mathbf{A}$ on the range of an idempotent polynomial have a particularly simple characterization for their fundamental operations. Namely, if $\mathbf{e}$ is idempotent for $\mathbf{A}$ and $U = \mathbf{e}(A)$, then the fundamental operations of $\mathbf{A}|_U$ consist of all polynomials of the form $\mathbf{ep}|_U$ where $\mathbf{p}$ ranges over all polynomials of $\mathbf{A}$. The collection of all idempotent polynomials for $\mathbf{A}$ is denoted $E(\mathbf{A})$.

Let $\alpha < \beta$ in the congruence lattice of a finite algebra $\mathbf{A}$. By $\mathrm{U}_{\mathbf{A}}(\alpha, \beta)$ we denote all sets of the form $\mathbf{f}(A)$, with at least two elements, where $\mathbf{f} \in \mathrm{Pol}_1\mathbf{A}$ and $\mathbf{f}(\beta) \nsubseteq \alpha$. Minimal members of $\mathrm{U}_{\mathbf{A}}(\alpha, \beta)$, that is, minimal when ordered by inclusion, are called *(α, β)-minimal sets of* $\mathbf{A}$. The set of all (α, β)-minimal sets of $\mathbf{A}$ is denoted $\mathrm{M}_{\mathbf{A}}(\alpha, \beta)$.

In a finite algebra $\mathbf{A}$ a quotient (α, β) in $\mathsf{Con}\,\mathbf{A}$ is called *tame* if there exist $V \in \mathrm{M}_{\mathbf{A}}(\alpha, \beta)$ and $\mathbf{e} \in E(\mathbf{A})$ such that $\mathbf{e}(A) = V$ and for all $\gamma \in \mathsf{Con}\,\mathbf{A}$ if $\alpha < \gamma < \beta$, then $\gamma|_V \neq \alpha|_V$ and $\gamma|_V \neq \beta|_V$. Note that every prime quotient is tame. A basic result in tame congruence theory is that if (α, β) is a tame quotient, then all (α, β)-minimal sets of $\mathbf{A}$ are polynomially isomorphic. If (α, β) is tame and $U \in \mathrm{M}_{\mathbf{A}}(\alpha, \beta)$, then any set of the form $a/\beta \cap U$ that is not of the form $a/\alpha \cap U$ is called a *trace* of U and an *(α, β)-trace* of $\mathbf{A}$. The union of all (α, β)-traces of U is called the *body* of U and those elements of U not in the body of U form the *tail* of U. If N is a trace for U, then $\alpha|_N$ denotes $\alpha \cap N^2$, and $\alpha|_N$ is a congruence on the nonindexed algebra $\mathbf{A}|_N$.

The interest in tame congruence theory in tame quotients and their minimal sets and traces arises from the fact that the local behavior of a tame quotient falls into one of five distinct situations. More specifically, for any finite algebra $\mathbf{A}$, for any tame quotient (α, β), and for any trace N of $U \in \mathrm{M}_{\mathbf{A}}(\alpha, \beta)$, the quotient algebra $(\mathbf{A}|_N)/(\alpha|_N)$ must be polynomially equivalent to one of the following five types of algebras:

1. a G-set,
2. a finite dimensional vector space over a finite field,
3. a 2-element Boolean algebra,
4. a 2-element distributive lattice,
5. a 2-element semilattice.

Moreover, the particular type **1**, **2**, **3**, **4**, or **5** is independent of the choice of U and N. This is called the *type* of the tame quotient (α, β) and is denoted $\mathrm{typ}(\alpha, \beta)$.

The type of a tame quotient in a finite algebra has significant consequences for local behavior and for the algebraic structure of the algebra and the quotient. For example, it is known that for a tame quotient (α, β):

- $\text{typ}(\alpha, \beta) = \mathbf{1}$ if and only if β is strongly Abelian over α, and
- $\text{typ}(\alpha, \beta) = \mathbf{2}$ if and only if β is Abelian but not strongly Abelian over α.

Because of this, types **1** and **2** are referred to as the *Abelian types* and types **3**, **4**, and **5** are the *non-Abelian types*.

In our work the tame quotients that we consider are usually prime quotients. The following terminology is used in connection with the set of types of prime quotients in finite algebra.

For $\gamma < \delta$ in $\mathsf{Con}\,\mathbf{A}$ the set of all types $\text{typ}(\alpha, \beta)$ for $\gamma \leqslant \alpha \prec \beta \leqslant \delta$ is denoted $\text{typ}\{\gamma, \delta\}$. The *type set of a finite algebra* $\mathbf{A}$, denoted $\text{typ}\{\mathbf{A}\}$, is $\text{typ}\{0_A, 1_A\}$. The *type set of a class* $\mathcal{K}$ of algebras consists of the union of the type sets of the finite algebras in $\mathcal{K}$ and is denoted $\text{typ}\{\mathcal{K}\}$.

Two preservation theorems involving the calculus of types that we will frequently use are that type is preserved under homomorphism and that projective prime quotients have the same type, that is, for a finite algebra $\mathbf{A}$,

- if $\delta \leqslant \alpha \prec \beta$ in $\mathsf{Con}\,\mathbf{A}$, then $\text{typ}(\alpha, \beta) = \text{typ}(\alpha/\delta, \beta/\delta)$ in $\mathbf{A}/\delta$,
- if $\alpha_1 \prec \beta_1$ and $\alpha_2 \prec \beta_2$ are projective prime quotients in $\mathsf{Con}\,\mathbf{A}$, then $\text{typ}(\alpha_1, \beta_1) = \text{typ}(\alpha_2, \beta_2)$.

These two results show that if $\mathbf{i} \in \text{typ}\{\mathbf{A}\}$, then there is a subdirectly irreducible algebra $\mathbf{B}$ with monolith μ such that $\mathbf{B}$ is a homomorphic image of $\mathbf{A}$ and $\text{typ}(0_B, \mu) = \mathbf{i}$. Thus, in a variety $\mathcal{V}$, every type that appears in $\text{typ}\{\mathcal{V}\}$ is of the form $\text{typ}(0, \mu)$ for some monolith μ of a finite subdirectly irreducible algebra in $\mathcal{V}$. Many of our arguments involve an analysis of $(0, \mu)$-minimal sets and traces for such a monolith μ.

We next summarize some of the algebraic properties that are consequences of a prime quotient having a particular type. We formulate these results for the situation that the prime quotient consists of the diagonal congruence relation and an atom in the congruence lattice.

Consider an arbitrary finite algebra $\mathbf{A}$ with $0_A \prec \beta$. Let $U \in M_{\mathbf{A}}(0_A, \beta)$ and let N be a $(0_A, \beta)$-trace contained in U. Suppose $\text{typ}(0_A, \beta) = \mathbf{3}, \mathbf{4}$ or $\mathbf{5}$. For these three types, it is known that N is the unique $(0_A, \beta)$-trace contained in U. The algebra $\mathbf{A}|_N$ is polynomially equivalent to a 2-element Boolean algebra, distributive lattice, or semilattice. For each of these non-Abelian types there is a binary polynomial $\wedge \in \text{Pol}_2\mathbf{A}$ such that $\wedge|_U$ is a *pseudo-meet*, [**32**, Definition 4.16]. This means that we can label the two elements of N with 0 and 1 so that $1 \wedge 1 = 1$, and $x \wedge 1 = 1 \wedge x = x \wedge x = x$ for all $x \in U - \{1\}$. Thus $\langle \{0,1\}, \wedge|_N \rangle$ is a meet-semilattice with order $0 < 1$. If $\text{typ}(0_A, \beta) = \mathbf{4}$, then there are two binary polynomials $\wedge$ and $\vee$ such that $\wedge|_U$ is a pseudo-meet and $\vee|_U$ is a *pseudo-join*. Here the behavior of the pseudo-join is just the dual to that of the pseudo-meet. So $\langle \{0,1\}, \wedge|_N, \vee|_N \rangle$ is a distributive lattice with $0 < 1$. Thus, every n-ary operation on $N = \{0, 1\}$ that preserves this order is of the form $\mathbf{p}|_N$ for some $\mathbf{p} \in \text{Pol}_n\mathbf{A}$. If $\text{typ}(0_A, \beta) = \mathbf{3}$, then in addition to the binary polynomials $\wedge$ and $\vee$ that we have in the type **4** case, there is also a unary polynomial $'$ such that $0' = 1, 1' = 0$ and $A' = U$. The algebra $\langle \{0,1\}, \wedge|_N, \vee|_N, '|_N \rangle$ is a Boolean algebra and thus every n-ary operation on N is the restriction to N of some n-ary polynomial on $\mathbf{A}$ that can be built using $\wedge, \vee$, and $'$.

Furthermore, if $\text{typ}(0_A, \beta) = \mathbf{3}, \mathbf{4}$, or $\mathbf{5}$, then there exists a congruence δ that is the *pseudo-complement of* β, that is, δ is the largest congruence relation θ of $\mathbf{A}$ for which $\beta \wedge \theta = 0_A$. (See [**32**, Lemma 5.12].)

If $\text{typ}(0_A, \beta) = \mathbf{2}$, then there may be more than one trace contained in U. Let B be the body of U. A useful result that applies to this type $\mathbf{2}$ case is that there is a $\mathbf{d} \in \text{Pol}_3\mathbf{A}$ such that

(1) $\mathbf{d}(x, x, x) = x$ for all $x \in U$.
(2) $\mathbf{d}(x, x, y) = y = \mathbf{d}(y, x, x)$ for all $x \in B$ and $y \in U$.
(3) For every $a, b \in B$, the unary polynomials given by $\mathbf{d}(x, a, b), \mathbf{d}(a, x, b)$, and $\mathbf{d}(a, b, x)$ are permutations of U
(4) B is closed under $\mathbf{d}$, that is, $\mathbf{d}(a, b, c) \in B$ for all $a, b, c \in B$.

The polynomial $\mathbf{d}$ is called a *pseudo-Maltsev operation* for U.

We conclude our discussion of tame congruence theory by citing some results that connect it with the theory of the generalized commutator.

An ordered pair (a, b) of distinct elements in an algebra $\mathbf{A}$ is called

- a 1-*snag* if there exists $\mathbf{p}(x, y) \in \text{Pol}_2\mathbf{A}$ for which $\mathbf{p}(a, b) = \mathbf{p}(b, a) = a$ and $\mathbf{p}(b, b) = b$,
- a 2-*snag* if there exists $\mathbf{p}(x, y) \in \text{Pol}_2\mathbf{A}$ for which $\mathbf{p}(a, b) = \mathbf{p}(b, a) = \mathbf{p}(a, a) = a$, and $\mathbf{p}(b, b) = b$,
- a 4-*snag* if there exits $\mathbf{p}(x, y) \in \text{Pol}_2\mathbf{A}$ for which $\mathbf{p}(a, b) = \mathbf{p}(b, a) = \mathbf{p}(a, a) = b$, and $\mathbf{p}(b, b) = a$.

The existence of 1-snags or 2-snags in a finite algebra $\mathbf{A}$ provides information about the set of types appearing in $\mathbf{A}$. Moreover, 1-snags and 2-snags provide links to the notions of strongly solvable and solvable congruence quotients. The following Theorem is [**32**, Theorem 7.2] and provides some of these connections.

THEOREM 2.5. *Let $\mathbf{A}$ be a finite algebra with $\delta \leqslant \gamma$ in* $\mathsf{Con}\,\mathbf{A}$. *The following are equivalent:*

(1) *The congruence γ is (strongly) solvable over δ.*
(2) *There are no 2-snags (1-snags) in $\gamma - \delta$.*
(3) *For all $\delta \leqslant \alpha \prec \beta \leqslant \gamma$, the prime quotient (α, β) is (strongly) Abelian.*
(4) *The type set of (γ, δ) is contained in $\{\mathbf{1}, \mathbf{2}\}$ (in $\{\mathbf{1}\}$).*

Some connections at the varietal level between the Abelian types and notions of solvability are given by the next theorem, which is [**32**, Theorem 7.11].

THEOREM 2.6. *Let $\mathcal{V}$ be a locally finite variety.*

(1) *$typ\{\mathcal{V}\} \subseteq \{\mathbf{1}\}$ if and only if every algebra in $\mathcal{V}$ is locally strongly solvable.*
(2) $\text{typ}\{\mathcal{V}\} \subseteq \{\mathbf{1}, \mathbf{2}\}$ *if and only if every algebra in $\mathcal{V}$ is locally solvable.*
(3) $\text{typ}\{\mathcal{V}\} \subseteq \{\mathbf{2}\}$ *if and only if $\mathcal{V}$ is congruence permutable and every algebra in $\mathcal{V}$ is locally solvable.*

The varietal conditions given in item (3) of Theorem 2.6 will be of special interest in our work. A variety $\mathcal{V}$ is called *affine* if it is congruence modular and Abelian. It can be argued that if $\mathcal{V}$ is affine then it is also congruence permutable. The properties of affine varieties are developed in [**22**]. Each affine variety $\mathcal{V}$ has a corresponding ring $\mathbf{R}$ with unit such that every algebra in $\mathcal{V}$ is polynomially equivalent to an $\mathbf{R}$-module and conversely every $\mathbf{R}$-module is polynomially equivalent to an algebra in $\mathcal{V}$. Throughout this work our modules will always be unitary left modules over a ring with an identity element. This polynomial equivalence between algebras in $\mathcal{V}$ and $\mathbf{R}$-modules is such that $\mathcal{V}$ is locally finite if and only if the ring $\mathbf{R}$ is finite.

The three non-Abelian types, **3**, **4** and **5**, are linked to the notion of congruence meet semi-distributivity. A lattice $\mathbf{L}$ is *meet semi-distributive* if and only if whenever $a, b, c \in L$ are such that $a \wedge b = a \wedge c$, then $a \wedge b = a \wedge (b \vee c)$ also holds. A variety $\mathcal{V}$ is said to be *congruence meet semi-distributive* if every algebra in $\mathcal{V}$ has a congruence lattice that is meet semi-distributive. A collection of conditions equivalent to congruence meet semi-distributivity is given in Theorem 9.10 of [**32**]. We mention, in particular, the following.

THEOREM 2.7. *A locally finite variety $\mathcal{V}$ is congruence meet semi-distributive if and only if* $\operatorname{typ}\{\mathcal{V}\} \subseteq \{\mathbf{3}, \mathbf{4}, \mathbf{5}\}$.

We next describe an important class of congruence meet semi-distributive varieties. A *discriminator operation* on a set A is a function $d : A^3 \to A$ given by $d(a, a, c) = c$ and $d(a, b, c) = a$ if $a \neq b$. An algebra $\mathbf{A}$ is said to have a *discriminator term* if $\operatorname{Clo} \mathbf{A}$ contains a discriminator operation. It is not hard to see that every nontrivial algebra that has a discriminator term is simple.

Let $\mathcal{C}$ by a class of algebras of the same similarity type and suppose $\mathbf{d}$ is a common discriminator term for the algebras in $\mathcal{C}$. The variety $\mathcal{V}$ generated by $\mathcal{C}$ is called a *discriminator variety*. Proofs of the following facts about discriminator varieties may be found in [**17**].

THEOREM 2.8. *Let $\mathcal{C}$ be a class of algebras of the same similarity type having a common discriminator term.*

(1) *The discriminator variety $\mathcal{V}(\mathcal{C})$ is both congruence distributive and congruence permutable, i.e., it is an arithmetical variety.*
(2) *Every directly indecomposable algebra in $\mathcal{V}(\mathcal{C})$ is simple.*
(3) *The simple algebras in $\mathcal{V}$ are the nontrivial members of* $\mathsf{SP_U}(\mathcal{C})$.

In any variety the subdirectly irreducible algebras are directly indecomposable, so in a discriminator variety the subdirectly irreducible algebras are simple, that is, every discriminator variety is semisimple. If the algebras in $\mathcal{C}$ have a common discriminator term and $\mathcal{C}$ is the class of all algebras that satisfy a set of universal sentences, then $\mathcal{C} = \mathsf{SP_U}(\mathcal{C})$ and the simple algebras in the discriminator variety $\mathcal{V}(\mathcal{C})$ are precisely the nontrivial members of $\mathcal{C}$. Another observation is that $\operatorname{typ}\{\mathcal{V}\} = \{\mathbf{3}\}$ for any locally finite discriminator variety $\mathcal{V}$, and more generally for any locally finite arithmetical variety.

We will use some elementary facts about partitions of a set and partitions of an integer. Let $\operatorname{Bell}(n)$ denote the n-th Bell number, which is the number of equivalence relations on a set of n elements. Asymptotic estimates for $\operatorname{Bell}(n)$ are known, e.g., [**48**] and [**57**], but for our needs the coarse inequalities $2^n \leqslant \operatorname{Bell}(n) \leqslant n! \leqslant 2^{n \log n}$ that hold for all but finitely many n will suffice. Thus, in our terminology, $\operatorname{Bell}(n)$ is both at least and at most exponential in n, and so is of exponential complexity. Two equivalence relations R_1 and R_2 are isomorphic, written $R_1 \simeq R_2$, if they are isomorphic as binary relations. For R_1 an equivalence relation on a finite set of n elements, we have $R_1 \simeq R_2$ if and only if the partition of the integer n given by the cardinalities of the equivalence classes of R_1 is the same as that for R_2. So the number of pairwise nonisomorphic equivalence relations on a set of size n is $\Pi(n)$, the number of partitions of n.

It is known (e.g., [**2**, p.70]) that

$$\Pi(n) \sim \frac{1}{4n\sqrt{3}} e^{\left(\pi\sqrt{\frac{2n}{3}}\right)}.$$

So $\Pi(n)$ is neither at most polynomial nor at least exponential as a function of n.

Some of our constructions of varieties with many models will be based on the following fact.

PROPOSITION 2.9. *The number of pairwise nonisomorphic equivalence relations on a set of size 2^k is at least doubly exponential as a function of k.*

PROOF. The formula for $\Pi(n)$ shows that there exist positive constants b and c such that

$$\Pi(2^k) \sim \frac{b}{2^k} 2^{(c2^{k/2})}.$$

□

Part 1

Introducing Generative Complexity

In this Part we consider G-spectra for varieties of algebras. We study how algebraic properties influence the generative complexity. Some of our results give sharp upper and lower bounds on the G-spectrum so as to place a particular variety or class of varieties at a precise level in our exponential hierarchy.

If there are no restrictions on $\mathcal{V}$ then $G_{\mathcal{V}}(k)$ can grow arbitrarily fast, faster than any preassigned function. Even with some fairly restrictive conditions varieties with arbitrarily large G-spectra can be found. Thus, in Example 5.12 we describe a method of constructing for any $f : \omega \to \omega$ a locally finite discriminator variety $\mathcal{V}$ in a language consisting of just two operation symbols for which $G_{\mathcal{V}}(k) \geqslant f(k)$ for all k.

In any locally finite variety $\mathcal{V}$ the behavior of the G-spectrum is related to that of the free spectrum of $\mathcal{V}$. In our exponential hierarchy, the G-spectrum can be at most one level higher than the free spectrum. In Chapter 3 we observe that for varieties generated by a finite algebra, the free spectrum can be at most one exponential level higher than the G-spectrum. Some of our other results reveal how algebraic properties can force a closer relationship between the free and G-spectra of varieties. For example, in Theorem 6.9 we show that for a locally finite, congruence uniform variety if the free spectrum is at most m-fold exponential, then so is the G-spectrum.

For a variety $\mathcal{V}$ generated by a finite algebra the complexity of the G-spectrum can range from polynomial to at most triply exponential. In Chapters 3 – 6 we present some general results for finitely generated varieties that place their G-spectra more precisely in this exponential hierarchy and we give many specific examples of varieties that have generative complexity at each of these levels.

For some varieties such as those in Example 1.1 or any variety of Abelian groups, of monadic algebras, or of modular ortholattices we can explicitly determine the G-spectra. For others, such as the variety $\mathcal{S}$ of semilattices, an exact formula for the G-spectrum is out of reach but we do argue in Chapter 4 that $G_{\mathcal{S}}(k) = 2^{\binom{k}{\lfloor k/2 \rfloor}(1+o(1))}$. Likewise, for the variety of distributive lattices $\mathcal{D}$ we prove that $G_{\mathcal{D}}(k) = 2^{2^k(1+o(1))}$. For other less tractable varieties the best estimate we can provide on their generative complexity is to place them at a particular level in the exponential hierarchy.

We are mainly interested in the interplay between varietal properties and the bounding, from above or below, of generative complexity. Two easily stated results of this type are:

- Any variety in a finite language that has a finite residual bound (i.e., a finite bound on the size of subdirectly irreducible algebras) is of at most doubly exponential generative complexity. (Corollary 6.2)
- Every finitely generated congruence modular variety has generative complexity that is at most doubly exponential. (Theorem 6.14)

The generative complexity of a variety is an invariant of the variety and is thus independent of the language or the particular set of identities used to present the variety. That is, so-called equivalent varieties have the same G-spectra. The position of the generative complexity of a variety in the exponential hierarchy that we consider is fairly robust in that it is preserved by broader types of equivalence. Thus in Chapter 7 we show if $\mathcal{V}$ and $\mathcal{W}$ are two varieties that are equivalent as categories, then there exist positive constants b and c such that $G_{\mathcal{V}}(k) \leqslant G_{\mathcal{W}}(bk)$ and $G_{\mathcal{W}}(k) \leqslant G_{\mathcal{V}}(ck)$ for all k.

CHAPTER 3

Definitions and Examples

Let $\mathcal{V}$ be an arbitrary nontrivial locally finite variety. What bounds can be provided for the size of $\mathrm{G}_{\mathcal{V}}(k)$? The algebras $\mathbf{F}_{\mathcal{V}}(1), \mathbf{F}_{\mathcal{V}}(2), \ldots, \mathbf{F}_{\mathcal{V}}(k)$ show that $k \leqslant \mathrm{G}_{\mathcal{V}}(k)$. As observed in Example 1.1 this lower bound is achieved when $\mathcal{V}$ is the variety of sets.

We may give an upper bound based on the free spectrum of $\mathcal{V}$. Every k-generated member of $\mathcal{V}$ is a homomorphic image of $\mathbf{F}_{\mathcal{V}}(k)$ and so we have $\mathrm{G}_{\mathcal{V}}(k) \leqslant |\mathrm{Con}(\mathbf{F}_{\mathcal{V}}(k))|$. Thus, for a finite algebra $\mathbf{A}$, the inequality $|\mathsf{Con}\,\mathbf{A}| \leqslant \mathrm{Bell}(|A|)$ always holds. This gives the following.

PROPOSITION 3.1. *If the free spectrum of a locally finite variety $\mathcal{V}$ is at most m-fold exponential, then $\mathrm{G}_{\mathcal{V}}(k)$ is at most $(m+1)$-fold exponential.* □

Thus any variety $\mathcal{V}$ with polynomial free spectrum has G-spectrum that is at most exponential. It is known that a variety $\mathcal{V}$ has free spectrum that is at most polynomial if and only if $\mathcal{V}$ has free spectrum that is a polynomial if and only if $\mathcal{V}$ is locally finite and there is a finite bound on the essential arities of the term operations of algebras in $\mathcal{V}$. For example, any variety generated by a finite, strongly Abelian algebra has a bound on the essential arities of its terms and thus has generative complexity that is at most exponential. Finite algebras in which there is a finite bound on the essential arities of the term operations of the algebra are investigated in [**45**].

We may also give a lower bound for $\mathrm{G}_{\mathcal{V}}(k)$ in terms of the free spectrum of $\mathcal{V}$ in the case that $\mathcal{V}$ is finitely generated. Clearly, $\mathrm{G}_{\mathcal{V}}(k)$ is bounded below by the number of elements in a maximal chain in the congruence lattice of $\mathbf{F}_{\mathcal{V}}(k)$.

PROPOSITION 3.2. *If the free spectrum of a finitely generated variety $\mathcal{V}$ is at least m-fold exponential for an $m \geqslant 1$, then $\mathrm{G}_{\mathcal{V}}(k)$ is at least $(m-1)$-fold exponential.*

PROOF. Let $\mathbf{A}$ generate $\mathcal{V}$ with $|A| = a$. We may consider $\mathbf{F} = \mathbf{F}_{\mathcal{V}}(k)$ as a subalgebra of $\mathbf{A}^{a^k}$. Consider the congruences $\pi_0 \geqslant \pi_1 \geqslant \ldots \geqslant \pi_{a^k}$ on $\mathbf{F}_{\mathcal{V}}(k)$ where $(s,t) \in \pi_i$ if and only if $s_j = t_j$ for all $j \leqslant i$. Thus, $\pi_0 = 1_F$ and $\pi_{a^k} = 0_F$. Moreover, $|\mathbf{F}/\pi_{i+1}| \leqslant |A||\mathbf{F}/\pi_i|$. From $|\mathbf{F}/\pi_0| = 1$ and $|\mathbf{F}/\pi_{a^k}| = |F|$, it follows that the number of strict inequalities $\pi_i < \pi_{i+1}$ is at least $(m-1)$-fold exponential as a function of k. Thus, the cardinality of a maximal chain of congruences of $\mathbf{F}_{\mathcal{V}}(k)$ is at least $(m-1)$-fold exponential. □

Note that if $\mathcal{V}$ is finitely generated, then in Proposition 3.2 the only possible values for m are 1 or 2. We do not know if the hypothesis that $\mathcal{V}$ be finitely generated is necessary in Proposition 3.2. However, Example 5.16 shows that for any function $h : \omega \to \omega$ there exists a variety $\mathcal{V}$ for which $|\mathrm{F}_{\mathcal{V}}(k)| \geqslant h(\mathrm{G}_{\mathcal{V}}(k))$ for all finite k.

Propositions 3.1 and 3.2 show that if a finitely generated variety $\mathcal{V}$ has a free spectrum of m-fold exponential complexity for $m \leqslant 2$, then $\mathrm{G}_{\mathcal{V}}(k)$ is of at least $(m-1)$-fold exponential complexity and of at most $(m+1)$-fold exponential complexity . Examples exist to show that for nonnegative integers m and r with $m \leqslant 2$ and $m-1 \leqslant r \leqslant m+1$ there exist finitely generated varieties with free spectra of m-fold exponential complexity and G-spectra of r-fold exponential complexity. See Table 5 in Chapter 20.

PROPOSITION 3.3. *If $\mathcal{V}$ is a locally finite variety with* **3** *or* **4** *in* $\mathrm{typ}\{\mathcal{V}\}$, *then* $\mathrm{G}_{\mathcal{V}}(k)$ *is at least exponential.*

PROOF. From Theorem 12.3 of [**32**] we know that there is a finitely generated subvariety of $\mathcal{V}$ whose free spectrum is at least doubly exponential. So Proposition 3.2 applies. □

This proposition also holds if type $\mathbf{5} \in \mathrm{typ}\{\mathcal{V}\}$, but in Theorem 8.1 we obtain the stronger result that a locally finite variety with **4** or **5** in its type set has many models. The lower bound given in Proposition 3.3 is obtained with, say, the variety of Boolean algebras.

We often look for collections of pairwise nonisomorphic k-generated algebras in a variety $\mathcal{V}$ by selecting certain congruences on $\mathbf{F}_{\mathcal{V}}(k)$. Our next definition and the subsequent material provide a method for finding such collections.

DEFINITION 3.4. *An algebra* $\mathbf{A}$ *is* uniquely generated *if there exists a set* G *that generates* $\mathbf{A}$ *and every generating set for* $\mathbf{A}$ *contains* G.

For example, a free semilattice is uniquely generated by its free generators as is any free lattice in a variety of lattices.

For an algebra $\mathbf{A}$ generated by a set G we are especially interested in congruences on $\mathbf{A}$ for which each element of G is in a singleton congruence class. To this end, for an algebra $\mathbf{A}$ and a set $X \subseteq A$, let

$$\mathrm{C_s}(\mathbf{A}, X) = \{\gamma \in \mathsf{Con}\,\mathbf{A} : x/\gamma = \{x\} \quad \text{for all } x \in X\}.$$

It is not hard to see that if $\alpha, \beta \in \mathsf{Con}\,\mathbf{A}$ and $\alpha \leqslant \beta \in \mathrm{C_s}(\mathbf{A}, X)$, then $\alpha \in \mathrm{C_s}(\mathbf{A}, X)$. The join of two congruences in $\mathrm{C_s}(\mathbf{A}, X)$ is also in $\mathrm{C_s}(\mathbf{A}, X)$. Thus, $\mathrm{C_s}(\mathbf{A}, X)$ is a lattice ideal of $\mathsf{Con}\,\mathbf{A}$. If $\mathbf{A}$ is nontrivial and X is nonempty, then $0_A \in \mathrm{C_s}(\mathbf{A}, X)$ but $1_A \notin \mathrm{C_s}(\mathbf{A}, X)$.

LEMMA 3.5. *Let* $\mathbf{A}$ *be uniquely generated by the set* G. *If* $\gamma \in \mathrm{C_s}(\mathbf{A}, G)$, *then* $\mathbf{A}/\gamma$ *is uniquely generated by* $\{g/\gamma : g \in G\}$.

PROOF. For $S \subseteq A$ let S/γ denote $\{a/\gamma : a \in S\}$. Clearly the set G/γ generates $\mathbf{A}/\gamma$. Let $H \subseteq A$ be such that H/γ generates $\mathbf{A}/\gamma$. For each $g \in G$ there is a term $\mathbf{t}_g$ and $h_1, \ldots, h_n \in H$ for which $\{g\} = g/\gamma = \mathbf{t}_g(h_1/\gamma, \ldots, h_n/\gamma)$. In $\mathbf{A}$ we have $\mathbf{t}_g(h_1, \ldots, h_n) = g$. So H generates $\mathbf{A}$ and thus $G \subseteq H$. Therefore $G/\gamma \subseteq H/\gamma$ as desired. □

LEMMA 3.6. *Let* $\mathbf{A}$ *be uniquely generated by the set* $G = \{g_1, \ldots, g_k\}$. *Suppose* $\gamma, \gamma' \in \mathrm{C_s}(\mathbf{A}, G)$ *and there exists an isomorphism* f *from* $\mathbf{A}/\gamma$ *onto* $\mathbf{A}/\gamma'$. *Then*

(1) *there is a permutation* σ *of* $\{1, \ldots, k\}$ *for which* $f(g_i/\gamma) = g_{\sigma(i)}/\gamma'$;
(2) *the congruence* γ' *is uniquely determined by* γ *and* σ.

PROOF. Both $\mathbf{A}/\gamma$ and $\mathbf{A}/\gamma'$ are uniquely generated by virtue of Lemma 3.5. Since f is an isomorphism it must be a bijection of G/γ onto G/γ' and thus the first claim of the Lemma holds. Let $a, b \in A$ be arbitrary and suppose $\mathbf{s}$ and $\mathbf{t}$ are terms for which $a = \mathbf{s}(g_{\sigma(1)}, \ldots, g_{\sigma(k)})$ and $b = \mathbf{t}(g_{\sigma(1)}, \ldots, g_{\sigma(k)})$. Then $(a, b) \in \gamma'$ if and only if

$$\mathbf{s}(g_{\sigma(1)}/\gamma', \ldots, g_{\sigma(k)}/\gamma') = \mathbf{t}(g_{\sigma(1)}/\gamma', \ldots, g_{\sigma(k)}/\gamma')$$

if and only if

$$\mathbf{s}(g_1/\gamma, \ldots, g_k/\gamma) = \mathbf{t}(g_1/\gamma, \ldots, g_k/\gamma)$$

if and only if

$$(\mathbf{s}(g_1, \ldots, g_k), \mathbf{t}(g_1, \ldots, g_k)) \in \gamma.$$

Thus, γ' is determined by γ and σ. □

COROLLARY 3.7. *Let* $\mathbf{A}$ *be uniquely generated by a set* G *of* k *elements. Then the number of pairwise nonisomorphic, homomorphic images of the algebra* $\mathbf{A}$ *is at least* $|\mathrm{C_s}(\mathbf{A}, G)|/k!$.

PROOF. From Lemma 3.6, for $\gamma \in \mathrm{C_s}(\mathbf{A}, G)$ the number of $\gamma' \in \mathrm{C_s}(\mathbf{A}, G)$ for which $\mathbf{A}/\gamma \simeq \mathbf{A}/\gamma'$ is at most $k!$. □

COROLLARY 3.8. *If* $\mathcal{V}$ *is a variety for which* $\mathbf{F} = \mathbf{F}_{\mathcal{V}}(k)$ *is uniquely generated by* X, *then*

$$\frac{|\mathrm{C_s}(\mathbf{F}, X)|}{k!} \leqslant \mathrm{G}_{\mathcal{V}}(k) \leqslant |\mathsf{Con}\, \mathbf{F}_{\mathcal{V}}(k)|.$$

As an example of the use of these definitions and techniques we consider the generative complexity of multiunary algebras

EXAMPLE 3.9. Let $\mathcal{V}$ an arbitrary, locally finite variety of multiunary algebras. We argue that $\mathrm{G}_{\mathcal{V}}(k)$ is at most polynomial if and only if every unary term operation of the free algebra $\mathbf{F}_{\mathcal{V}}(1)$ is a constant or a permutation. Moreover, if $\mathrm{G}_{\mathcal{V}}(k)$ is not at most polynomial, then $\mathrm{G}_{\mathcal{V}}(k)$ is bounded below by $\Pi(k)$, the number of partitions of the integer k.

First suppose every unary operation on $\mathbf{F}_{\mathcal{V}}(1)$ is a permutation or constant. We list the universe of $\mathbf{F}_{\mathcal{V}}(\{x_1\})$ as

$$\{\mathbf{p}_0(x_1), \mathbf{p}_1(x_1), \ldots, \mathbf{p}_r(x_1), \mathbf{c}_1(x_1), \ldots, \mathbf{c}_s(x_1)\}$$

where $\mathbf{p}_0(x_1) = x_1$, the $\mathbf{p}_i$ are permutations, and the $\mathbf{c}_j$ are constant operations. Note that $s = 0$ is possible. Let $x_1, \ldots, x_k$ be the free generators of $\mathbf{F}_{\mathcal{V}}(k)$. We view $\mathbf{F}_{\mathcal{V}}(\{x_i\})$ as a subuniverse of $\mathbf{F}_{\mathcal{V}}(k)$. The set of constants in $\mathbf{F}_{\mathcal{V}}(k)$ is denoted by $\mathrm{F}_{\mathcal{V}}(0)$.

Let $a \in \mathrm{F}_{\mathcal{V}}(\{x_i\}) - \mathrm{F}_{\mathcal{V}}(0)$ and $b \in \mathrm{F}_{\mathcal{V}}(\{x_j\}) - \mathrm{F}_{\mathcal{V}}(0)$ with $i \neq j$. Let γ be the congruence relation on $\mathbf{F}_{\mathcal{V}}(k)$ generated by the pair (a, b). Consider $\Gamma = \{(\mathbf{q}(a), \mathbf{q}(b)) : \mathbf{q} \in \{\mathbf{p}_0, \mathbf{p}_1, \ldots, \mathbf{p}_r, \mathbf{c}_1, \ldots, \mathbf{c}_s\}\}$. Since the $\mathbf{p}_i$ are permutations and the $\mathbf{c}_i$ are constants, Γ is a bijection from $\mathbf{F}_{\mathcal{V}}(\{x_i\})$ to $\mathbf{F}_{\mathcal{V}}(\{x_j\})$. The symmetric reflexive closure of Γ is a congruence relation, which must be γ. Thus $\mathbf{F}_{\mathcal{V}}(k)/\gamma$ is isomorphic to $\mathbf{F}_{\mathcal{V}}(k-1)$.

Now consider any congruence relation τ of $\mathbf{F}_{\mathcal{V}}(k)$. Suppose there exist $(a, b) \in \tau$ as in the previous paragraph and $\gamma = \mathrm{Cg}^{\mathbf{F}_{\mathcal{V}}(k)}(a, b)$. Then

$$\mathbf{F}_{\mathcal{V}}(k)/\tau \simeq (\mathbf{F}_{\mathcal{V}}(k)/\gamma)/(\tau/\gamma).$$

So we may replace τ by a congruence τ' on $\mathbf{F}_{\mathcal{V}}(k-1)$

$$\mathbf{F}_{\mathcal{V}}(k)/\tau \simeq \mathbf{F}_{\mathcal{V}}(k-1)/\tau'.$$

Continuing in this way we need only consider $\mathbf{F}_{\mathcal{V}}(m)$ for $m \leqslant k$ and congruences τ on $\mathbf{F}_{\mathcal{V}}(m)$ such that $(a,b) \in \tau$ implies that both a and b are in the same subalgebra of $\mathbf{F}_{\mathcal{V}}(m)$ generated by some generator x_i. Therefore $\mathrm{G}_{\mathcal{V}}(k)$ is bounded above by the number of such congruences. For such a congruence τ on $\mathbf{F}_{\mathcal{V}}(m)$ we have that

$$\tau = \bigcup_{1 \leqslant i \leqslant m} \tau|_{\mathbf{F}_{\mathcal{V}}(\{x_i\})}$$

and that $\tau|_{\mathbf{F}_{\mathcal{V}}(\{x_i\})}$ is a congruence relation on $\mathbf{F}_{\mathcal{V}}(\{x_i\})$. Let $\{\theta_1, \theta_2, \ldots, \theta_t\}$ be a list of all the congruence relations on $\mathbf{F}_{\mathcal{V}}(1)$. The number of nonisomorphic algebras that arise as quotients of $\mathbf{F}_{\mathcal{V}}(m)$ by the congruences τ is bounded above by the number of nonnegative integral solutions of $z_1 + z_2 + \cdots + z_t = m$, where z_j is number of x_i for which $\tau|_{\mathbf{F}_{\mathcal{V}}(\{x_i\})}$ is θ_j. The number of such solutions is $\binom{m+t-1}{t-1}$. So

$$\mathrm{G}_{\mathcal{V}}(k) \leqslant \sum_{m=0}^{k} \binom{m+t-1}{t-1},$$

which is a polynomial in k of degree $t-1$.

We now prove the converse and the "Moreover" part of the Example. Suppose $\mathbf{F}_{\mathcal{V}}(\{x\})$ contains an element $\mathbf{q}_1$ that is neither constant nor a permutation. The subuniverse $Q = \{\mathbf{q}_1(x), \mathbf{q}_2(x), \ldots, \mathbf{q}_s(x)\}$ generated by this element is proper and is not $\mathrm{F}_{\mathcal{V}}(0)$. Note that no $\mathbf{q}_i$ is a permutation. Let $X = \{x_1, \ldots, x_k\}$ be the free generators of $\mathbf{F} := \mathbf{F}_{\mathcal{V}}(k)$. Let R be an arbitrary equivalence relation on $\{x_1, \ldots, x_k\}$. We associate with each such R a quotient algebra $\mathbf{A}_R$ of $\mathbf{F}_{\mathcal{V}}(k)$ so that nonisomorphic equivalence relations give rise to nonisomorphic members of $\mathcal{V}$. Since the number of nonisomorphic equivalence relations on a k-element set is $\Pi(k)$, this construction will justify our claim.

Let $B_1, B_2, \ldots, B_r$ be the equivalence classes of R. For each free generator x_u let Q_u denote the set $\{\mathbf{q}_1(x_u), \mathbf{q}_2(x_u), \ldots, \mathbf{q}_s(x_u)\}$. For each B_j define the congruence relation

$$\gamma_j := \mathrm{Cg}^{\mathbf{F}}(\{(\mathbf{q}_i(x_u), \mathbf{q}_i(x_w)) : x_u, x_w \in B_j, 1 \leqslant i \leqslant s\}).$$

The nontrivial congruence classes of γ_j are contained in $\bigcup_{x_u \in B_j} Q_u$ and are precisely sets of the form $\{\mathbf{q}_i(x_u) : x_u \in B_j\}$ for each equivalence class B_j and nonconstant $\mathbf{q}_i$.

Define

$$\gamma = \bigvee_{1 \leqslant j \leqslant r} \gamma_j \qquad \text{and} \qquad \mathbf{A}_R = \mathbf{F}_{\mathcal{V}}(k)/\gamma.$$

Note that $x_u/\gamma = \{x_u\}$, that is, $\gamma \in C_s(\mathbf{F}, X)$, but $\mathbf{F}$ need not be uniquely generated. Each nontrivial congruence class of γ is a congruence class of a unique γ_j. The algebra $\mathbf{A} := \mathbf{A}_R$ is k-generated but not $(k-1)$-generated. Choose any generating set for $\mathbf{A}$, say, $\{g_1/\gamma, \ldots, g_k/\gamma\}$. We wish to recover R from this set. In general, we cannot assume $\{g_1, \ldots, g_k\} = \{x_1, \ldots, x_k\}$. However, since $x_u/\gamma = \{x_u\}$ each x_u must be in the subalgebra of $\mathbf{F}$ generated by a g_j. This implies that $g_i \notin Q_w$ for any x_w and therefore $g_i/\gamma = \{g_i\}$. Since $\mathbf{F}$ is k-generated and not $(k-1)$-generated we see that for every g_j there is a unique x_u such that $x_u \in \mathrm{Sg}^{\mathbf{F}}(\{g_j\})$. We renumber the g_i so that $x_i \in \mathrm{Sg}^{\mathbf{F}}(\{g_i\})$.

Therefore, if $(x_i, x_j) \in R$, then

$$|\mathrm{Sg}^{\mathbf{A}}(\{g_i/\gamma\}) \cap \mathrm{Sg}^{\mathbf{A}}(\{g_j/\gamma\})| \geqslant |Q|,$$

while if $(x_i, x_j) \notin R$, then the cardinality of this intersection is $|\mathrm{F}_{\mathcal{V}}(0)|$. Since $|Q| > |\mathrm{F}_{\mathcal{V}}(0)|$ we can distinguish these two cases. Thus, using only the algebraic structure of $\mathbf{A}_R$, a unique equivalence relation on the generating set $\{g_1/\gamma, \ldots, g_k/\gamma\}$ can be found that is isomorphic to the original equivalence relation R on X.

EXAMPLE 3.10. We present a finitely generated variety $\mathcal{V}$ having G-spectrum that is neither at most polynomial nor at least exponential. Let $\mathcal{V}$ be the variety generated by the unary algebra $\mathbf{A} = \langle\{0, 1, 2\}, \mathbf{f}\rangle$ in which $\mathbf{f}(2) = 1$ and $\mathbf{f}(1) = \mathbf{f}(0) = 0$. We show that

$$\mathrm{G}_{\mathcal{V}}(k) = \sum_{0 \leqslant r \leqslant k} \Pi(r)(k + 1 - r),$$

and hence

$$\Pi(k) \leqslant \mathrm{G}_{\mathcal{V}}(k) \leqslant (k+1)^2 \Pi(k).$$

As we observed in Chapter 2 the function $\Pi(k)$ is neither at most polynomial nor at least exponential.

The variety $\mathcal{V}$ has a constant 0 and satisfies the identities $\mathbf{f}^2(y) \approx \mathbf{f}(0) \approx 0$. The free algebra for $\mathcal{V}$ on k free generators has $2k+1$ elements: the free generators $x_1, \ldots, x_k$, their images $\mathbf{f}(x_1), \ldots, \mathbf{f}(x_k)$, and the constant 0. It is easily checked that every k-generated algebra in $\mathcal{V}$ is of the following form. There exist nonnegative integers r and s with $r + s \leqslant k$ and generators $g_1, \ldots, g_r$ and $g_{r+1}, \ldots, g_{r+s}$; these generators are not in the range of $\mathbf{f}$; for each $i \leqslant r$ we have $\mathbf{f}(g_i) \neq 0$ but $\mathbf{f}(g_i) = 0$ for $i > r$. An equivalence relation can be defined on $g_1, \ldots, g_r$ by placing two generators in the same equivalence class if and only if they have the same image under the fundamental operation $\mathbf{f}$. Clearly, if two algebras have the same r and s and have isomorphic equivalence relations on the elements not mapped to 0 by $\mathbf{f}$, then they are isomorphic as algebras. For a given choice of r there are $k + 1 - r$ choices for s. Since the number of nonisomorphic equivalence relations on a set of r elements is $\Pi(r)$, we obtain the desired expression for $\mathrm{G}_{\mathcal{V}}(k)$.

CHAPTER 4

Semilattices and Lattices

In this Chapter we consider the generative complexity of the varieties of semilattices and distributive lattices. We show that they both have at least doubly exponentially many models and we are able to provide sharp estimates on their G-spectra. These results are crucial for our later analysis in Chapter 8 that shows if a variety has **4** or **5** in its set of types, then it has many models.

THEOREM 4.1. *The variety $\mathcal{S}$ of semilattices has many models. In fact,*

$$\mathrm{G}_{\mathcal{S}}(k) = 2^{\binom{k}{\lfloor k/2 \rfloor}(1+o(1))}.$$

PROOF. If X is a set, then let $\mathcal{P}^+(X)$ denote the set of all nonempty subsets of X. For a positive integer k a concrete representation of the free algebra $\mathbf{F} = \mathbf{F}_{\mathcal{S}}(k)$ is the algebra $\langle \mathcal{P}^+(X), \cup \rangle$. Here the free generators are the k singletons $\{x\}$ for $x \in X$. The algebra $\mathbf{F}$ is uniquely generated by this set of singleton sets. Thus by Corollary 3.8 we have

$$\frac{|\mathrm{C_s}(\mathbf{F}, X)|}{k!} \leqslant \mathrm{G}_{\mathcal{S}}(k) \leqslant |\mathsf{Con}\,\mathbf{F}|.$$

Our proof proceeds by giving sharp estimates for the cardinalities of $\mathrm{C_s}(\mathbf{F}, X)$ and $\mathsf{Con}\,\mathbf{F}$.

Let U be any upwardly closed subset of $\mathcal{P}^+(X)$. Consider the equivalence relation θ_U on $\mathcal{P}^+(X)$ whose only nonsingleton equivalence class is U. Clearly, θ_U is a congruence relation for $\mathbf{F}$. If U contains no singleton $\{x\}$, then $\theta_U \in \mathrm{C_s}(\mathbf{F}, X)$. If we choose $U \subseteq \mathcal{P}^+(X)$ to be all sets above a given collection of subsets of X, each of size $\lfloor k/2 \rfloor$, then there are $2^{\binom{k}{\lfloor k/2 \rfloor}}$ ways to select such a U. Hence

$$2^{\binom{k}{\lfloor k/2 \rfloor}} \leqslant |\mathrm{C_s}(\mathbf{F}, X)|,$$

and from Corollary 3.8 we get

$$\frac{2^{\binom{k}{\lfloor k/2 \rfloor}}}{k!} \leqslant \mathrm{G}_{\mathcal{S}}(k).$$

If $S \subseteq X$ and $\gamma \in \mathsf{Con}\,\mathbf{F}$, then the class S/γ is closed under union and thus contains a largest member, say $\overline{S}$. Then it is easy to check that the map $S \mapsto \overline{S}$ satisfies the three axioms for a closure operator on X:

(1) $S \subseteq \overline{S}$,
(2) $\overline{\overline{S}} = \overline{S}$,
(3) $S \subseteq T \subseteq X \Rightarrow \overline{S} \subseteq \overline{T}$.

If we define $\overline{\emptyset} = \emptyset$, then the congruence γ induces a closure operator $S \mapsto \overline{S}$ on X, and distinct congruence relations give rise to different closure operators under this mapping. Conversely, if $S \mapsto \overline{S}$ is a closure operator on X for which $\overline{\emptyset} = \emptyset$, then it can be checked that the relation γ on $\mathcal{P}^+(X)$ given by $(S,T) \in \gamma$ if and only if $\overline{S} = \overline{T}$ is a congruence relation for $\mathbf{F}$. Thus, there is a bijection between the set of closure operators on X for which $\emptyset$ is closed and $\mathsf{Con}\,\mathbf{F}$.

Let $c(k)$ denote the number of closure operators on a set of size k and let $c_\emptyset(k)$ be the number for which $\overline{\emptyset} = \emptyset$. These two functions are related by

$$c(k) = \sum_{i=0}^{k} \binom{k}{k-i} c_\emptyset(i),$$

and thus

$$c_\emptyset(k) \leqslant c(k) \leqslant 2^k c_\emptyset(k).$$

The function $c(k)$ has received attention in recent years since it is the number of functional dependencies involving k attributes and the number of intersection families on a set of size k. The articles [**10**] and [**11**] discuss $c(k)$ and $c_\emptyset(k)$ and give estimates for their values. Alexseev [**1**] has shown that both $c(k)$ and $c_\emptyset(k)$ are of size

$$2^{\binom{k}{\lfloor k/2 \rfloor}(1+o(1))}.$$

Therefore

$$|\mathsf{Con}\,\mathbf{F}| = 2^{\binom{k}{\lfloor k/2 \rfloor}(1+o(1))}.$$

Since our lower bound for $\mathrm{G}_{\mathcal{S}}(k)$ is also same as that in Alexseev's estimate we conclude that

$$\mathrm{G}_{\mathcal{S}}(k) = 2^{\binom{k}{\lfloor k/2 \rfloor}(1+o(1))}.$$

□

A similar analysis, resulting in the same estimate, applies to semilattices with 0 or 1.

We next consider the G-spectra of varieties of lattices.

THEOREM 4.2. *The variety $\mathcal{D}$ of distributive lattices has many models. In fact,*

$$\mathrm{G}_{\mathcal{D}}(k) = 2^{2^k(1+o(1))}.$$

PROOF. If $\mathbf{L}$ is a finite member of the variety $\mathcal{D}$, then it is known (e.g., [**55**, Theorem 2.75]) that $\mathsf{Con}\,\mathbf{L}$ is a Boolean lattice whose atoms are congruence relations of the form $\mathrm{Cg}^{\mathbf{L}}(a,b)$ for a covering pair $a \prec b$ of $\mathbf{L}$. Moreover, every covering pair of $\mathbf{L}$ determines an atom in this way. The number of atoms in $\mathsf{Con}\,\mathbf{L}$ is equal to the length of the lattice $\mathbf{L}$.

Let $\mathbf{F} = \mathbf{F}_{\mathcal{D}}(k)$. The length of $\mathbf{F}$ is 2^k. Thus $|\mathsf{Con}\,\mathbf{F}| = 2^{2^k}$. For any variety of lattices, the generators of a free lattice for the variety are always meet and join irreducible. Therefore $\mathbf{F}$ is uniquely generated by the set of its free generators. Let $X = \{x_1, \ldots, x_k\}$ generate $\mathbf{F}$ and for each x_i let x_i^+ and x_i^- denote the unique elements covering x_i and covered by x_i, respectively. If $\gamma \in \mathsf{Con}\,\mathbf{F}$ is such that $x_i/\gamma \neq \{x_i\}$, then either (x_i^-, x_i) or (x_i^+, x_i) are in γ. So there are at most $2k$ atoms of $\mathsf{Con}\,\mathbf{F}$ that are not in $\mathrm{C_s}(\mathbf{F}, X)$. The largest member of the ideal $\mathrm{C_s}(\mathbf{F}, X)$ is therefore the join of at least $2^k - 2k$ atoms, and so $2^{2^k-2k} \leqslant |\mathrm{C_s}(\mathbf{F}, X)|$. From Corollary 3.8 we get

$$\frac{2^{2^k-2k}}{k!} \leqslant \mathrm{G}_{\mathcal{D}}(k) \leqslant 2^{2^k}$$

and so,

$$G_{\mathcal{D}}(k) = 2^{2^k(1+o(1))},$$

as claimed. □

If we consider the variety of distributive lattices instead of bounded distributive lattices, then the same argument works, except the length of $\mathbf{F}$ is $2^k - 2$.

COROLLARY 4.3. *Every nontrivial variety of lattices has many models.*

PROOF. Every nontrivial variety of lattices contains distributive lattices as a subvariety. □

We note that although the free algebras on k free generators for the variety $\mathcal{B}$ of Boolean algebras and the variety $\mathcal{D}$ of bounded distributive lattices have isomorphic congruence lattices, $\mathcal{B}$ has few models while $\mathcal{D}$ has many models. Likewise, although the free Boolean group on k free generators and the free semilattice with 1 on k free generators each have 2^k elements, the variety of Boolean groups has very few models while semilattices with 1 have many models. Other examples comparing free spectra and G-spectra are given in Table 5 in Chapter 20.

The varieties of semilattices, distributive lattices, Boolean algebras, and Boolean groups are each generated by 2-element algebras. An analysis of the generative complexity of every variety generated by a 2-element algebra along with a discussion of the free spectra and fine spectra of these varieties is given in [**6**].

CHAPTER 5

Varieties with a Large Number of Models

We provide several examples of locally finite varieties having large generative complexity. In addition to being of interest in their own right, these examples serve to illustrate the difficulties that might arise in attempting to extend the classification theorems in Part II.

We first focus on finitely generated varieties and observe that their G-spectra are at most triply exponential. We provide some examples to show this upper bound is obtained, even when restricting to strongly solvable varieties. These examples run counter to the expectation that a variety of type **1** would have small generative complexity and that the free algebras in such varieties would have a great deal of homogeneity. However, if we restrict to finitely generated strongly Abelian varieties, then the generative complexity is at most exponential, and we provide examples for which this upper bound is obtained. We observe that every locally finite Abelian variety has a G-spectrum that is at most doubly exponential and we provide an example of a finitely generated Abelian variety of doubly exponential generative complexity. We next exhibit examples of locally finite varieties with arbitrarily large G-spectra. We show how such varieties can be found even when restricting to discriminator varieties or locally strongly solvable varieties.

If a variety $\mathcal{V}$ is finitely generated, then $|\mathrm{F}_{\mathcal{V}}(k)|$ is an at most doubly exponential function of k. Indeed, if $\mathcal{V}$ is generated by an algebra $\mathbf{A}$, then $\mathbf{F}_{\mathcal{V}}(k)$ is a subalgebra of $\mathbf{A}^{|A|^k}$. From Proposition 3.1 we obtain the following observation.

PROPOSITION 5.1. *If $\mathcal{V}$ is a finitely generated variety, then $\mathrm{G}_{\mathcal{V}}(k)$ is at most triply exponential.* □

As we shall see, most of the familiar finitely generated varieties (including all congruence modular ones, see Theorem 6.14) are of at most doubly exponential generative complexity. We now present a construction that yields finitely generated varieties of finite similarity type having large G-spectra.

PROPOSITION 5.2. *Let $\mathbf{A} = \langle A, (\mathbf{f}_i)_{i\in I}\rangle$ be an arbitrary finite algebra. There exists an algebra $\mathbf{A}' = \langle A', \mathbf{h}, (\mathbf{f}'_i)_{i\in I}\rangle$ in which $\mathbf{h}$ is a unary operation, the operations $\mathbf{f}_i$ and $\mathbf{f}'_i$ are of the same arity, $|A'| = 2|A|$, $\mathrm{typ}\{\mathbf{A}'\} = \mathrm{typ}\{\mathbf{A}\} \cup \{\mathbf{1}\}$, and if $\mathcal{V}$ and $\mathcal{V}'$ are the varieties generated by $\mathbf{A}$ and $\mathbf{A}'$, then*

$$\mathrm{G}_{\mathcal{V}'}(k) \geqslant \frac{\mathrm{Bell}(|\mathrm{F}_{\mathcal{V}}(k)|)}{k!}.$$

PROOF. Let $A = \{1, 2, \ldots, m\}$ and form $A' = \{1, 2, \ldots, 2m\}$. For each fundamental operation $\mathbf{f}_i$ on $\mathbf{A}$ of arity n_i define $\mathbf{f}'_i$ on A' by

$$\mathbf{f}'_i(\bar{b}) = \begin{cases} \mathbf{f}_i(\bar{b}), & \text{if } \bar{b} \in A^{n_i}, \\ 1, & \text{otherwise.} \end{cases}$$

Let $\mathbf{h} : A' \to A'$ be given by

$$\mathbf{h}(b) = \begin{cases} b+m, & \text{if } 1 \leqslant b \leqslant m, \\ b, & \text{otherwise.} \end{cases}$$

We form the algebra $\mathbf{A}' = \langle A', \mathbf{h}, (\mathbf{f}'_i)_{i\in I}\rangle$. Note that no fundamental operation of $\mathbf{A}'$ has all of A' in its range. Let $\mathcal{V}$ and $\mathcal{V}'$ be the varieties generated by $\mathbf{A}$ and $\mathbf{A}'$, respectively. The free algebras $\mathbf{F}_{\mathcal{V}}(k)$ and $\mathbf{F}_{\mathcal{V}'}(k)$ are each generated by $X = \{x_1, \ldots, x_k\}$ and $\mathbf{F}_{\mathcal{V}'}(k)$ is uniquely generated by X since no term operation other than a variable has all of the set A' in its range.

For each term $\mathbf{t}$ in the language of $\mathbf{A}$ that is not a variable, let $\mathbf{t}'$ be the corresponding term for $\mathbf{A}'$ in which each instance of $\mathbf{f}_i$ is replaced by $\mathbf{f}'_i$ for all $i \in I$. If $\mathbf{t}$ is the variable x_j, then let $\mathbf{t}'$ denote x_j. The algebra $\mathbf{F}_{\mathcal{V}'}(k)$ includes all $\mathbf{t}'$ and all $\mathbf{h}(\mathbf{t}')$ for every $\mathbf{t} \in \mathrm{F}_{\mathcal{V}}(k)$. Moreover, all of these elements are pairwise distinct. So $|\mathrm{F}_{\mathcal{V}'}(k)| \geqslant 2|\mathrm{F}_{\mathcal{V}}(k)|$.

Let γ be any equivalence relation on $\mathbf{F}_{\mathcal{V}'}(k)$ for which the only nontrivial blocks are subsets of $\{\mathbf{h}(\mathbf{t}') : \mathbf{t} \in \mathrm{F}_{\mathcal{V}}(k)\}$. We claim that γ is a congruence relation on $\mathbf{F}_{\mathcal{V}'}(k)$. For if $\mathbf{f}'_i$ is an n_i-ary fundamental term of $\mathcal{V}'$ and $(u_j, w_j) \in \gamma$ for $1 \leqslant j \leqslant n_i$, with, say, $u_1 \neq w_1$, and $u_1 = \mathbf{h}(\mathbf{s}')$ and $w_1 = \mathbf{h}(\mathbf{t}')$ for some $\mathbf{s}, \mathbf{t} \in \mathrm{F}_{\mathcal{V}}(k)$, then $\mathbf{f}'_i(\mathbf{h}(\mathbf{s}'), \ldots) = \mathbf{f}'_i(\mathbf{h}(\mathbf{t}'), \ldots) = 1$, so $\mathbf{f}'_i$ preserves γ. The term $\mathbf{h}$ also preserves γ since if $(\mathbf{h}(\mathbf{s}'), \mathbf{h}(\mathbf{t}')) \in \gamma$, then $(\mathbf{h}(\mathbf{h}(\mathbf{s}')), \mathbf{h}(\mathbf{h}(\mathbf{t}'))) = (\mathbf{h}(\mathbf{s}'), \mathbf{h}(\mathbf{t}')) \in \gamma$.

Therefore, each such γ is in $C_s(\mathbf{F}_{\mathcal{V}'}(k), X)$. An application of Theorem 3.8 gives

$$\mathrm{G}_{\mathcal{V}'}(k) \geqslant \frac{\mathrm{Bell}(|\mathrm{F}_{\mathcal{V}}(k)|)}{k!}.$$

We next show that $\mathrm{typ}\{\mathbf{A}'\} = \mathrm{typ}\{\mathbf{A}\} \cup \{\mathbf{1}\}$. To see this, note that if $A = \{1, \ldots, m\}$ and $A' = A \cup \{m+1, \ldots, 2m\}$, then every polynomial for $\mathbf{A}'$ other than a projection operation has range contained in A or in $\{m+1, \ldots, 2m\}$. If $\mathbf{A}$ has no fundamental operations, then $\mathrm{typ}\{\mathbf{A}'\} = \mathrm{typ}\{\mathbf{A}\} = \{\mathbf{1}\}$ and we are done. Otherwise, the element 1 is a constant for $\mathbf{A}'$ since it is given by $\mathbf{f}_1(\mathbf{h}(x), \ldots)$. Also, $\mathbf{A}'$ satisfies the identities $\mathbf{h}(\mathbf{h}(x)) \approx \mathbf{h}(x)$ and $\mathbf{f}_i(\ldots, \mathbf{h}(x), \ldots) \approx 1$. Therefore every term $\mathbf{t}(x_1, \ldots, x_n)$ for $\mathbf{A}'$ is equivalent to the constant 1 or to a term in which either $\mathbf{h}$ does not appear at all or to a term of the form $\mathbf{h}(\mathbf{s}(x_1, \ldots, x_n))$ in which $\mathbf{h}$ does not appear in $\mathbf{s}$. Note that if $\mathbf{s}$ is a term in which $\mathbf{h}$ does not appear and at least one $\mathbf{f}_i$ does appear, then for $b_2, \ldots b_n \in A'$ and $c, d > m$ we have $\mathbf{s}(c, b_2, \ldots, b_n) = \mathbf{s}(d, b_2, \ldots, b_n)$ Thus, if $\mathbf{p}(x_1, \ldots x_k)$ is a polynomial for $\mathbf{A}'$ in which at least one $\mathbf{f}_i$ appears, then $\mathbf{p}(\overline{c}) = \mathbf{p}(\overline{d})$ for all $\overline{c}, \overline{d} \in \{m+1, \ldots, 2m\}^k$. Also, every polynomial for $\mathbf{A}'$ can be written as a polynomial all of whose constants are elements of A. So if $\mathbf{p} \in \mathrm{Pol}\,\mathbf{A}'$ then $\mathbf{p}|_A$ is a polynomial for $\mathbf{A}$.

We show $\mathbf{1} \in \mathrm{typ}\{\mathbf{A}'\}$. If $|A| = 1$, then $\mathrm{typ}\{\mathbf{A}'\} = \mathbf{1}$, and we are done. If $|A| \geqslant 2$, then every polynomial operation of $\mathbf{A}'$ restricted to $\{m+1, m+2\}$ is constant or the identity operation. The only nontrivial congruence class of $\mathrm{Cg}^{\mathbf{A}'}(m+1, m+2)$ is $\{m+1, m+2\}$, and $\mathrm{Cg}^{\mathbf{A}'}(m+1, m+2)$ contains no 1-snags. So $\mathrm{typ}(0_{A'}, \mathrm{Cg}^{\mathbf{A}'}(m+1, m+2)) = \mathbf{1}$.

It remains to show that for $\mathbf{i} \neq \mathbf{1}$, we have $\mathbf{i} \in \mathrm{typ}\{\mathbf{A}\}$ if and only if $\mathbf{i} \in \mathrm{typ}\{\mathbf{A}'\}$. We argue using subtraces as presented in [**8**]. Let $\mathbf{B}$ be a finite algebra. For $c, d \in B$ let

$$S^{\mathbf{B}}(c, d) = \{(\mathbf{f}(c), \mathbf{f}(d)) : \mathbf{f} \in \mathrm{Pol}_1\mathbf{B} \;\&\; \forall \mathbf{g} \in \mathrm{Pol}_1\mathbf{B}\ \{\mathbf{gf}(c), \mathbf{gf}(d)\} \neq \{c, d\}\}.$$

Then $\{c,d\}$ is called a *subtrace* of $\mathbf{B}$ if (c,d) is not in the transitive closure of $S^{\mathbf{B}}(c,d)$. The *type of a subtrace* $\{c,d\}$, denoted $\text{typ}\{c,d\}$, is defined in terms of the 1-, 2- and 4-snag structure on (c,d) and (d,c) as follows:

- $\text{typ}\{c,d\} = \mathbf{5}$ if and only if exactly one of (c,d) and (d,c) is a 2-snag.
- $\text{typ}\{c,d\} = \mathbf{4}$ if and only if both (c,d) and (d,c) are 2-snags and neither is a 4-snag.
- $\text{typ}\{c,d\} = \mathbf{3}$ if and only if (c,d) is a 4-snag.
- $\text{typ}\{a,b\} = \mathbf{2}$ if and only if (c,d) or (d,c) is a 1-snag and neither is a 2-snag.
- $\text{typ}\{c,d\} = \mathbf{1}$ if and only if neither (c,d) nor (d,c) are 1-snags.

From this definition it follows that $\mathbf{i} \in \text{typ}\{\mathbf{B}\}$ if and only if there is subtrace $\{c,d\}$ of $\mathbf{B}$ with $\text{typ}\{c,d\} = \mathbf{i}$.

Now in the algebra $\mathbf{A}'$ any 1-, 2- or 4-snag (c,d) is witnessed by a binary polynomial $\mathbf{p}$ that depends on both variables and has c,d in its range. So $\mathbf{p}$ may be written with no instance of $\mathbf{h}$ and with the range of $\mathbf{p}$ contained in A. Hence (c,d) is a 1-, 2- or 4-snag on $\mathbf{A}$ if and only if (c,d) is a 1-, 2- or 4-snag on $\mathbf{A}'$. If $\mathbf{f}$ is a unary polynomial for $\mathbf{A}'$ that is not constant, not the identity, and has elements of $\mathbf{A}$ in its range, then $\mathbf{f}|_A \in \text{Pol}_1\mathbf{A}$. So if $a,b \in A$, then $\{a,b\}$ is a subtrace of $\mathbf{A}'$ if and only if $\{a,b\}$ is a subtrace of $\mathbf{A}$. The 1-, 2-, or 4-snag structure induced on $\{a,b\}$ in $\mathbf{A}$ is the same as that induced on $\{a,b\}$ in $\mathbf{A}'$. Therefore $\text{typ}\{a,b\}$ in $\mathbf{A}$ is the same as $\text{typ}\{a,b\}$ in $\mathbf{A}'$. □

EXAMPLE 5.3. Let $\mathbf{A}$ be a 2-element Boolean algebra and let $\mathbf{A}'$ be the 4-element algebra as in Proposition 5.2. Consider $\mathcal{B}$ the variety of Boolean algebras and $\mathcal{B}'$ the variety generated by $\mathbf{A}'$. From $|\mathrm{F}_{\mathcal{B}}(k)| = 2^{2^k}$ we see that

$$\mathrm{G}_{\mathcal{B}'}(k) \geqslant \frac{\text{Bell}(2^{2^k})}{k!},$$

and thus $\mathrm{G}_{\mathcal{B}'}(k)$ is at least triply exponential.

The next example shows that even in the strongly solvable setting there are finitely generated varieties of triply exponential generative complexity.

EXAMPLE 5.4. There exists a finitely generated variety $\mathcal{V}$ with $\text{typ}\{\mathcal{V}\} = \{\mathbf{1}\}$ for which $\mathrm{G}_{\mathcal{V}}(k)$ of triply exponential complexity.

To construct this variety we first note that if a variety $\mathcal{V}$ is generated by a nontrivial finite algebra all of whose prime quotients are of type $\mathbf{1}$, then $\text{typ}\{\mathcal{V}\} = \{\mathbf{1}\}$ (see [**32**, Corollary 7.6 and Theorem 7.11]). So if a finite algebra $\mathbf{A}$ has $\text{typ}\{\mathbf{A}\} = \{\mathbf{1}\}$ and the variety $\mathcal{V}$ generated by $\mathbf{A}$ has free spectrum that is at least doubly exponential, then the variety $\mathcal{V}'$ obtained from $\mathcal{V}$ by the construction in Proposition 5.2 would have triply exponential generative complexity. Such algebras $\mathbf{A}$ do exist. For example, Keith Kearnes has pointed out to us that the 4-element algebra $\mathbf{A}$ given as Example 1 on page 505 of [**42**] is an algebra with type set $\{\mathbf{1}\}$ and for which the twin monoid is a group that is not nilpotent. Therefore, by [**43**], the free spectrum of the variety generated by $\mathbf{A}$ is at least doubly exponential.

The following is a corollary of Proposition 3.1 and the fact that locally finite strongly Abelian varieties have polynomial free spectra [**50**].

COROLLARY 5.5. *The* G*-spectrum of a locally finite strongly Abelian variety is at most singly exponential.* □

Our next example shows that the upper bound of Corollary 5.5 can be achieved even among multiunary varieties. This example is obtained by applying Proposition 5.2 to Example 3.10

EXAMPLE 5.6. Consider the unary algebra $\mathbf{A} = \langle\{1,2,3\},\mathbf{f}\rangle$ with $\mathbf{f}(3) = 2$ and $\mathbf{f}(2) = \mathbf{f}(1) = 1$. Form $\mathbf{A}' = \langle\{1,2,\ldots,6\},\mathbf{h},\mathbf{f}'\rangle$ as in Proposition 5.2. Let $\mathcal{V}$ and $\mathcal{V}'$ be the varieties generated by $\mathbf{A}$ and $\mathbf{A}'$. We have $|\mathrm{F}_{\mathcal{V}}(k)| = 2k+1$ and so by the construction

$$\mathrm{G}_{\mathcal{V}'}(k) \geqslant \frac{\mathrm{Bell}(2k+1)}{k!}.$$

Now the number of partitions of a set of $2k+1$ elements into k blocks each of two elements and one block of with one element is

$$\frac{(2k+1)!}{k!2^k}.$$

Thus,

$$\frac{\mathrm{Bell}(2k+1)}{k!} \geqslant \frac{(2k+1)!}{k!k!2^k} \geqslant \frac{\binom{2k+1}{k}}{2^k},$$

and thus $\mathrm{G}_{\mathcal{V}'}(k)$ is at least exponential as a function of k.

The size of $\mathrm{F}_{\mathcal{V}'}(k)$ is linear in k as is always the case for free algebras in any finitely generated multiunary variety. Proposition 3.1 shows that $\mathrm{G}_{\mathcal{V}'}(k)$ is at most exponential. So the multiunary variety $\mathcal{V}'$ is of exponential generative complexity.

In Corollary 6.13 locally finite congruence modular Abelian varieties are shown to be of at most exponential generative complexity. The next example, which is based on an algebra in [**46**] and was suggested to us by Ross Willard, shows that this singly exponential upper bound does not hold in an arbitrary finitely generated Abelian variety. Note that by [**47**] every locally finite Abelian variety is finitely generated, and therefore by [**7**] has an at most singly exponential free spectrum. So the G-spectrum of such a variety is at most doubly exponential.

EXAMPLE 5.7. There exists a finitely generated Abelian variety of doubly exponential generative complexity.

The variety is generated by the algebra $\mathbf{A} = \langle\{0,1,2\};\oplus,\mathbf{f}\rangle$ where $\oplus$ is a binary operation for which $i \oplus j = 0$ except $0\oplus 1 = 0 \oplus 2 = 1 \oplus 0 = 2 \oplus 0 = 1$ and $\mathbf{f}$ is unary with $\mathbf{f}(0) = 0, \mathbf{f}(1) = \mathbf{f}(2) = 2$.

The operation $\oplus$ is associative and commutative. (In the language of semigroup theory $\langle A;\oplus\rangle$ is the *inflation* of the two-element Boolean group by the element 2.) The element 0 is definable as $x \oplus x$. The algebra $\mathbf{A}$ and hence the variety $\mathcal{V}$ generated by $\mathbf{A}$ satisfy the following identities:

$$x \oplus y \oplus 0 \approx x \oplus y \tag{1}$$
$$\mathbf{f}(\mathbf{f}(x)) \approx \mathbf{f}(x) \tag{2}$$
$$x \oplus \mathbf{f}(y) \approx x \oplus y. \tag{3}$$

Let $\mathbf{F}$ denote $\mathbf{F}_{\mathcal{V}}(X)$ for $X = \{x_1,\ldots,x_k\}$. Every nonzero element of $\mathbf{F}$ is equal to one of the following terms:

$0 \oplus x_i$,
$x_{i_1} \oplus \cdots \oplus x_{i_n}$ with $1 \leqslant i_1 < \cdots < i_n \leqslant k$,
$\mathbf{f}(x_{i_1} \oplus \cdots \oplus x_{i_n})$ with $1 \leqslant i_1 < \cdots < i_n \leqslant k$.

The free spectrum of $\mathcal{V}$ is $2^{k+1}+k-1$ since all of these terms are pairwise distinct. Therefore $\mathrm{G}_{\mathcal{V}}(k)$ is at most doubly exponential by Proposition 3.1. The algebra $\mathbf{F}$ is uniquely generated since no free generator is in the range of either $\oplus$ or $\mathbf{f}$. For each $\mathbf{t} \in F$ with $\mathbf{t} \neq 0$, let $\gamma_{\mathbf{t}} = \mathrm{Cg}^{\mathbf{F}}(\mathbf{t}, \mathbf{f}(\mathbf{t}))$. Note that $\{\mathbf{t}, \mathbf{f}(\mathbf{t})\}$ is the only nonsingleton block of $\gamma_{\mathbf{t}}$. For $S \subseteq F$ with $S \neq \emptyset$ and $0, x_1, \ldots, x_k \notin S$, form $\gamma_S = \bigvee_{\mathbf{t} \in S} \gamma_{\mathbf{t}}$. Each γ_S is in $\mathrm{C_s}(\mathbf{F}, X)$ and there are $2^{2^{k+1}-2}-1$ such congruence relations. Corollary 3.8 applies to show that $\mathrm{G}_{\mathcal{V}}(k)$ is at least doubly exponential. Thus $\mathcal{V}$ is of doubly exponential generative complexity.

We argue that $\mathcal{V}$ is Abelian. Suppose that $\mathbf{t}$ is a term and $\mathbf{t}(u, \overline{y}) = \mathbf{t}(u, \overline{z})$ in an algebra $\mathbf{C} \in \mathcal{V}$. We wish to show $\mathbf{t}(v, \overline{y}) = \mathbf{t}(v, \overline{z})$. If $\mathbf{t}$ is unary, then there is nothing to do. So we may assume that $\mathbf{t}(u, \overline{y})$ is $u \oplus y_1 \oplus \cdots \oplus y_m$ or $\mathbf{f}(u \oplus y_1 \oplus \cdots \oplus y_m)$ for some $m \geqslant 1$. In the first case, if we add u to $u \oplus y_1 \oplus \cdots \oplus y_m = u \oplus z_1 \oplus \cdots \oplus z_m$ we get $0 \oplus y_1 \oplus \cdots \oplus y_m = 0 \oplus z_1 \oplus \cdots \oplus z_m$ since $u \oplus u = 0$. Adding v and applying (1) we obtain $v \oplus y_1 \oplus \cdots \oplus y_m = v \oplus z_1 \oplus \cdots \oplus z_m$ as desired. In the latter case, we add u to both sides and by means of (3) get $u \oplus u \oplus y_1 \oplus \cdots \oplus y_m = u \oplus u \oplus z_1 \oplus \cdots \oplus z_m$. Now we argue as in the former case and apply $\mathbf{f}$.

Our next examples show that in a locally finite setting the G-spectra of varieties can exceed any pre-assigned function.

EXAMPLE 5.8. For any sequence $p_0, p_1, p_2, \ldots$ of positive integers there exists a locally finite variety $\mathcal{V}$ such that $\mathrm{G}_{\mathcal{V}}(k) \geqslant p_k$ for all $k \geqslant 2$. For this example we use a construction of [**24**] as presented in [**25**, p. 69] that produces a variety $\mathcal{V}$ that has each p_n as its number of essentially n-ary operations. For such a $\mathcal{V}$ it is known that $|\mathrm{F}_{\mathcal{V}}(k)| = \sum_{i \leqslant k} \binom{k}{i} p_i$.

The fundamental terms for $\mathcal{V}$ are the constants $\mathbf{f}_1, \ldots, \mathbf{f}_{p_0}$, the unary terms $\mathbf{f}_{i1}$ for $1 \leqslant i \leqslant p_1 - 1$, and for each $k \geqslant 2$ the k-ary terms $\mathbf{f}_{1k}, \ldots, \mathbf{f}_{p_k k}$. An algebra $\mathbf{A}$ that generates $\mathcal{V}$ is constructed as follows. For $i < \omega$ let A_i be a collection of pairwise disjoint sets such that $A_0 = \{c_0, c_1, \ldots, c_{p_0}\}$ and A_i is countably infinite for $i \geqslant 1$. The algebra $\mathbf{A}$ has universe $\bigcup_{i \geqslant 0} A_i$ and fundamental term operations given by

$$\mathbf{f}_{ik}(a_1, \ldots, a_k) = \begin{cases} c_0, & \text{if } a_1, \ldots, a_k \in A_i \text{ and } |\{a_1, \ldots, a_k\}| = k, \\ c_1, & \text{otherwise,} \end{cases}$$

and $\mathbf{f}_i = c_i$ for $1 \leqslant i \leqslant p_0$.

In this variety, any substitution of $\mathbf{f}_i$ or $\mathbf{f}_{im}$ in a $\mathbf{f}_{jn}$ yields a constant. For each $m \leqslant p_k$ let γ_m be the congruence relation on $\mathbf{F}_{\mathcal{V}}(k)$ generated by identifying each of $\mathbf{f}_{1k}, \ldots, \mathbf{f}_{mk}$ with the constant $\mathbf{f}_1$. Since every composition of the $\mathbf{f}_{ij}$ produces a constant function it follows that $\gamma_1 < \gamma_2 < \cdots < \gamma_{p_k}$ in $\mathsf{Con}\,(\mathbf{F}_{\mathcal{V}}(k))$. So the quotients $\mathbf{F}_{\mathcal{V}}(k)/\gamma_m$ are pairwise nonisomorphic for $1 \leqslant m \leqslant p_k$.

In Example 5.8 we take a known construction of locally finite varieties that have arbitrarily large free spectra and note that these varieties also have large G-spectra. The varieties constructed have infinitely many fundamental operations. Our next example, a recent theorem of Ralph McKenzie, significantly sharpens this by working with varieties of groupoids, that is, algebras in which there is only one fundamental operation and it is binary.

EXAMPLE 5.9 (R. McKenzie [**54**]). *Let f be any function from integers to integers. There exists a locally finite variety $\mathcal{V}$ of groupoids such that $|\mathrm{F}_{\mathcal{V}}(k)| \geqslant f(k)$ and $\mathrm{G}_{\mathcal{V}}(k) \geqslant f(k)$ for all $k > 0$.*

We next show that in the hierarchy of m-fold exponential generative complexity every level is nonempty. For each level we exhibit a locally finite variety of finite similarity type. Moreover, in our construction each variety has type set $\{\mathbf{1}\}$.

EXAMPLE 5.10. A construction that for each integer $m \geqslant 2$ creates a variety in a finite language that is locally finite, locally strongly solvable and of m-fold generative complexity.

We start with a variety $\mathcal{V}$ having an associative and commutative binary operation $\oplus$ and a constant 0 so that $\mathcal{V}$ satisfies the identities $x \oplus 0 \approx 0$ and $x \oplus x \approx 0$. The elements of $\mathbf{F}_{\mathcal{V}}(n)$ other than 0 can be written as $x_{i_1} \oplus x_{i_2} \oplus \cdots \oplus x_{i_j}$ with $1 \leqslant i_1 < i_2 < \cdots < i_j \leqslant n$. Thus there is a natural bijection between the elements of $\mathbf{F}_{\mathcal{V}}(n)$ and the subsets of $X = \{x_1, \ldots, x_n\}$ with 0 corresponding to $\emptyset$. So $|\mathrm{F}_{\mathcal{V}}(n)| = 2^n$. A crucial observation is that if $\mathbf{t}_1$ and $\mathbf{t}_2$ are elements of $\mathrm{F}_{\mathcal{V}}(n)$ and $\mathbf{t} = \mathbf{t}_1 \oplus \mathbf{t}_2$ is not 0, then $\mathbf{t}$ is more complex than either $\mathbf{t}_1$ or $\mathbf{t}_2$, and in particular, $\mathbf{t}$ cannot be a free generator of $\mathbf{F}_{\mathcal{V}}(n)$. So $\mathbf{F}_{\mathcal{V}}(n)$ is uniquely generated by X. An argument similar to the one given for semilattices in the proof of Theorem 4.1 shows that $\mathbf{F}_{\mathcal{V}}(n)$ has at least $2^{\binom{n}{\lfloor n/2 \rfloor}}$ congruence relations in which each free generator is in a singleton congruence class, that is, $\mathrm{C_s}(\mathbf{F}_{\mathcal{V}}(n), X)$ is at least doubly exponential as a function of n. The algebra $\mathbf{F}_{\mathcal{V}}(n)$ has no 1-snags since if $\mathbf{s}(a, a) = a$ for a binary polynomial $\mathbf{s}$, then $a = 0$. But $\mathbf{s}(0, b) = 0$ for all choices of b, so no 1-snags are possible.

Next, let $\mathcal{W}$ be an arbitrary locally finite variety with a constant term 0 such that the identity $\mathbf{f}(x_1, \ldots, x_{i-1}, 0, x_{i+1}, \ldots, x_r) \approx 0$ holds for every fundamental operation $\mathbf{f}$ and every $1 \leqslant i \leqslant r$. For a binary operation symbol $\oplus$ not in the language of $\mathcal{W}$ we construct from $\mathcal{W}$ a new variety $\mathcal{W}^{\oplus}$ by adding $\oplus$ to the language of $\mathcal{W}$ and requiring that all the identities of $\mathcal{W}$ hold for $\mathcal{W}^{\oplus}$, all the identities of $\mathcal{V}$ hold for $\oplus$ in $\mathcal{W}^{\oplus}$, and the identities $\mathbf{f}(x_1, \ldots, x_{i-1}, y \oplus z, x_{i+1}, \ldots, x_r) \approx 0$ hold for every fundamental operation $\mathbf{f}$ of $\mathcal{W}$ and every $1 \leqslant i \leqslant r$.

Let $\mathbf{F}$ denote the free algebra on k free generators for $\mathcal{W}^{\oplus}$ and let c be the cardinality of the free algebra $\mathbf{F}_{\mathcal{W}}(k)$ with say, $\mathrm{F}_{\mathcal{W}}(k) = \{0, \mathbf{t}_1, \ldots, \mathbf{t}_{c-1}\}$. Every element of $\mathbf{F}$ other than 0 can be written in the form $\mathbf{t}_{i_1} \oplus \mathbf{t}_{i_2} \oplus \cdots \oplus \mathbf{t}_{i_n}$ for $1 \leqslant i_1 < i_2 < \cdots < i_n \leqslant c - 1$. Thus, $\mathbf{F}$ has at most 2^{c-1} elements. However, if we take the algebra $\mathbf{F}_{\mathcal{W}}(k)$ and introduce a new binary operation $\oplus$ and form the free $\mathcal{V}$ algebra freely generated by the nonzero elements of $\mathrm{F}_{\mathcal{W}}(k)$ and extend the fundamental operations of $\mathcal{W}$ in the obvious way, then we obtain an algebra in $\mathcal{W}^{\oplus}$ of cardinality 2^{c-1}. So this algebra must be the free algebra $\mathbf{F}$. That is, $\mathbf{F}$ can be constructed by forming the free algebra for $\mathcal{V}$ using the nonzero elements of $\mathbf{F}_{\mathcal{V}}(k)$ as free generators. It follows that if the free spectrum of $\mathcal{W}$ is m-fold exponential, then the free spectrum of $\mathcal{W}^{\oplus}$ is $(m+1)$-fold exponential. It is easily checked that if $\mathbf{F}_{\mathcal{W}}(k)$ is uniquely generated, then so is $\mathbf{F}$. We also note that if $\mathbf{F}_{\mathcal{W}}(k)$ has no 1-snags, then $\mathbf{F}$ also has no 1-snags. To see this, suppose $\langle b, a \rangle$ is a 1-snag in $\mathbf{F}$. So there exists $\mathbf{s}(x, y) \in \mathrm{Pol}_2\mathbf{F}$ such that $\mathbf{s}(a, b) = \mathbf{s}(b, a) = b$ and $\mathbf{s}(a, a) = a$. The element a cannot be 0 since $\mathbf{s}(0, b) = 0$. The conditions $\mathbf{s}(a, b) = b$ and $\mathbf{s}(a, a) = a$ show that a and b can be obtained from the free generators without the use of $\oplus$. So $\langle b, a \rangle$ is a 1-snag in $\mathbf{F}_{\mathcal{W}}(k)$.

Finally, for every positive m we construct a locally finite, locally strongly solvable variety $\mathcal{V}_m$ of $(m+1)$-fold generative complexity whose fundamental operations consist of the constant 0 and m binary operations $\oplus_1, \ldots, \oplus_m$. We let $\mathcal{V}_1$ be the variety $\mathcal{V}$ but with the fundamental binary operation written as $\oplus_1$. For $m \geqslant 2$ we define $\mathcal{V}_m$ to be $\mathcal{V}_{m-1}^{\oplus_m}$. Let $\mathbf{F}$ denote the free algebra for $\mathcal{V}_m$ freely generated by $X = \{x_1, \ldots, x_k\}$. We may view the elements of $\mathbf{F}$ as being in the universe of a free $\mathcal{V}$ algebra in which the binary operation is denoted $\oplus_m$ and the free generators are the elements of $\mathbf{F}_{\mathcal{V}_{m-1}}(k)$. The free spectrum of $\mathcal{V}_1$ is 2^k and from the construction of each $\mathcal{V}_i$ from $\mathcal{V}_{i-1}$ we have that $\mathcal{V}_m$ is locally finite with a free spectrum that is m-fold exponential as a function of k. From Proposition 3.1 it follows that $\mathcal{V}_m$ has generative complexity that is at most $(m+1)$-fold exponential. If c denotes the cardinality of $\mathbf{F}_{\mathcal{V}_{m-1}}(k)$, then the cardinality of $\mathrm{C_s}(\mathbf{F}, X)$ is at least $2^{\binom{c}{\lfloor c/2 \rfloor}}$ and is therefore at least doubly exponential as a function of c. Since c is $(m-1)$-fold exponential as a function of k we may conclude from Corollary 3.8 that the G-spectrum of $\mathcal{V}_m$ is at least a doubly exponential function of an $(m-1)$-fold exponential function, that is, it is at least $(m+1)$-fold exponential. Thus $\mathcal{V}_m$ is of $(m+1)$-fold exponential generative complexity. Since for all k the free algebra $\mathbf{F}_{\mathcal{V}}(k)$ has no 1-snags and for all i the 1-snags in $\mathbf{F}_{\mathcal{V}_i}(k)$ must be 1-snags in $\mathbf{F}_{\mathcal{V}_{i-1}}(k)$, it follows that $\mathrm{typ}\{\mathcal{V}_m\} = \{\mathbf{1}\}$, so $\mathcal{V}_m$ is locally strongly solvable.

We next consider discriminator varieties. In Chapter 6 we show that every finitely generated discriminator variety is of exponential generative complexity. Discriminator varieties belong to the wider class of congruence meet semi-distributive varieties. A characterization of finitely generated congruence meet-semi-distributive varieties having few models is given in Corollary 10.3. However, for discriminator varieties that are not finitely generated but still locally finite the situation can be quite different. In our next example we present a construction method that allows us to build locally finite discriminator varieties with large generative complexity well beyond the 1-fold exponential level. The examples obtained by this method show that neither Corollary 10.3 nor our main Theorem 17.2 can be relaxed from finitely generated to locally finite.

EXAMPLE 5.11. For a class $\mathcal{C}$ of algebras let $\mathcal{C}^{\mathbf{d}}$ denote the variety generated by the class of all finite algebras of the form $\mathbf{A}^{\mathbf{d}}$, where $\mathbf{A}^{\mathbf{d}}$ is obtained from $\mathbf{A} \in \mathcal{C}$ by augmenting it with a new ternary operation $\mathbf{d}(x, y, z)$ that acts as a discriminator function as in Theorem 2.8, i.e., $\mathbf{d}(x, x, z) = z$ and $\mathbf{d}(x, y, z) = x$ if $x \neq y$. If $\mathcal{C}$ is a locally finite, universal class of algebras with $\mathrm{G}_{\mathcal{C}}(k)$ finite for all finite k, then we show that $\mathcal{C}^{\mathbf{d}}$ is a locally finite discriminator variety and $\mathrm{G}_{\mathcal{C}^{\mathbf{d}}}(k) \geqslant 2^{\mathrm{G}_{\mathcal{C}}(k)-1}$.

The variety $\mathcal{C}^{\mathbf{d}}$ is a discriminator variety since $\mathbf{d}$ is a common discriminator term for all the algebras in $\mathcal{C}$. As observed in Theorem 2.8, $\mathcal{C}^{\mathbf{d}}$ is a semisimple arithmetical variety and since $\mathcal{C}$ is assumed to be a universal class, the simple algebras as well as the directly indecomposable algebras in $\mathcal{C}^{\mathbf{d}}$ are precisely the algebras $\mathbf{A}^{\mathbf{d}}$ where $\mathbf{A}$ is a finite member of $\mathcal{C}$ having at least two elements. Note that since $\mathbf{d}(x, y, z) \in \{x, z\}$, an algebra $\mathbf{A}$ in $\mathcal{C}$ is k-generated if and only if $\mathbf{A}^{\mathbf{d}}$ is. So the number $g(k)$ of k-generated simple algebras in $\mathcal{C}^{\mathbf{d}}$ is $\mathrm{G}_{\mathcal{C}}(k)$ or $\mathrm{G}_{\mathcal{C}}(k) - 1$, depending on whether $\mathcal{C}$ contains a trivial algebra. Our assumptions on $\mathcal{C}$ guarantee that $g(k)$ is finite for all finite k. Let $\mathbf{B}_1, \ldots, \mathbf{B}_{g(k)}$ be a transversal of the k-generated simple algebras in $\mathcal{C}^{\mathbf{d}}$ and $u(k)$ be the largest cardinality of the $\mathbf{B}_i$. By $\mathbf{F}$ we denote the free algebra for $\mathcal{C}^{\mathbf{d}}$ freely generated by a set X, with $|X| = k$. Now $\mathbf{F}$ is a subdirect product of k-generated subdirectly irreducible algebras. Thus $\mathbf{F}$ is a subalgebra of a product of algebras of the form $\mathbf{B}_i^{\alpha_i}$ where $1 \leqslant i \leqslant g(k)$ and α_i is bounded above by the number of distinct mappings h of X to B_i for which $h(X)$ generates all of $\mathbf{B}_i$. Each α_i is finite, in fact $\alpha_i \leqslant |B_i|^k \leqslant u(k)^k$, so it follows that $\mathbf{F}$ is finite. Thus it is a direct product of directly indecomposable algebras and we may write

$$\mathbf{F} = \prod_{i=1}^{g(k)} \mathbf{B}_i^{\alpha_i}. \tag{4}$$

The congruence lattice of $\mathbf{F}$ is a Boolean lattice and all congruences permute, so every congruence of $\mathbf{F}$ is a factor congruence. For all choices of $0 \leqslant \beta_i \leqslant \alpha_i$ we have that every algebra of the form

$$\mathbf{B} = \prod_{i=1}^{g(k)} \mathbf{B}_i^{\beta_i} \tag{5}$$

is a homomorphic image of $\mathbf{F}$ and is therefore k-generated. Conversely, every k-generated algebra in $\mathcal{C}^{\mathbf{d}}$ is of this form. Since finite algebras in discriminator varieties are uniquely factorable, it follows that each k-generated algebra in $\mathcal{C}^{\mathbf{d}}$ is uniquely determined by the tuple $(\beta_1, \dots, \beta_{g(k)})$ of exponents in (5). So the G-spectrum of $\mathcal{C}^{\mathbf{d}}$ is given by the formula

$$\mathrm{G}_{\mathcal{C}^{\mathbf{d}}}(k) = (1+\alpha_1)(1+\alpha_2)\dots(1+\alpha_{g(k)}). \tag{6}$$

Since $1 \leqslant \alpha_i \leqslant u(k)^k$ for every i we see that for all k,

$$2^{\mathrm{G}_{\mathcal{C}}(k)-1} \leqslant \mathrm{G}_{\mathcal{C}^{\mathbf{d}}}(k) \leqslant (1+u(k)^k)^{\mathrm{G}_{\mathcal{C}}(k)}. \tag{7}$$

Note that in this construction, since we are dealing with a discriminator variety we have $\mathrm{typ}\{\mathcal{C}^{\mathbf{d}}\} = \{\mathbf{3}\}$. Another observation is that if the class $\mathcal{C}$ has arbitrarily large finite algebras, then by Jónsson's Lemma the variety $\mathcal{C}^{\mathbf{d}}$ is not finitely generated. Finally, we note that if $\mathcal{C}$ is class of algebras in a finite language and $\mathcal{C}$ is uniformly locally finite (that is, there is a function $u : \omega \longrightarrow \omega$ such that every k–generated algebra in $\mathcal{C}$ has at most $u(k)$ elements), then the hypothesis that $\mathrm{G}_{\mathcal{C}}(k)$ is finite is satisfied.

We apply this construction to create specific examples, some of which will be used later in the book.

EXAMPLE 5.12. If we apply the construction of Example 5.11 using for $\mathcal{C}$ the varieties of Example 5.9, then for any $f : \omega \longrightarrow \omega$ we see that there exists a locally finite discriminator variety $\mathcal{V}$ in a language consisting of one binary and one ternary operation such that $\mathrm{G}_{\mathcal{V}}(k) \geqslant f(k)$ for all k.

EXAMPLE 5.13. If in the construction of Example 5.11 we let $\mathcal{C}$ be the variety of sets, then $\mathrm{G}_{\mathcal{C}}(k) = k$, the variety $\mathcal{V} = \mathcal{C}^{\mathbf{d}}$ is the variety of pure discriminator algebras whose directly indecomposable algebras are $\mathbf{B}_2, \dots, \mathbf{B}_k$, where the algebra $\mathbf{B}_i$ is of cardinality i and has the discriminator operation as its only fundamental operation. For the free algebra $\mathbf{F}_{\mathcal{V}}(k)$ each α_i in (4) is $S(k,i)$, the Stirling number of the second kind, that is, the number of ways to partition a set of k elements into i nonempty subsets. (Details are contained in [**64**].) Therefore the free spectrum of $\mathcal{V}$ is at least doubly exponential. However, $\mathrm{Bell}(k) = S(k,1) + \cdots + S(k,k)$, so $|\mathrm{F}_{\mathcal{V}}(k)| \leqslant k^{\mathrm{Bell}(k)}$. From (6) we have

$$\mathrm{G}_{\mathcal{V}}(k) = \prod_{i=2}^{k}(1+S(k,i)),$$

for all $k > 1$. Note that $\mathrm{G}_{\mathcal{V}}(k)$ eventually dominates 2^{ck} for any choice of a constant c but $\mathrm{G}_{\mathcal{V}}(k) \leqslant \mathrm{Bell}(k)^k$.

EXAMPLE 5.14. An interesting application is obtained by letting $\mathcal{C}$ in Example 5.11 be the variety of Boolean algebras. In this case $\mathcal{C}^{\mathbf{d}}$ is the variety $\mathcal{M}$ of monadic algebras. The function $g(k)$ is 2^k and for each $1 \leqslant i \leqslant 2^k$ we may let the simple algebra $\mathbf{B}_i$ be obtained from the Boolean algebra with i atoms. For a given k the precise value of the α_i in (4) is known to be $\binom{2^k}{i}$, see e.g., [**59**]. We may apply (4) to show that the free spectrum of $\mathcal{M}$ is of triply exponential complexity, specifically,

$$|\mathrm{F}_{\mathcal{M}}(k)| = 2^{2^k 2^{(2^k-1)}},$$

a result originally obtained in [**3**]. The variety of monadic algebras has many models as can be seen from (6) or (7).

Example 5.10 contains examples of locally finite varieties of m-fold exponential generative complexity for every $m \geqslant 2$. Each variety has $\{\mathbf{1}\}$ as its set of types. We now present for every positive integer m a locally finite discriminator variety of finite similarity type having G-spectrum of m-fold exponential complexity. Since these are discriminator varieties, each has $\{\mathbf{3}\}$ as its set of types. The example is due to Ralph McKenzie.

EXAMPLE 5.15 (R. McKenzie). For each $m > 0$ we construct a locally finite discriminator variety in a language consisting of m ternary operations, in which the free spectrum is $(m+1)$-fold exponential and the G-spectrum is m-fold exponential.

Let $\mathcal{V}_1$ be the variety of pure discriminator algebras of Example 5.13. As we have observed the free spectrum of this variety is doubly exponential but the G-spectrum is singly exponential. For $m > 1$ we obtain $\mathcal{V}_m$ from $\mathcal{V}_{m-1}$ by applying the construction in Example 5.11 using $\mathcal{V}_{m-1}$ for $\mathcal{C}$. From (7) and the induction hypothesis we see that the G-spectrum of $\mathcal{V}_m$ is bounded below by an m-fold exponential function and bounded above by an m-fold exponential function raised to an $(m-1)$-fold exponential function, which therefore gives an m-fold exponential upper bound.

Our final example, which is a theorem of Ralph McKenzie, shows that the free spectrum can exceed the G-spectrum by an arbitrary amount.

EXAMPLE 5.16 (R. McKenzie [**54**]). *Let h be any function from integers to integers. There exists a locally finite discriminator variety $\mathcal{V}$ of finite type such that $|\mathrm{F}_{\mathcal{V}}(n)| \geqslant h(\mathrm{G}_{\mathcal{V}}(n))$ for all $n > 1$.*

CHAPTER 6

Upper Bounds

In this Chapter we show how additional hypotheses on a variety can provide upper bounds for the G-spectra of locally finite varieties that are sharper than those provided by the general theory, and in particular, by Proposition 3.1. Several of the results presented in this Chapter play a crucial role in the proofs of the characterization theorems in Part II.

Throughout this Chapter $\mathcal{C}$ is a class of algebras all of the same similarity type. For each positive integer k we choose $\mathcal{T}_k$ to be a collection of pairwise nonisomorphic, k-generated members of $\mathcal{C}$ such that every k-generated algebra in $\mathcal{C}$ is isomorphic to a subalgebra of a product of members of $\mathcal{T}_k$.

For example, if $\mathcal{C}$ is a locally finite variety, then we may let $\mathcal{T}_k$ be a transversal of either the k-generated subdirectly irreducible members of $\mathcal{C}$ or of the k-generated directly indecomposable algebras in $\mathcal{C}$. If $\mathcal{C}$ is a quasivariety generated by an algebra $\mathbf{A}$, then one choice for $\mathcal{T}_k$ is a transversal of the k-generated subalgebras of $\mathbf{A}$.

We are interested in $\mathcal{C}$ and $\mathcal{T}_k$ for which $\mathcal{T}_k$ is a finite set of finite algebras. In such a case we let $t(k) = |\mathcal{T}_k|$ and let $m(k)$ be the maximal cardinality of a member of $\mathcal{T}_k$. If $\mathcal{C}$ is a locally finite variety, then for any choice of $\mathcal{T}_k$ we have $t(k) \leqslant \mathrm{G}_{\mathcal{V}}(k)$ and $m(k) \leqslant |\mathrm{F}_{\mathcal{V}}(k)|$. Likewise if $\mathcal{C}$ is a quasivariety generated by a finite set of finite algebras, then every $\mathcal{T}_k$ is a finite set of finite algebras.

THEOREM 6.1. *Let $\mathcal{C}$ be a class of algebras and for each positive integer k let $\mathcal{T}_k$ be a finite set of finite k-generated algebras in $\mathcal{C}$ such that every k-generated member of $\mathcal{C}$ is a subalgebra of a finite product of algebras in $\mathcal{T}_k$. If $t(k) = |\mathcal{T}_k|$ and $m(k)$ is the cardinality of a maximal-sized member of $\mathcal{T}_k$, then*

$$\mathrm{G}_{\mathcal{C}}(k) \leqslant 2^{t(k)m(k)^k}.$$

PROOF. Let k be arbitrary and let t and m denote $t(k)$ and $m(k)$. Suppose $\mathcal{T}_k = \{\mathbf{A}_1, \dots, \mathbf{A}_t\}$. Let $\mathbf{B}$ be an arbitrary k-generated member of $\mathcal{C}$, generated by $g_1, \dots, g_k$. There exist integers $n_1, \dots, n_t$ such that $B \subseteq A_1^{n_1} \times \cdots \times A_t^{n_t}$, and we assume that this representation is irredundant in that $n = n_1 + \cdots + n_t$ is as small as possible. Every element b of B is an n-tuple $(b(1), \dots, b(n))$. For each $1 \leqslant j \leqslant n$ consider the vector $(g_1(j), \dots, g_k(j))$. These vectors are all distinct since the representation of $\mathbf{B}$ is irredundant. For a given $\mathbf{A}_i$ there are at most $|A_i|^k$ such vectors and $2^{|A_i|^k}$ sets of such vectors. Thus, the number of possible k-generated algebras $\mathbf{B}$ is bounded above by

$$2^{|A_1|^k} 2^{|A_2|^k} \cdots 2^{|A_t|^k} \leqslant 2^{tm^k},$$

which proves the claim. □

We present a number of corollaries of this theorem. Recall that a cardinal κ is a *residual bound* for a variety $\mathcal{V}$ if every subdirectly irreducible algebra in $\mathcal{V}$ has cardinality less than κ.

COROLLARY 6.2. *If $\mathcal{V}$ is a variety of finite similarity type with a finite residual bound, then there is a constant c such that*

$$\mathrm{G}_{\mathcal{V}}(k) \leqslant 2^{2^{ck}}.$$

PROOF. First observe that $\mathcal{V}$ is locally finite. If we apply Theorem 6.1 with $\mathcal{T}_k$ a transversal of the k-generated subdirectly irreducible algebras in $\mathcal{V}$, then eventually $t(k)$ and $m(k)$ are constant. □

COROLLARY 6.3. *If $\mathcal{Q}$ is a finitely generated quasivariety, then there exists a constant c for which*

$$\mathrm{G}_{\mathcal{Q}}(k) \leqslant 2^{2^{ck}}.$$

PROOF. If $\mathcal{Q}$ is generated by a finite set of finite algebras $\mathbf{A}_1, \ldots, \mathbf{A}_r$, then we may choose $\mathcal{T}_k$ to be all k-generated subalgebras of the $\mathbf{A}_i$. Then $t(k)$ and $m(k)$ are eventually constant and we argue as in the previous corollary. □

In the event that $\mathcal{T}_k$ in Theorem 6.1 is a transversal of the k-generated directly indecomposable algebras of $\mathcal{C}$, then a sharper estimate on $\mathrm{G}_{\mathcal{C}}(k)$ is sometimes possible. Namely, if $\mathcal{T}_k = \{\mathbf{A}_1, \ldots, \mathbf{A}_t\}$ and $\mathbf{B} \in \mathcal{C}$ is k-generated, then there exist integers $n_1, \ldots, n_t$ such that $\mathbf{B} \simeq \mathbf{A}_1^{n_1} \times \cdots \times \mathbf{A}_t^{n_t}$. So $\mathbf{B}$ is uniquely determined by the t-tuple $(n_1, \ldots, n_t)$. We write m for $m(k)$ and use the bounds $0 \leqslant n_i \leqslant m^k$ given in the proof of Theorem 6.1 to obtain

$$\mathrm{G}_{\mathcal{C}}(k) \leqslant (1 + m^k)^t \leqslant m^{(k+1)t}. \tag{8}$$

We apply (8) to give sufficient conditions for a variety to have few models. The bounds given by the next theorem are a key ingredient in the proofs of the characterization theorems 17.2 and 18.2.

THEOREM 6.4. *Let $\mathcal{C}$ be a class of algebras and $\mathcal{T}_k$ a transversal of the finite k-generated directly indecomposable members of $\mathcal{C}$ such that every k-generated member of $\mathcal{C}$ is isomorphic to a direct product of members of $\mathcal{T}_k$. Suppose there exist polynomials $p(k)$ and $q(k)$ such that $t(k) \leqslant p(k)$ and $m(k) \leqslant 2^{q(k)}$. Then*

$$\mathrm{G}_{\mathcal{C}}(k) \leqslant (2^{q(k)})^{(k+1)p(k)}.$$

□

The following is an application of Theorem 6.4.

EXAMPLE 6.5. Let $\mathcal{V}$ be the variety generated by the ring (with unit in the similarity type) $\mathbf{Z}_{p^2}$ for a prime p. We determine $t(k)$ and $m(k)$ for this variety.

Consider the following unary terms:

$$\begin{aligned}
\mathbf{q}(x) &= x^{p^2-p}, \\
\mathbf{e}(x) &= -\prod_{i=0}^{p-1}(x-i), \\
\mathbf{e}'(x) &= x - \mathbf{e}(x).
\end{aligned}$$

By Euler's theorem we know that $x^{p^2-p} \equiv 1 (\text{mod } p^2)$ whenever $p \not| x$. On the other hand, $p^2 - p$ is an even number so in $\mathbf{Z}_{p^2}$ we have

$$\mathbf{q}(x) = \begin{cases} 1, & \text{if } p \not| x, \\ 0, & \text{if } p | x. \end{cases}$$

The reader can check that $\mathbf{e}(x)$ is maximal in the set $\{z \leqslant x : p \text{ divides } z\}$.

Note that the only subdirectly irreducible rings in $\mathcal{V}$ are $\mathbf{Z}_{p^2}$ and $\mathbf{Z}_p$. Now let $\mathbf{D}$ be a finite directly indecomposable ring in $\mathcal{V}$. Then $\mathbf{D}$ is a subalgebra of $\mathbf{Z}_{p^2}^X \times \mathbf{Z}_p^Y$ for some finite sets X, Y. Note that $\mathbf{D}(0,1) = \{\overline{0}, \overline{1}\}$, as otherwise for $a \in \mathbf{D}(0,1) - \{\overline{0}, \overline{1}\}$ we get a proper decomposition $\mathbf{D} = \mathbf{D}/D \cdot a \times \mathbf{D}/D \cdot (\overline{1} - a)$. In particular for $x \in D$ we have $\mathbf{e}'(x) \in \{\overline{0}, \overline{1}, \dots, \overline{p-1}\}$. Indeed, $\mathbf{e}'(x) \in \mathbf{D}(0, 1, \dots, p-1)$ so that applying the term $\mathbf{q}$ to $\mathbf{e}'(x), \mathbf{e}'(x) - \overline{1}, \dots, \mathbf{e}'(x) - \overline{p-1}$ we get that $\mathbf{e}'(x)$ is a constant function.

Now assume that $\mathbf{D} \in \mathcal{V}$ is a directly indecomposable ring generated by the set $\{a_1, \dots, a_k\}$. Then $a_i = \mathbf{e}(a_i) + \mathbf{e}'(a_i)$ together with the fact that $\mathbf{e}'(a_i)$ is a constant give that $\mathbf{D}$ is generated by $\{\mathbf{e}(a_1), \dots, \mathbf{e}(a_k)\}$. Note that for all i and j we have $\mathbf{e}(a_i) \cdot \mathbf{e}(a_j) = \overline{0}$ so that the $(+, \cdot)$-subalgebra of $\mathbf{D}$ generated by the $\mathbf{e}(x_i)$'s is simply a subgroup of $\langle \mathbf{D}(0, p, 2p, \dots, (p-1)p); + \rangle$ generated by $\{\overline{p}, \mathbf{e}(a_1), \dots, \mathbf{e}(a_k)\}$. Since in the variety of groups generated by $\langle Z_p; + \rangle$ there are at most $k+2$ nonisomorphic algebras that are $(k+1)$-generated it follows that in $\mathcal{V}$ we have $t(k) \leqslant k+2$. Moreover, the maximal size of a $(k+1)$-generated $\mathbf{Z}_p$-group is p^{k+1} so the maximal possible cardinality of a k-generated, directly indecomposable ring in $\mathcal{V}$ is $p \cdot p^{k+1}$ and hence $m(k) \leqslant p^{k+2}$. So in Theorem 6.4 the functions $p(k)$ and $q(k)$ are both linear, and thus $\mathcal{V}$ has few models.

In Corollary 18.3 we generalize this example and prove that a cyclic ring $\mathbf{Z}_m$ generates a variety with few models if and only if m is cube free.

A variety is called *directly representable* if it is generated by a finite algebra and has, up to isomorphism, only finitely many finite, directly indecomposable members. The paper [**52**] is the main source of results about directly representable varieties. The book [**17**] also discusses them. The directly representable varieties of groups are precisely the finitely generated Abelian varieties. Also, every finitely generated, semisimple, congruence permutable variety is directly representable.

If $\mathcal{V}$ is a directly representable variety, then Theorem 6.4 may be applied since both $m(k)$ and $t(k)$ are bounded by constants. So there is a constant c such that $\mathrm{G}_{\mathcal{V}}(k) \leqslant 2^{ck}$.

In [**52**] is the important result that if $\mathcal{V}$ is directly representable, then $\mathcal{V}$ is congruence permutable, and hence $\mathrm{typ}\{\mathcal{V}\} \subseteq \{\mathbf{2}, \mathbf{3}\}$. Also, [**52**] contains a proof that a directly representable variety is either affine or contains a non-Abelian simple algebra.

In the case that $\mathcal{V}$ is an affine directly representable variety (i.e., $\mathrm{typ}\{\mathcal{V}\} \subseteq \{\mathbf{2}\}$), then every directly indecomposable algebra in $\mathcal{V}$ is Abelian. Such a $\mathcal{V}$ is generated by a finite Abelian algebra and by [**7**] there is an integer d such that $|\mathrm{F}_{\mathcal{V}}(k)| \leqslant 2^{dk}$. Let $\mathbf{D}_1, \dots, \mathbf{D}_t$ be a transversal of the directly indecomposable algebras in $\mathcal{V}$. If $\mathbf{A}$ is a finite member of $\mathcal{V}$, then there exist nonnegative integers $n_1, \dots, n_t$ such that $\mathbf{A} \simeq \mathbf{D}_1^{n_1} \times \cdots \times \mathbf{D}_t^{n_t}$. If $\mathbf{A}$ is k-generated, we obtain the inequality $n_1 + \cdots + n_t \leqslant dk$ since $|A| \leqslant 2^{dk}$ and $2 \leqslant |D_i|$ for all i. The number of nonnegative integral solutions of this inequality is $\binom{dk+t}{t}$, which is a polynomial of degree t in k.

If instead $\mathcal{V}$ contains a non-Abelian simple algebra, then $\mathbf{3} \in \operatorname{typ}\{\mathcal{V}\}$, and thus $\mathrm{G}_{\mathcal{V}}(k)$ is at least exponential by Proposition 3.3.

We summarize these observations.

COROLLARY 6.6. *Let $\mathcal{V}$ be a directly representable variety having, up to isomorphism, t directly indecomposable members. If $\mathcal{V}$ is affine and d is such that $|\mathrm{F}_{\mathcal{V}}(k)| \leqslant 2^{dk}$, then*

$$\mathrm{G}_{\mathcal{V}}(k) \leqslant \binom{dk+t}{t}$$

and $\mathcal{V}$ has very few models. If $\mathcal{V}$ is not affine, then there exist constants b and c for which

$$2^{bk} \leqslant \mathrm{G}_{\mathcal{V}}(k) \leqslant 2^{ck}$$

and $\mathcal{V}$ is of exponential generative complexity. □

The affine case in Corollary 6.6 is used in Theorem 13.4 for the characterization of locally finite varieties omitting type **1** that have very few models. The non-affine case is used in Corollary 10.3 for the characterization of finitely generated, congruence meet semi-distributive varieties that do not have many models. Both cases appear in the proof of the main Theorem 17.2.

If $\mathcal{V}$ is a directly representable variety for which a transversal of directly indecomposable algebras consists of exactly one algebra $\mathbf{A}$, then it is easily seen that $\mathbf{F}_{\mathcal{V}}(k) \simeq \mathbf{A}^{\mathrm{G}_{\mathcal{V}}(k)-1}$. In Chapter 5 of [**65**] there is a careful analysis of $\mathrm{G}_{\mathcal{V}}(k)$ for such $\mathcal{V}$.

EXAMPLE 6.7. We illustrate our analysis of Abelian, directly representable varieties by considering the G-spectra of varieties of Abelian groups. Any variety $\mathcal{V}$ of Abelian groups is generated by $\mathbf{F}_{\mathcal{V}}(1)$. If this free group is isomorphic to $\mathbf{Z}$, then $\mathcal{V}$ is not locally finite, and $\mathrm{G}_{\mathcal{V}}(k) = \omega$ for all finite k. If $\mathbf{F}_{\mathcal{V}}(1)$ is not isomorphic to $\mathbf{Z}$, then $\mathcal{V}$ is locally finite, with say, $|\mathrm{F}_{\mathcal{V}}(1)| = m$. Then $\mathcal{V}$ is generated by the cyclic group $\mathbf{Z}_m$. We have $\mathbf{F}_{\mathcal{V}}(k) \simeq \mathbf{Z}_m^k$. Let m have prime decomposition $p_1^{e_1} \dots p_r^{e_r}$. The directly indecomposables in $\mathcal{V}$ are the $\mathbf{Z}_{p_i^d}$ for $1 \leqslant d \leqslant e_i$. So a transversal of directly indecomposables for $\mathcal{V}$ contains $t = e_1 + \cdots + e_r$ members. The free k-generated algebra for $\mathcal{V}$ is

$$\mathbf{F}_{\mathcal{V}}(k) = \prod_{i=1}^{r} (\mathbf{Z}_{p_i^{e_i}})^k.$$

A typical k-generated member of $\mathcal{V}$ is a homomorphic image of $\mathbf{F}_{\mathcal{V}}(k)$ and is isomorphic to a group of the form

$$\prod_{i=1}^{r} \prod_{j=1}^{e_i} (\mathbf{Z}_{p_i^j})^{n_{ij}}$$

where $n_{i1} + \cdots + n_{ie_i} \leqslant k$ for all $1 \leqslant i \leqslant r$. Moreover, distinct sequences of n_{ij} give rise to nonisomorphic groups. Since the number of solutions of each such inequality is $\binom{k+e_i}{e_i}$, we see that

$$\mathrm{G}_{\mathcal{V}}(k) = \prod_{i=1}^{r} \binom{k+e_i}{e_i},$$

which is a polynomial of degree t in k.

Thus, every locally finite variety of Abelian groups has very few models. We use this result in Theorem 18.1 to characterize finitely generated varieties of groups with very few models.

An algebra $\mathbf{A}$ is *congruence uniform* if for every congruence θ of $\mathbf{A}$ all the θ-blocks have the same cardinality. A variety is congruence uniform if every algebra in the variety is. We wish to investigate the behavior of $\mathrm{G}_{\mathcal{V}}(k)$ when $\mathcal{V}$ is a locally finite, congruence uniform variety. Note that [**52**] contains the theorem that every directly representable variety is congruence uniform.

LEMMA 6.8. *If* $\mathbf{A}$ *is a finite, congruence uniform algebra, then*

$$|\mathsf{Con}\,\mathbf{A}| \leqslant |A|^{2\log_2 |A|}.$$

PROOF. If α is covered by β in $\mathsf{Con}\,\mathbf{A}$, then the cardinality of each α-block is at most 1/2 the cardinality of a β-block. From this it follows that if ℓ is the height of the lattice $\mathsf{Con}\,\mathbf{A}$, then $|A| \geqslant 2^\ell$. Therefore, every congruence in $\mathsf{Con}\,\mathbf{A}$ can be expressed as the join of at most ℓ principal congruences of $\mathbf{A}$. So

$$|\mathsf{Con}\,\mathbf{A}| \leqslant \binom{|A|}{2}^{\ell} \leqslant |A|^{2\ell} \leqslant |A|^{2\log_2 |A|}.$$

□

THEOREM 6.9. *Let* $\mathcal{V}$ *be a locally finite, congruence uniform variety. If the free spectrum of* $\mathcal{V}$ *is at most* m*-fold exponential, then* $\mathrm{G}_{\mathcal{V}}(k)$ *is also at most* m*-fold exponential.*

PROOF. $\mathrm{G}_{\mathcal{V}}(k) \leqslant |\mathrm{Con}(\mathrm{F}_{\mathcal{V}}(k))| \leqslant |\mathrm{F}_{\mathcal{V}}(k)|^{2\log_2 |\mathrm{F}_{\mathcal{V}}(k)|} = 2^{2(\log_2 |\mathrm{F}_{\mathcal{V}}(k)|)^2}$. □

COROLLARY 6.10. *Let* $\mathcal{V}$ *be a locally finite variety omitting types* **1** *and* **5**. *If the free spectrum of* $\mathcal{V}$ *is at most exponential, then* $\mathrm{G}_{\mathcal{V}}(k)$ *is at most exponential.*

PROOF. The hypotheses together with [**32**, Theorem 12.5] force $\mathcal{V}$ to be congruence permutable and nilpotent. Thus $\mathcal{V}$ is congruence uniform by [**22**, Corollary 7.5]. An application of Theorem 6.9 completes the proof. □

Note that the hypotheses in Corollary 6.10 that $\mathcal{V}$ omit types **1** and **5** and have a free spectrum that is at most exponential together imply that $\mathrm{typ}\{\mathcal{V}\} = \{\mathbf{2}\}$.

COROLLARY 6.11. *Let* $\mathbf{A}$ *be a finite nilpotent algebra in a congruence modular variety of finite type. If* $\mathbf{A}$ *is a direct product of algebras of prime power order, then it generates a variety with few models.*

PROOF. By Theorem 2 in [**5**] the free spectrum of the variety $\mathcal{V}$ generated by $\mathbf{A}$ is at most exponential. Since $\mathcal{V}$ must be congruence uniform, Theorem 6.9 gives the desired conclusion. □

Corollary 6.11 and Theorem 6.4 are used in the final step of our proof of Theorem 17.2 to provide an exponential upper bound for the G-spectra of the varieties considered there.

A converse to Corollary 6.11 is contained in Theorem 14.6. There we prove that for $\mathcal{V}$ a finitely generated nilpotent variety, $\mathcal{V}$ has few models if and only if the clone of $\mathcal{V}$ is finitely generated and $\mathcal{V}$ itself is generated by a finite algebra that factors as a direct product of algebras of prime power order.

Our next two results are applications of Theorem 6.9 to some specific nilpotent varieties.

COROLLARY 6.12. *Every locally finite variety of nilpotent groups has few models.*

PROOF. A locally finite variety of nilpotent groups must have a bound r on the nilpotence class of its members since every variety is closed under the formation of ultraproducts. From [**31**] or [**58**] we know that if $\mathcal{V}$ is a locally finite variety of nilpotent groups of class r, then $\log |\mathrm{F}_{\mathcal{V}}(k)|$ is a polynomial of degree r. Theorem 6.9 applies to give $\mathrm{G}_{\mathcal{V}}(k) \leqslant 2^{p(k)}$ for a polynomial $p(k)$ of degree $2r$. □

COROLLARY 6.13. *Every locally finite affine variety has few models. In particular, every locally finite variety of modules has few models.*

PROOF. If $\mathcal{V}$ is a locally finite affine variety, then $\mathcal{V}$ is generated by the finite algebra $\mathbf{F}_{\mathcal{V}}(2)$ and there are finite constants r and c such that $|\mathrm{F}_{\mathcal{V}}(k)| = cr^k$, (see [**22**, Theorem 9.16 (6),(7)]). The variety $\mathcal{V}$ is also congruence uniform since it is congruence modular and nilpotent. Thus, Theorem 6.9 applies. □

By the proof of Theorem 6.9 it follows that the G-spectrum of a locally finite affine variety is bounded above by 2^{ck^2} for some $c > 0$. In [**36**] there is a construction that shows this bound can be obtained.

Our next result also uses the height of the congruence lattice of a free algebra to bound the generative complexity of a variety.

THEOREM 6.14. *If $\mathcal{V}$ is a finitely generated, congruence modular variety, then $\mathrm{G}_{\mathcal{V}}(k)$ is at most doubly exponential.*

PROOF. Let $\mathbf{L}$ be a finite lattice. The height of $\mathbf{L}$ is denoted $h(\mathbf{L})$ and for $a \leqslant b$ in L the interval sublattice from a to b in $\mathbf{L}$ is $\mathbf{I}[a, b]$. In any finite modular lattice $\mathbf{L}$ and for any $a_1, a_2 \in L$,

$$h(\mathbf{I}[a_1, 1]) \geqslant h(\mathbf{I}[a_1, a_1 \vee a_2]) = h(\mathbf{I}[a_1 \wedge a_2, a_2]).$$

An easy induction shows that for $a_1, \ldots, a_m \in L$

$$\sum_{i=1}^{m} h(\mathbf{I}[a_i, 1]) \geqslant h(\mathbf{I}[a_1 \wedge \cdots \wedge a_m, 1]).$$

Now suppose that $\mathbf{A}$ is a finite algebra that generates a congruence modular variety $\mathcal{V}$. Let k be arbitrary. The free algebra $\mathbf{F} = \mathbf{F}_{\mathcal{V}}(k)$ is a subdirect product of m factors, where each factor is a subalgebra of $\mathbf{A}$ and $m \leqslant |A|^k$. Let ℓ be the maximal height of the congruence lattices of these subdirect factors. Note that for sufficiently large k the value of ℓ depends only on $|A|$. Suppose $\gamma_1, \ldots, \gamma_m$ are the congruence relations on $\mathbf{C} = \mathsf{Con}\,\mathbf{F}$ corresponding to the projections of $\mathbf{F}$ onto each of these factors. So $\gamma_1 \wedge \cdots \wedge \gamma_m = 0_F$. Then, since $\mathbf{C}$ is a modular lattice,

$$m \cdot \ell \geqslant \sum_{i=1}^{m} h(\mathbf{I}[\gamma_i, 1_F]) \geqslant h(\mathbf{I}[\bigwedge_{i=1}^{m} \gamma_i, 1_F]) = h(\mathbf{C}),$$

and so the height of $\mathbf{C}$ is an at most exponential function of k. We argue as in the proof of Lemma 6.8 that every congruence relation on $\mathbf{F}$ is the join of at most $h(\mathbf{C})$ principal congruences. Thus,

$$|C| \leqslant |F|^{2h(\mathbf{C})} \leqslant |A|^{2|A|^k h(\mathbf{C})} \leqslant |A|^{2\ell|A|^{2k}}.$$

Since $\mathrm{G}_{\mathcal{V}}(k) \leqslant |C|$ we see that $\mathrm{G}_{\mathcal{V}}(k)$ is at most doubly exponential as a function of k. □

This doubly exponential upper bound on $\mathrm{G}_{\mathcal{V}}(k)$ for finitely generated, congruence modular varieties allows us in Chapter 18, where we consider the G-spectrum of groups and rings, to restrict the generative complexity hierarchy to the at most doubly exponential level.

We conclude this Chapter with a study of $\mathrm{G}_{\mathcal{V}}(k)$ for semisimple arithmetical varieties. If $\mathcal{V}$ is a semisimple arithmetical variety generated by a finite algebra $\mathbf{A}$, then every finitely generated directly indecomposable algebra in $\mathcal{V}$ is a simple algebra, and by Jónsson's Lemma, is a homomorphic image of a subalgebra of $\mathbf{A}$. So $\mathcal{V}$ is a non-Abelian directly representable variety and $\operatorname{typ}\{\mathcal{V}\} \subseteq \{\mathbf{3}\}$. Therefore, if $\mathcal{V}$ is nontrivial, finitely generated, semisimple arithmetical variety, then $\mathcal{V}$ has the exponential generative complexity given by Corollary 6.6.

We consider relaxing the hypothesis that $\mathcal{V}$ is finitely generated in this situation by replacing it with an at most doubly exponential bound on the free spectrum of $\mathcal{V}$. We show in Theorem 6.17 that such a $\mathcal{V}$ has few models.

First, we require the following general fact.

LEMMA 6.15. *Suppose $\mathcal{V}$ is a locally finite variety, $\mathbf{A} \in \mathcal{V}$ is k-generated, and $m \geqslant k$. Then the number of distinct congruence relations α of $\mathbf{F}_{\mathcal{V}}(m)$ for which $\mathbf{F}_{\mathcal{V}}(m)/\alpha \simeq \mathbf{A}$ is at least $|A|^{m-k}$.*

PROOF. Let $x_1, \ldots, x_m$ generate $\mathbf{F}_{\mathcal{V}}(m)$. We denote by $\mathbf{F}$ the subalgebra of $\mathbf{F}_{\mathcal{V}}(m)$ generated by $x_1, \ldots, x_k$. So $\mathbf{F}$ is $\mathbf{F}_{\mathcal{V}}(k)$. Let h be any homomorphism of $\mathbf{F}$ onto $\mathbf{A}$ and let $\overline{h}$ be any homomorphism of $\mathbf{F}_{\mathcal{V}}(m)$ onto $\mathbf{A}$ that extends h. For each x_i, with $k+1 \leqslant i \leqslant m$, there exists $\mathbf{t}(x_1, \ldots, x_k) \in F$ such that $\overline{h}(x_i) = \overline{h}(\mathbf{t}(x_1, \ldots, x_k)) = h(\mathbf{t}(x_1, \ldots, x_k))$. In particular, $x_i \in \mathbf{t}(x_1, \ldots, x_k)/\ker(\overline{h}) \supseteq \mathbf{t}(x_1, \ldots, x_k)/\ker(h)$. Thus, $\ker(\overline{h})$ is determined by the assignment of the variables $x_{k+1}, \ldots, x_m$ to the $|A|$ congruence classes of $\ker(h)$. There are $|A|^{m-k}$ such assignments, each resulting in a different congruence α of $\mathbf{F}_{\mathcal{V}}(m)$ for which $\mathbf{F}_{\mathcal{V}}(m)/\alpha \simeq \mathbf{A}$. □

LEMMA 6.16. *Suppose $\mathcal{V}$ is a nontrivial, semisimple, arithmetical, and locally finite variety. If there exists a polynomial $f(k)$ of degree δ for which $|\mathrm{F}_{\mathcal{V}}(k)| \leqslant 2^{2^{f(k)}}$ for all k, then the size of a k-generated, directly indecomposable algebra in $\mathcal{V}$ is at most $2^{g(k)}$, for some polynomial $g(k)$ of degree $\delta - 1$.*

PROOF. The hypotheses guarantee that $\delta \geqslant 1$ and that every finite $\mathbf{B} \in \mathcal{V}$ is the direct product of simple algebras,

$$\mathbf{B} = \prod_{i=1}^{n} \mathbf{B}/\alpha_i$$

where $\alpha_1, \ldots, \alpha_n$ are all the coatoms of $\operatorname{Con} \mathbf{B}$. Let $\mathcal{T}_k = \{\mathbf{S}_1, \ldots, \mathbf{S}_{t(k)}\}$ be a transversal of the k-generated directly indecomposable (i.e., simple) algebras in $\mathcal{V}$ and, as usual, let $m(k)$ be the cardinality of a largest member of $\mathcal{T}_k$. We have

$$\mathbf{F}_{\mathcal{V}}(2k) = \prod_{i=1}^{t(2k)} \mathbf{S}_i^{n_i}$$

where the $\mathbf{S}_i$ range over $\mathcal{T}_{2k}$ and n_i is the number of coatoms α of $\operatorname{Con} \mathbf{F}_{\mathcal{V}}(2k)$ for which $\mathbf{F}_{\mathcal{V}}(2k)/\alpha \simeq \mathbf{S}_i$. By Lemma 6.15 we have $n_i \geqslant |S_i|^k$ for every $\mathbf{S}_i \in \mathcal{T}_k$. So

$$2^{2^{f(2k)}} \geqslant |\mathrm{F}_{\mathcal{V}}(2k)| \geqslant m(k)^{(m(k)^k)}$$

and therefore

$$f(2k) \geqslant k\log_2(m(k))$$

for all sufficiently large k. This gives $2^{f(2k)/k} \geqslant m(k)$ for all large k. Let $g(k)$ be any polynomial of degree $\delta - 1$ that dominates $f(2k)/k$ for all k and for which $2^{g(k)} \geqslant m(k)$ for small k. □

An interesting special case of this lemma is when there is a constant a for which $|\mathrm{F}_{\mathcal{V}}(k)| \leqslant a^{a^k}$, that is, $\delta = 1$, and consequently $m(k)$ is bounded by a constant.

THEOREM 6.17. *Suppose $\mathcal{V}$ is a nontrivial, semisimple, arithmetical, and locally finite variety for which there exists a polynomial $f(k)$ with $|\mathrm{F}_{\mathcal{V}}(k)| \leqslant 2^{2^{f(k)}}$ for all k. If there is a polynomial $p(k)$ that dominates the size of a transversal of the k-generated simple algebras in $\mathcal{V}$, then $\mathrm{G}_{\mathcal{V}}(k) \leqslant 2^{r(k)}$, where $r(k)$ is a polynomial whose degree is the sum of the degrees of p and f.*

PROOF. By Lemma 6.16 we know $m(k) \leqslant 2^{q(k)}$ for some polynomial $q(k)$ of degree one less than that of $f(k)$. From Theorem 6.4 we have

$$\mathrm{G}_{\mathcal{V}}(k) \leqslant (2^{q(k)})^{(k+1)p(k)}.$$

We let $r(k) = q(k)(k+1)p(k)$. □

The final example in this Chapter is a special semisimple arithmetical variety that is not finitely generated but is of singly exponential generative complexity as is every nontrivial finitely generated semisimple arithmetical variety. This example motivates Problem 7 of Chapter 19.

EXAMPLE 6.18. For each $2 \leqslant n \leqslant \omega$ let $\mathbf{MO}_n$ be the ortholattice consisting of $2n$ pairwise incomparable elements and the two lattice bounds. We denote by $\mathcal{MO}_n$ the variety of modular ortholattices generated by $\mathbf{MO}_n$. It is known [**41**] that the variety $\mathcal{MO}_\omega$ is semisimple, arithmetical, and locally finite, but is not finitely generated. The k-generated directly indecomposable members of $\mathcal{MO}_\omega$ consist of the 2-element ortholattice $\mathbf{MO}_0$ and the $\mathbf{MO}_n$ for each $2 \leqslant n \leqslant k$. Thus, $t(k) = k$ and $m(k) = 2k + 2$ for all finite $k \geqslant 1$. Theorem 6.4 applies, so $\mathcal{MO}_\omega$ has few models.

For this variety we can actually determine $\mathrm{G}_{\mathcal{MO}_\omega}(k)$ explicitly. The free algebra $\mathbf{F}$ for $\mathcal{MO}_\omega$ freely generated by k free generators is isomorphic to

$$\mathbf{MO}_0^{f_0} \times \prod_{i=2}^{k} \mathbf{MO}_i^{f_i}$$

where f_i is the number of coatoms α in the congruence lattice of $\mathbf{F}$ for which $\mathbf{F}/\alpha \simeq \mathbf{MO}_i$. Every k-generated member of $\mathcal{MO}_\omega$ is isomorphic to

$$\mathbf{MO}_0^{e_0} \times \prod_{i=2}^{k} \mathbf{MO}_i^{e_i}$$

where $0 \leqslant e_i \leqslant f_i$. Moreover, distinct sequences $(e_0, e_2, \ldots, e_k)$ correspond to nonisomorphic members of $\mathcal{MO}_\omega$. Thus, for $k > 1$

$$\mathrm{G}_{\mathcal{MO}_\omega}(k) = (1 + f_0)(1 + f_2) \ \ldots \ (1 + f_k).$$

It remains to determine the f_i. Each coatom α of $\mathsf{Con}\,\mathbf{F}$ corresponds to a map h of the k generators of $\mathbf{F}$ to $\mathbf{MO}_i$ for which the image of h is a generating set. There are 2^k such maps for $i = 0$. For $i \geqslant 2$, let ℓ be the number of generators mapped by h to the bounds $\{0,1\}$. The choices for ℓ range from 0 to $k-i$ and for each ℓ there are $\binom{k}{\ell}$ ways to select the generators that h sends to $\{0,1\}$. This leaves $k-\ell$ generators to be partitioned into i nonempty blocks, each block consisting of the generators mapped to a pair of complementary atoms of $\mathbf{MO}_i$. There are $S(k-\ell,i)$ such partitions, where $S(n,b)$ denotes the number of ways to partition a set of n elements into b nonempty blocks — a Stirling number of the second kind. Because of the automorphism structure of $\mathbf{MO}_i$ we may choose one element in each of the i blocks and send it arbitrarily to one of the $2i$ atoms, making sure that all i complementary pairs of atoms are represented. Having done this, there are 2^{k-i} distinct ways to map the remaining $k-i$ generators of $\mathbf{F}$ so as to preserve the block structure. This gives

$$f_i = 2^{k-i}\sum_{\ell=o}^{k-i}\binom{k}{\ell}S(k-\ell,i),$$

for all $2 \leqslant i \leqslant k$. A description of the free objects in $\mathcal{MO}_\omega$, obtained using duality theory, is given in [**30**]. We note that $\mathrm{G}_{\mathcal{MO}_\omega}(k)$ eventually dominates 2^{ck} for all choices of c, and thus, the bound given by Corollary 6.6 does not hold.

CHAPTER 7

Categorical Invariants

In this Chapter we consider different ways to associate another variety with a given variety and how the generative complexities of the two varieties are related. For a variety $\mathcal{V}$ we consider the generative complexity of $\mathcal{W}$ in the cases where $\mathcal{W}$ is a variety equivalent to $\mathcal{V}$, a subvariety of $\mathcal{V}$, a varietal power of $\mathcal{V}$, a matrix power of $\mathcal{V}$ and a variety categorically equivalent to $\mathcal{V}$. Our main result is that if $\mathcal{V}$ is a variety that is equivalent as a category to a variety $\mathcal{W}$, then there are positive constants b and c such that $\mathrm{G}_{\mathcal{V}}(k) \leqslant \mathrm{G}_{\mathcal{W}}(bk)$ and $\mathrm{G}_{\mathcal{W}}(k) \leqslant \mathrm{G}_{\mathcal{V}}(ck)$ for all k. Thus, if $\mathrm{G}_{\mathcal{V}}(k)$ is a polynomial of degree n, then so is $\mathrm{G}_{\mathcal{W}}(k)$ and if $\mathcal{V}$ is of m-fold exponential generative complexity, then so is $\mathcal{W}$.

Two algebras are said to be *term equivalent* if they share the same universe and have the same set of term operations. Algebras $\mathbf{A}$ and $\mathbf{B}$ are *weakly isomorphic* if there is an algebra $\mathbf{C}$ such that $\mathbf{A}$ is term equivalent to $\mathbf{C}$ and $\mathbf{C}$ is isomorphic to $\mathbf{B}$. Two varieties $\mathcal{V}$ and $\mathcal{W}$ are called *equivalent* if for every n the free algebras $\mathbf{F}_{\mathcal{V}}(n)$ and $\mathbf{F}_{\mathcal{W}}(n)$ are weakly isomorphic. For example, the variety of Boolean algebras is equivalent to the variety of Boolean rings. If $\mathcal{V}$ and $\mathcal{W}$ are equivalent, then there is a natural bijection between the algebras in the two varieties and this bijection preserves cardinality and generating sets. Thus, equivalent varieties have the same G-spectra.

If $\mathcal{W}$ is a subvariety of a variety $\mathcal{V}$, then of course $\mathrm{G}_{\mathcal{W}}(k) \leqslant \mathrm{G}_{\mathcal{V}}(k)$ for all k. If $\mathcal{W}$ is a proper subvariety of $\mathcal{V}$, then there is an integer n for which $\mathbf{F}_{\mathcal{V}}(n)$ and $\mathbf{F}_{\mathcal{W}}(n)$ are not isomorphic. For such an n we have $\mathrm{G}_{\mathcal{W}}(n)+1 \leqslant \mathrm{G}_{\mathcal{V}}(n)$. Thus, if $\mathcal{V}$ is a locally finite variety with $\mathcal{W}$ is a subvariety of $\mathcal{V}$, then $\mathcal{W}$ is a proper subvariety of $\mathcal{V}$ if and only if there is an n such that $\mathrm{G}_{\mathcal{W}}(k) < \mathrm{G}_{\mathcal{V}}(k)$ for all $k \geqslant n$.

Given two varieties $\mathcal{V}_1$ and $\mathcal{V}_2$ the varietal product of $\mathcal{V}_1$ and $\mathcal{V}_2$, denoted $\mathcal{V}_1 \otimes \mathcal{V}_2$, consists of all algebras $\mathbf{A}_1 \otimes \mathbf{A}_2$ for $\mathbf{A}_1 \in \mathcal{V}_1, \mathbf{A}_2 \in \mathcal{V}_2$, with universe $A_1 \times A_2$ and whose n-ary fundamental operations are all $\mathbf{t}_1 \times \mathbf{t}_2$ where the $\mathbf{t}_i$ range over all n-ary terms of $\mathcal{V}_i$ and $\mathbf{t}_1 \times \mathbf{t}_2$ acts coordinatewise on $A_1 \times A_2$. It is easily seen that $\mathrm{G}_{\mathcal{V}_1 \otimes \mathcal{V}_2}(k) = \mathrm{G}_{\mathcal{V}_1}(k) \times \mathrm{G}_{\mathcal{V}_2}(k)$. For example, if $\mathcal{V}$ has a linear G-spectrum, then the n-th varietal power $\mathcal{V} \otimes \mathcal{V} \otimes \cdots \otimes \mathcal{V}$ of $\mathcal{V}$ has a G-spectrum that is a polynomial of degree n. Thus, for every positive integer n there exists a variety whose G-spectrum is a polynomial of degree n.

A class $\mathcal{C}$ of algebras may be viewed as a category whose objects are the algebras in $\mathcal{C}$ and whose maps are the homomorphisms between these algebras. Two classes of algebras $\mathcal{C}_1$ and $\mathcal{C}_2$ are called *categorically equivalent* if there are two functors $F_1 : \mathcal{C}_1 \to \mathcal{C}_2$ and $F_2 : \mathcal{C}_2 \to \mathcal{C}_1$ such that the composition functors $F_1 \circ F_2$ and $F_2 \circ F_1$ are naturally equivalent to the identity maps on $\mathcal{C}_2$ and $\mathcal{C}_1$, respectively. For example, if $\mathcal{B}$ is the variety of Boolean algebras and $\mathcal{V}$ is a variety categorically equivalent to $\mathcal{B}$, then it is known, e.g., [**33**], that $\mathcal{V}$ must be a variety generated by a primal algebra.

For an algebra $\mathbf{A}$ and a positive integer m the m-th *matrix power* of $\mathbf{A}$, denoted $\mathbf{A}^{[m]}$, is an algebra with universe A^m whose fundamental k-ary operations consist of all m-tuples $(\mathbf{p}_1, \ldots, \mathbf{p}_m)$ where each $\mathbf{p}_i$ is a km-ary term operation of $\mathbf{A}$ and $(\mathbf{p}_1, \ldots, \mathbf{p}_m)$ maps $(A^m)^k$ to A^m by sending $(\bar{a}_1, \ldots, \bar{a}_k)$ to $(\mathbf{p}_1(\bar{a}_1, \ldots, \bar{a}_k), \ldots, \mathbf{p}_m(\bar{a}_1, \ldots, \bar{a}_k))$. For a class $\mathcal{C}$ of algebras $\mathcal{C}^{[m]}$ denotes all m-th matrix powers of algebras in $\mathcal{C}$. It is known that if $\mathcal{V}$ is a variety, then so is $\mathcal{V}^{[m]}$ and the correspondence $\mathbf{A} \mapsto \mathbf{A}^{[m]}$ is a categorical equivalence of the varieties $\mathcal{V}$ and $\mathcal{V}^{[m]}$.

A unary term σ for a variety $\mathcal{V}$ is *idempotent* if $\mathcal{V} \models \sigma(\sigma(x)) \approx \sigma(x)$. If σ is an idempotent term for $\mathcal{V}$ and $\mathbf{A} \in \mathcal{V}$, then $\mathbf{A}(\sigma)$ denotes the algebra with universe $\sigma(A)$ whose fundamental operations are of the form $\sigma\mathbf{t}|_{\sigma(A)}$ where $\mathbf{t}$ ranges over all terms for $\mathcal{V}$. The class of all algebras $\mathbf{A}(\sigma)$ for $\mathbf{A} \in V$ is denoted $\mathcal{V}(\sigma)$. It is known that if $\mathcal{V}$ is a variety, then so is $\mathcal{V}(\sigma)$. A unary term σ for $\mathcal{V}$ is called *invertible* if there exist an r-ary term $\mathbf{t}$ and unary terms $\mathbf{t}_1, \ldots, \mathbf{t}_r$ such that $\mathcal{V} \models x \approx \mathbf{t}(\sigma(\mathbf{t}_1(x)), \ldots, \sigma(\mathbf{t}_r(x)))$. The r-ary term $\mathbf{t}$ is called an *inverting term* for σ. In [**53**] it is shown that if σ is an invertible idempotent term for a variety $\mathcal{V}$, then the correspondence $\mathbf{A} \mapsto \mathbf{A}(\sigma)$ is a categorical equivalence between $\mathcal{V}$ and $\mathcal{V}(\sigma)$.

The following characterization of categorical equivalence for varieties of algebras by means of matrix powers and invertible idempotents is due to Ralph McKenzie.

THEOREM 7.1 (R. McKenzie [**53**]). *Two varieties $\mathcal{V}$ and $\mathcal{W}$ are categorically equivalent if and only if $\mathcal{W}$ is equivalent to $\mathcal{V}^{[m]}(\sigma)$ for some positive integer m and some invertible idempotent σ for $\mathcal{V}^{[m]}$.*

We use this Theorem to compare the G-spectra of categorically equivalent varieties.

LEMMA 7.2. *Let $\mathbf{A}$ be a k-generated algebra, m a positive integer, and σ an invertible idempotent term for $\mathbf{A}$. Then*

(1) *The matrix power $\mathbf{A}^{[m]}$ is $\lceil k/m \rceil$ generated.*
(2) *The algebra $\mathbf{A}(\sigma)$ is kr-generated where r is the arity of the inverting term for σ.*

PROOF. Suppose $\{g_1, \ldots, g_k\}$ generates $\mathbf{A}$. For each $a \in A$ let $\mathbf{h}_a$ be a k-ary term for $\mathbf{A}$ such that $a = \mathbf{h}_a(g_1, \ldots, g_k)$. To simplify notation we write q for $\lceil k/m \rceil$. Let $b_1, b_2, \ldots, b_q \in A^m$ be any collection of m-tuples for which each g_i appears at least once among the b_j. For an arbitrary $(a_1, \ldots, a_m) \in A^m$ consider the q-ary term $(\mathbf{p}_1, \ldots, \mathbf{p}_m)$ for $\mathbf{A}^{[m]}$ where $\mathbf{p}_i(b_1, b_2, \ldots, b_q) = \mathbf{h}_{a_i}(g_1, g_2, \ldots, g_k)$. Thus, $(\mathbf{p}_1, \ldots, \mathbf{p}_m)(b_1, \ldots, b_q) = (a_1, \ldots, a_m)$, that is, the set $\{b_1, \ldots, b_q\}$ generates $\mathbf{A}^{[m]}$.

For the second claim, suppose $\mathbf{A} \models x \approx \mathbf{t}(\sigma\mathbf{t}_1(x), \ldots, \sigma\mathbf{t}_r(x))$. We show that the algebra $\mathbf{A}(\sigma)$ is generated by the elements $\sigma\mathbf{t}_i(g_j)$ for $1 \leqslant i \leqslant r$ and $1 \leqslant j \leqslant k$. Clearly each $\sigma\mathbf{t}_i(g_j) \in \sigma(A)$. Let $a \in \sigma(A)$ be arbitrary. We have $a = \sigma(a) = \sigma\mathbf{h}_a(g_1, \ldots, g_k)$. If in this equality each g_j is replaced by $\mathbf{t}(\sigma\mathbf{t}_1(g_j), \ldots, \sigma\mathbf{t}_r(g_j))$, the result is

$$a = \sigma\mathbf{h}_a(\mathbf{t}(\sigma\mathbf{t}_1(g_1), \ldots, \sigma\mathbf{t}_r(g_1)), \ldots, \mathbf{t}(\sigma\mathbf{t}_1(g_k), \ldots, \sigma\mathbf{t}_r(g_k))).$$

Thus, $\sigma\mathbf{h}_a(\mathbf{t}(x_{11}, \ldots, x_{r1}), \ldots, \mathbf{t}(x_{1k}, \ldots, x_{rk}))$ is a term for $\mathbf{A}(\sigma)$ that has value a when the $\sigma\mathbf{t}_i(g_j)$ are substituted for the variables x_{ij}. □

Suppose $\mathcal{V}$ is a variety and σ is an invertible idempotent term for $\mathcal{V}^{[m]}$. From Lemma 7.2 there exists a positive constant b such that if $\mathbf{A} \in \mathcal{V}$ is k-generated, then the algebra $\mathbf{A}^{[m]}(\sigma)$ is generated by at most bk elements. Thus $\mathrm{G}_{\mathcal{V}}(k) \leqslant \mathrm{G}_{\mathcal{V}^{[k]}(\sigma)}(bk)$. This observation together with Theorem 7.1 give the following.

THEOREM 7.3. *If $\mathcal{V}$ and $\mathcal{W}$ are two categorically equivalent varieties, then there are positive constants b and c such that $\mathrm{G}_{\mathcal{V}}(k) \leqslant \mathrm{G}_{\mathcal{W}}(bk)$ and $\mathrm{G}_{\mathcal{W}}(k) \leqslant \mathrm{G}_{\mathcal{V}}(ck)$ for all k.* □

For example, let $\langle L, \leqslant\rangle$ be a finite lattice order and let $\mathbf{A}$ be the algebra with universe L and fundamental operations consisting of all operations that preserve $\leqslant$. If $\mathcal{V}$ is the variety generated by $\mathbf{A}$ and $\mathcal{D}$ is the variety of bounded distributive lattices, then it is known ([**20**], [**62**]) that $\mathcal{V}$ and $\mathcal{D}$ are categorically equivalent. Note that $\mathcal{D}$ is equivalent to the variety generated by the algebra with universe $\{0,1\}$ and fundamental term operations consisting of all operations that preserve the relation $0 \leqslant 1$. In fact, if the lattice L has $c_1, \dots, c_r$ as its join irreducible elements, then it is possible to find an invertible idempotent σ on $\mathbf{A}$ such that $\sigma(L) = \{0,1\}$ by defining $\sigma(0) = 0, \sigma(a) = 1$ if $a \neq 0$ with the inverting term for σ the r-ary term $\mathbf{t}(x_1, \dots, x_r) = (x_1 \wedge c_1) \vee \cdots \vee (x_r \wedge c_r)$ and the $\mathbf{t}_i$ given by $\mathbf{t}_i(x) = c_i$ if $c_i \leqslant x$ and $\mathbf{t}_i(x) = 0$ otherwise. Then $\mathbf{A} \models x \approx \mathbf{t}(\sigma(\mathbf{t}_1(x)), \dots, \sigma(\mathbf{t}_r(x))$ and we have by Lemma 7.2 and Theorem 4.2 the bound $\mathrm{G}_{\mathcal{V}}(k) \leqslant \mathrm{G}_{\mathcal{D}}(rk) = 2^{2^{rk}(1+o(1))}$.

As another application of Theorem 7.3 we consider Example 3.9 in which we show that if $\mathcal{V}$ is a locally finite variety of multiunary algebras, then $\mathcal{V}$ is of polynomial generative complexity if and only if every unary term operation of $\mathbf{F}_{\mathcal{V}}(1)$ is a permutation or a constant. By Theorem 7.3 if the variety $\mathcal{W}$ is categorically equivalent to $\mathcal{V}$, then $\mathcal{W}$ is also of polynomial generative complexity. We use McKenzie's Theorem 7.1 and a suggestion of Keith Kearnes to provide a characterization of such varieties. Let $\mathcal{W}$ be term equivalent to $\mathcal{V}^{[m]}(\sigma)$ for some positive integer m and some invertible idempotent unary term σ for $\mathcal{V}^{[m]}$. For every $\mathbf{A} \in \mathcal{V}$, every term operation is essentially unary and is either a permutation of A or is a constant. The unary term σ for $\mathcal{V}^{[m]}$ is of the form $\sigma(x_1, \dots, x_m) = (\mathbf{t}_1(x_{h(1)}), \dots, \mathbf{t}_m(x_{h(m)}))$ where each $\mathbf{t}_i$ is a unary term for $\mathcal{V}$ and $h : \{1, \dots, m\} \to \{1, \dots, m\}$. Suppose r of the $\mathbf{t}_i$ give rise to permutations and $m - r$ of them correspond to constants. Without loss of generality $\sigma(x_1, \dots, x_m) = (\mathbf{t}_1(x_{h(1)}), \dots, \mathbf{t}_r(x_{h(r)}), c_{r+1}, \dots, c_m)$ where the c_j are constant terms. Now σ is idempotent so for each $1 \leqslant i \leqslant r$ we have $\mathbf{t}_i(x_{h(i)}) = \mathbf{t}_i(\mathbf{t}_{h(i)}(x_{h(h(i))}))$. The $\mathbf{t}_i$ are permutations so we may cancel to obtain $x_{h(i)} = \mathbf{t}_{h(i)}(x_{h(h(i))})$. This implies $h(i) = h(h(i))$ and $\mathbf{t}_{h(i)}$ is the identity permutation. Thus if q is the number of coordinates for which $h(i) = i$, then $|\sigma(A)| = |A|^q$ and the projection map from $\mathbf{A}^{[m]}(\sigma)$ to $\mathbf{A}^{[q]}$ obtained by projecting onto these coordinates is an isomorphism. Thus $\mathcal{W}$ is term equivalent to $\mathcal{V}^{[q]}$.

Part 2

Varieties with Few Models

In this Part of the book we study locally finite varieties with few and very few models. One way to describe such varieties is to understand their finite subdirectly irreducible algebras. This analysis splits into five cases depending on the type of the monolith in the subdirectly irreducibles. First, in Chapter 8 we rule out types **4** and **5** by showing that they cannot occur in such varieties. Chapter 9 shows that subdirectly irreducible algebras with type **3** monolith have to be simple. Thus the minimal sets of type **3** have no tails. Also type **2** minimal sets are shown to have empty tails, but the proof of this is postponed to Chapter 11. In particular, we get that every locally finite variety that omits type **1** and has few models is congruence modular. Actually, in Chapter 10, we show that such a variety has to be congruence permutable.

In Chapters 11 and 12 a further study of subdirectly irreducible algebras with type **2** monolith is undertaken. It is shown that the centralizer of such a monolith is the solvable radical and is either the total congruence or a unique coatom in the congruence lattice.

With an additional assumption that there is no type **1** in the variety we show that the solvable radical is nilpotent. Chapter 14 puts further restrictions on this nilpotent part. We prove that every finite nilpotent algebra decomposes into a direct product of algebras of prime power order. In particular, all finite nilpotent subdirectly irreducibles are of prime power order. On the other hand, if a subdirectly irreducible is not nilpotent (or equivalently if it has type **3** prime quotient at the top) then the nilpotent radical is Abelian. Further restrictions on the Abelian part are inferred in Chapter 16. This detailed understanding of nilpotent congruences is possible due to the description of locally finite varieties that omit type **1** and have very few models. This description is obtained in Chapter 13.

The final Chapter 17 of this part of the book collects all necessary conditions that a locally finite variety omitting type **1** has to fulfill in order to have few models. We also prove there that this list of conditions is in fact complete in the case of finitely generated varieties.

CHAPTER 8

Types 4 or 5 Need Not Apply

THEOREM 8.1. *If $\mathcal{V}$ is a locally finite variety such that* $\operatorname{typ}\{\mathcal{V}\} \cap \{\mathbf{4}, \mathbf{5}\} \neq \emptyset$, *then $\mathcal{V}$ has many models.*

PROOF. Suppose that $\mathcal{V}$ is a locally finite variety that has **4** or **5** in its set of types. We code k-generated semilattices into certain k-generated algebras in $\mathcal{V}$ in such a way that from Theorem 4.1 we see that $\mathcal{V}$ has many models.

Let $k \geqslant 2$ be arbitrary. Consider the variety $\mathcal{S}_1$ of meet semilattices with unit 1. Let $\mathbf{F}$ be the free algebra for $\mathcal{S}_1$ that is freely generated by $X = \{x_1, \ldots, x_k\}$. Thus each x_i is covered by 1 in the order relation of $\mathbf{F}$, and 1 is the largest element in this order. We denote $x_1 \wedge \cdots \wedge x_k$ by 0. From the proof of Theorem 4.1 we know that the algebra $\mathbf{F}$ is uniquely generated by X and that $|\mathrm{C_s}(\mathbf{F}, X)| \geqslant 2^{\binom{k}{\lfloor k/2 \rfloor}}$. Let C be a maximal collection of pairwise nonisomorphic quotients $\mathbf{F}/\gamma$ where each $\gamma \in \mathrm{C_s}(\mathbf{F}, X)$. For such a $\mathbf{F}/\gamma$ we let g_i denote x_i/γ. Then $\mathbf{F}/\gamma$ is uniquely generated by the set $G = \{g_1, \ldots, g_k\}$ since Lemma 3.5 applies. We have $|C| \geqslant 2^{\binom{k}{\lfloor k/2 \rfloor}}/k!$ by Corollary 3.7.

Let $\mathbf{S} = \langle S, \wedge, 1 \rangle \in C$ be arbitrary. We embed $\mathbf{S}$ into $\{0,1\}^S$ with a map σ defined by

$$\sigma(t)_s = \begin{cases} 1, & \text{if } s \leqslant t, \\ 0, & \text{if } s \nleqslant t. \end{cases}$$

Thus, σ is a subdirect embedding, except the projection on coordinate $0 \in S$ is not onto $\{0, 1\}$.

We note the following properties of the mapping σ.

(1) For all $t \in S$, the element t is the least u in S for which $\sigma(u)_t = 1$.
(2) If $0 < u \in S$, then there exist $s, t \in S$, with $s < t$, for which $\sigma(s)_u = 0$ and $\sigma(t)_u = 1$.
(3) For $s \neq t \in S$ the set $\{(\sigma(u)_s, \sigma(u)_t) : u \in S\}$ includes both $(0, 1)$ and $(1, 0)$ if and only if s and t are incomparable in the semilattice order; and includes $(1, 0)$ but not $(0, 1)$ if $s < t$.
(4) $\sigma(1) = \overline{1}$ and $\sigma(0)$ is all 0's, except $\sigma(0)_0 = 1$.

Let $\mathbf{A} \in \mathcal{V}$ be a finite subdirectly irreducible algebra with monolith μ and $\operatorname{typ}(0_A, \mu) = \mathbf{4}$ or $\mathbf{5}$. There exists $U \in M_{\mathbf{A}}(0_A, \mu)$ with $N = \{0, 1\} \subseteq U$ a $(0_A, \mu)$-trace of $\mathbf{A}$. Let $\mathbf{e} \in E(\mathbf{A})$ be such that $\mathbf{e}(A) = U$. We have a pseudo-meet $\wedge$ that is a meet operation on $\{0, 1\}$ with $0 \wedge 1 = 0$. If $\operatorname{typ}(0_A, \mu) = \mathbf{4}$, then let $\vee$ denote the pseudo-join.

For $\mathbf{S} \in C$ let $\mathbf{D} = \mathbf{D_S}$ be the diagonal subalgebra of $\mathbf{A}^S$ generated by $\sigma(S) \cup \Delta$. The algebra $\mathbf{D}$ is actually generated by $\{\sigma(g_1), \ldots, \sigma(g_k)\} \cup \Delta$ since the pseudo-meet extends coordinatewise to $\mathbf{D}$. Every $f \in D$ is μ-constant since $(0, 1) \in \mu$.

In the type **4** case let $\mathbf{L}$ denote the distributive sublattice of $\langle \{0, 1\}, \wedge, \vee \rangle^S$ generated by $\sigma(S)$. The lattice $\mathbf{L}$ is k-generated since for $s, t \in S$ the greatest lower bound of $\sigma(s)$ and $\sigma(t)$ in $\mathbf{L}$ is $\sigma(s \wedge t)$. The join irreducible elements of $\mathbf{L}$ are all $\sigma(t)$ for $t \in S, t \neq 0$. Nonisomorphic $\mathbf{S}$ give rise to nonisomorphic $\mathbf{L}$.

We first show that $D(0,1)$ is well behaved.

CLAIM 1: If $\operatorname{typ}(0_A, \mu) = \mathbf{5}$, then $D(0,1) = \sigma(S) \cup \{\overline{0}\}$ and if $\operatorname{typ}(0_A, \mu) = \mathbf{4}$, then $D(0,1) = L \cup \{\overline{0}\}$.

Proof. Let $f \in D(0,1)$ with $f \neq \overline{0}$. Suppose $A = \{0, 1, \ldots, n-1\}$. Then there is a term $\mathbf{q}$ for which $\mathbf{q}^{\mathbf{D}}(\sigma(g_1), \ldots, \sigma(g_k), \overline{0}, \ldots, \overline{n-1}) = f$. Define $\mathbf{p}(y_1, \ldots, y_k) = \mathbf{q}^{\mathbf{A}}(y_1, \ldots, y_k, 0, \ldots, n-1)$. So $\mathbf{p}$ is a k-ary polynomial for $\mathbf{A}$ as is $\mathbf{ep}$. There is at least one k-tuple $\bar{a} \in \{0,1\}^k$ for which $\mathbf{ep}(\bar{a}) \in \{0,1\}$. Thus $\mathbf{ep} : \{0,1\}^k \to \{0,1\}$ since μ is a congruence of $\mathbf{A}$ and $0/\mu \cap \mathbf{e}(A) = \{0,1\}$. It follows that $\mathbf{ep}|_{\{0,1\}}$ is a polynomial of $\mathbf{A}|_{\{0,1\}}$. Thus $\mathbf{ep}|_{\{0,1\}} = \mathbf{p}'|_{\{0,1\}}$ for a $\mathbf{p}' \in \operatorname{Pol} \mathbf{A}$ such that $\mathbf{p}'$ can be built from the constants of A and the pseudo-meet if $\operatorname{typ}(0_A, \mu) = \mathbf{5}$ and from the constants, pseudo-meet and pseudo-join if $\operatorname{typ}(0_A, \mu) = \mathbf{4}$. We may extend $\mathbf{p}'$ coordinatewise to a polynomial $\overline{\mathbf{p}'}$ of $\mathbf{D}$ and we have $\overline{\mathbf{p}'}(\sigma(g_1), \ldots, \sigma(g_k)) = f$. Thus, $f \in \sigma(S)$ in the type $\mathbf{5}$ case and $f \in L$ in the type $\mathbf{4}$ case.

We wish to argue that if $\mathbf{S} \not\simeq \mathbf{S}'$, then $\mathbf{D_S} \not\simeq \mathbf{D_{S'}}$. If all constants of A are in the language of $\mathbf{A}$, then this would be fairly easy to do. But if not, then there can be an isomorphism between $\mathbf{D_S}$ and $\mathbf{D_{S'}}$ that does not send Δ to Δ. Our strategy is to use the atoms of $\mathsf{Con}\,\mathbf{D}$ to recover all projection kernels η_t for $t \in S$, and from these η_t recover the comparability relation on $\mathbf{S}$. Although the order relation on $\mathbf{S}$ is in general not recoverable from the comparability relation, the number of nonisomorphic $\mathbf{S}$ having isomorphic comparability relations is relatively small, and so we can conclude that $\mathcal{V}$ has many models.

It is immediate that $\overline{\mu} \geqslant \eta_t$ for if $(f,g) \in \eta_t$, then $(f_t, g_t) \in \mu$ and so $(f,g) \in \overline{\mu}$ since f and g are μ-constant. Moreover, $\overline{\mu} > \eta_t$ since $(\overline{0}, \overline{1}) \in \overline{\mu}$. We know that $\mathsf{Con}\,\mathbf{A} \simeq \mathsf{Con}\,(\mathbf{D}/\eta_t)$ and that $\mathsf{Con}\,(\mathbf{A}/\mu) \simeq \mathsf{Con}\,(\mathbf{D}/\overline{\mu})$. So $\overline{\mu} \succ \eta_t$ for all $t \in S$, and if $\delta > \eta_t$, then $\delta \geqslant \overline{\mu}$. Also, we have $\operatorname{typ}(\eta_t, \overline{\mu}) = \operatorname{typ}(0_A, \mu) \in \{\mathbf{4}, \mathbf{5}\}$.

CLAIM 2: For every $t \in S$ and $c, d \in D$ with $c_t \neq d_t$ there exists $f \in D(\{0,1\})$ for which $\sigma(t) > f$ and $(\sigma(t), f) \in \operatorname{Cg}^{\mathbf{D}}(c,d)$.

Proof. We have $(0,1) \in \mu \in \operatorname{Cg}^{\mathbf{A}}(c_t, d_t)$. Thus in $\mathbf{A}$ there is a Maltsev chain $0 = u_0, u_1, \ldots, u_m = 1$ and unary polynomials $\mathbf{p}_1, \mathbf{p}_2, \ldots, \mathbf{p}_m$ such that $\{u_{i-1}, u_i\} = \{\mathbf{p}_i(c_t), \mathbf{p}_i(d_t)\}$ for all $1 \leqslant i \leqslant m$ with m minimal. For $1 \leqslant i \leqslant m$ let $\epsilon_i \in \{0,1\}$ be given by $\epsilon_i = 0$ if $(\mathbf{p}_i(c_t), \mathbf{p}_i(d_t)) = (u_{i-1}, u_i)$ and $\epsilon_i = 1$ if $(\mathbf{p}_i(d_t), \mathbf{p}_i(c_t)) = (u_{i-1}, u_i)$. We may assume $\mathbf{p}_i : A \to U$ since $0, 1 \in U$ so $\mathbf{ep}_i$ can replace $\mathbf{p}_i$ for every i. Extend these $\mathbf{p}_i$ coordinatewise to $\overline{\mathbf{p}_i} \in \operatorname{Pol} \mathbf{D}$. We write w_i^0 for $\overline{\mathbf{p}_i}(c)$ and w_i^1 for $\overline{\mathbf{p}_i}(d)$. Thus, $w_i^0(t) = \mathbf{p}_i(c_t)$ and $w_i^1(t) = \mathbf{p}_i(d_t)$. We obtain the $w_i^0, w_i^1 \in U^S$ and $w_1^{\epsilon_1}(t) = 0, w_m^{1-\epsilon_m}(t) = 1, (w_i^{\epsilon_i}, w_i^{1-\epsilon_i}) \in \operatorname{Cg}^{\mathbf{D}}(c,d)$ and $(w_i^{1-\epsilon_i}, w_{i+1}^{\epsilon_{i+1}}) \in \eta_t$. Note that if $f \in U^S$ and $f_t \neq 0, 1$, then $f_s = f_t$ for all $s \in S$ since f is μ-constant, f_t is in the tail of U, and $0_A \prec \mu$. So $f = \bar{a}$ for some $a \in U$. On the other hand, if $f_t \in \{0,1\}$, then $f \in \{0,1\}^S$ for a similar reason. By the minimality of m we have for each i that $w_i^{1-\epsilon_i}(t) = w_{i+1}^{\epsilon_{i+1}}(t) \in U \setminus \{0,1\}$, and so $w_i^{1-\epsilon_i} = w_{i+1}^{\epsilon_{i+1}}$. From this we have $(w_1^{\epsilon_1}, w_m^{1-\epsilon_m}) \in \operatorname{Cg}^{\mathbf{D}}(c,d)$. Note that $\sigma(t) \wedge w_m^{1-\epsilon_m} = \sigma(t)$ since $\sigma(t)$ is the least element of $D \cap \{0,1\}^S$ whose t coordinate is 1. Define f to be $\sigma(t) \wedge w_1^{\epsilon_1}$. Then $\sigma(t) > f$ since $w_1^{\epsilon_1}(t) = 0$. So $(\sigma(t), f) \in \operatorname{Cg}^{\mathbf{D}}(c,d)$.

We apply this Claim to show that the atoms of $\mathsf{Con}\,\mathbf{D}$ can be used to code the elements of S.

CLAIM 3: Let α be an atom of $\mathsf{Con}\,\mathbf{D}$. There exist $t \in S$ and $f \in D(\{0,1\})$ such that $\sigma(t)$ and f differ only on coordinate t and $\alpha = \mathrm{Cg}^{\mathbf{D}}(\sigma(t), f)$.

Proof. If $(c,d) \in \alpha$ and $c_t \neq d_t$, then we may argue as in the previous Claim that $0_D < \mathrm{Cg}^{\mathbf{D}}(\sigma(t), f) \leqslant \alpha$ for some $\sigma(t) > f$. Thus $\alpha = \mathrm{Cg}^{\mathbf{D}}(\sigma(t), f)$ since α is an atom. If $\sigma(t)$ and f differ on some coordinate other than t, say r, then $\sigma(t)_r = 1$ and $f_r = 0$. From $\sigma(t)_r = 1$ we have $t > r$. Applying the previous Claim to coordinate r we obtain $h \in D(0,1)$ with $\alpha = \mathrm{Cg}^{\mathbf{D}}(\sigma(r), h)$ and $h < \sigma(r) < \sigma(t)$. But $\sigma(r)$ and h are both 0 on coordinate t, and so $(c,d) \notin \alpha$, a contradiction. So $\sigma(t)$ and f differ only at t.

Thus $\alpha \leqslant \overline{\mu}$ for any atom $\alpha \in \mathsf{Con}\,\mathbf{D}$ since $(\sigma(t), f) \in \overline{\mu}$. Also, every atom α determines a unique element $t \in S$. The two prime quotients $\langle 0_D, \alpha\rangle$ and $\langle \eta_t, \overline{\mu}\rangle$ are projective and so $\mathrm{typ}(0_D, \alpha) = \mathrm{typ}(0_A, \mu) \in \{\mathbf{4}, \mathbf{5}\}$. From [**32**, Lemma 5.12] there exists a pseudocomplement δ of α in $\mathsf{Con}\,\mathbf{D}$. So $\delta \geqslant \eta_t$ and since $\alpha \leqslant \overline{\mu}$ we have $\delta \not\geqslant \overline{\mu}$. This shows $\delta = \eta_t$. Thus the congruence $\overline{\mu}$ is recoverable from the isomorphism type of $\mathsf{Con}\,\mathbf{D}$ since $\overline{\mu}$ is the join of the pseudocomplements of any two atoms of $\mathsf{Con}\,\mathbf{D}$.

CLAIM 4: Every congruence covered by $\overline{\mu}$ in $\mathsf{Con}\,\mathbf{D}$ is of the form η_t for some $t \in S$.

Proof. Suppose not. Let $\tau \prec \overline{\mu}$ with $\tau \neq \eta_t$ for every $t \in S$. For each $t \in S$ there exists $(c,d) \in \tau$ with $c_t \neq d_t$. Start with t the top element of S so $\sigma(t) = \overline{1}$. There exists $f^1 \in \mathbf{D}(0,1)$ for which $(\overline{1}, f^1) \in \tau$ by Claim 2. Continue and find $f^2 \in \mathbf{D}(\{0,1\})$ with $(f^1, f^2) \in \tau$ and $f^1 > f^2$. Eventually we get $(f^m, \overline{0}) \in \tau$, and by transitivity we have $(\overline{1}, \overline{0}) \in \tau$. From this it follows that τ collapses $\overline{0}, \overline{1}$ and the k generators of $\sigma(\mathbf{S})$ into one congruence class. This gives $|\mathbf{D}/\tau| \leqslant |\mathbf{A}/\mu| = |\mathbf{D}/\overline{\mu}|$, which contradicts $\tau < \overline{\mu}$.

Thus, the elements of S correspond to subcovers of $\overline{\mu}$, and $\overline{\mu}$ is recoverable from the atoms of $\mathsf{Con}\,\mathbf{D}$. Next we recover the comparability relation on $(S, \leqslant)$, up to order isomorphism, from the isomorphism type of $\mathbf{D}$.

Let γ be the binary relation of comparability on $(S, \leqslant)$, i.e.,

$$\gamma = \{(s,t) : s \leqslant t \text{ or } t \leqslant s\}.$$

For $s \neq t$ consider $\eta_s \wedge \eta_t \in \mathsf{Con}\,\mathbf{D}$. We have $(f,g) \in \eta_s \wedge \eta_t$ if and only if $f_t = g_t$ and $f_s = g_s$. The number of elements in $\mathbf{D}/(\eta_s \wedge \eta_t)$ is the number of congruence classes of $\eta_s \wedge \eta_t$, which is the cardinality of $D_{st} = \{(f_s, f_t) : f \in D\}$. The algebra $\mathbf{D}$ is generated by $\sigma(S) \cup \Delta$, and so D_{st}, which is a subuniverse of $\mathbf{A}^2$, is generated by $G_{st} = \{(\sigma(u)_s, \sigma(u)_t) : u \in S\} \cup \{(a,a) : a \in A\}$. We have observed that if s and t are incomparable, then G_{st} includes both $(0,1)$ and $(1,0)$, while if s and t are comparable, then G_{st} includes precisely one of these pairs. All the algebras $\mathbf{D}/(\eta_s \wedge \eta_t)$ have the same cardinality if $(s,t) \in \gamma$ as do all such algebras if $(s,t) \notin \gamma$, with the former algebras having the strictly smaller cardinality. We use this difference in cardinality to recover γ.

Define a binary relation Γ on $\Pi = \{\eta_t : t \in S\}$ by $(\eta_s, \eta_t) \in \Gamma$ if and only if $\mathbf{D}/(\eta_s \wedge \eta_t)$ has this smaller cardinality. We have (S, γ) and (Π, Γ) are isomorphic relations. So, up to isomorphism, the relation γ on S can be recovered from $\mathsf{Con}\,\mathbf{D}$.

We next associate with each $\mathbf{S} \in C$ a set $\Lambda(\mathbf{S})$ of ordered sets such that

(1) **S** is order isomorphic to a member of $\Lambda(\mathbf{S})$;
(2) The comparability relation of each ordered set in $\Lambda(\mathbf{S})$ is isomorphic to the comparability relation γ of **S**;
(3) $|\Lambda(\mathbf{S})| \leqslant 2^{k^2}$.

Suppose we have such a $\Lambda(\mathbf{S})$ for each $\mathbf{S} \in C$. If $\mathbf{S} \in C$ corresponds to $\mathbf{D_S} \in \mathcal{V}$ and (S, γ) is isomorphic to (Π, Γ), then from property (2), the set $\Lambda(\mathbf{S})$ can be constructed using (Π, Γ), and the relation (Π, Γ) can be determined using only the isomorphism type of $\mathbf{D_S}$. So if $\mathbf{S}, \mathbf{S}' \in C$ with $\Lambda(\mathbf{S}) \neq \Lambda(\mathbf{S}')$, then $\mathbf{D_S} \not\cong \mathbf{D_{S'}}$. From properties (1) and (3) of $\Lambda(\mathbf{S})$ we see that for a given $\mathbf{S} \in C$ at most 2^{k^2} of the $\mathbf{S}' \in C$ have $\Lambda(\mathbf{S}) = \Lambda(\mathbf{S}')$. Since $|C| \geqslant 2^{\binom{k}{\lfloor k/2 \rfloor}}/k!$ there are at least

$$g(k) = \frac{2^{\binom{k}{\lfloor k/2 \rfloor}}}{2^{k^2}k!}$$

k-generated semilattices $\mathbf{S} \in C$ for which the sets $\Lambda(\mathbf{S})$ are pairwise distinct. Thus there are at least $g(k)$ pairwise nonisomorphic $(k+n)$-generated algebras in $\mathcal{V}$, where $n = |A|$. We conclude that $\mathrm{G}_{\mathcal{V}}(k) \geqslant g(k-n)$, so $\mathcal{V}$ has many models.

It remains to construct $\Lambda(\mathbf{S})$ for each $\mathbf{S} \in C$ using only (S, γ). The top and bottom elements of **S** are recoverable from γ since $t \in S$ is such an element if and only if $(s, t) \in \gamma$ for all $s \in S$. We work with S^-, which is S with these two elements removed. Let $M \subseteq S^-$ be the set of k maximal elements. For any set $Z \subseteq S^-$, the property that Z is an antichain in $(S, \leqslant)$ is definable in terms of γ. For each k-element antichain Z in S^- define the binary predicate P_Z by

$$P_Z(s,t) \;\text{ iff }\; \{z \in Z : (s,z) \in \gamma\} \supseteq \{z \in Z : (t,z) \in \gamma\}.$$

Note that for $Z = M$ we have $P_M(s,t)$ if and only if $s \leqslant t$ in $(S^-, \leqslant)$. Let $\Lambda(\mathbf{S})$ denote all ordered sets, up to order isomorphism, that can be obtained by adjoining a top and bottom element to (S^-, P_Z) when P_Z is an order relation on S^- and Z ranges over the k-element antichains of S^-. Then $|\Lambda(\mathbf{S})| \leqslant \binom{|S|}{k} \leqslant (2^k)^k = 2^{k^2}$. Thus the three desired conditions for $\Lambda(\mathbf{S})$ hold, and the proof of Theorem 8.1 is complete. □

EXAMPLE 8.2. Consider a pseudocomplemented lattice $\mathbf{L} = \langle L, \wedge, \vee, ', 0, 1\rangle$ or a pseudocomplemented semilattice $\mathbf{L} = \langle L, \wedge, ', 0, 1\rangle$. In such an algebra, the map $a \mapsto a''$ is known to be a homomorphism, e.g. [**23**]. The Glivenko congruence relation γ on **L** is the kernel of this homomorphism and is given by $\gamma := \{(a,b) : a' = b'\}$. Covering pairs in $\mathsf{Con}\,\mathbf{L}$ that are below γ all have type **4** or **5** since the pseudocomplement operation is constant on congruence classes of γ. Thus, if $\gamma \neq 0_L$, then by Theorem 8.1 any variety containing **L** has many models. The only pseudocomplemented lattices or semilattices in which the Glivenko congruence is trivial are Boolean algebras. Thus, every nontrivial variety of pseudocomplemented lattices other than the variety of Boolean algebras has many models. For pseudocomplemented semilattices it is known that the only nontrivial variety other than the variety of Boolean algebras is the entire variety of pseudocomplemented semilattices, and so this is the only variety of pseudocomplemented semilattices that has many models.

COROLLARY 8.3. *Every locally finite variety with very few models is locally solvable.*

PROOF. It follows from Theorem 8.1 and Proposition 3.3 that if $\mathcal{V}$ has very few models, then $\text{typ}\{\mathcal{V}\} \subseteq \{\mathbf{1}, \mathbf{2}\}$. So $\mathcal{V}$ is locally solvable by [**32**, Theorem 7.11]. □

CHAPTER 9

Semisimple May Apply

In this Chapter we prove that every locally finite, congruence meet semidistributive variety that does not have many models is semisimple and is, in fact, congruence distributive.

THEOREM 9.1. *If $\mathcal{V}$ is a locally finite variety that contains a finite, subdirectly irreducible algebra with monolith of type* **3** *that is not simple, then $\mathcal{V}$ has many models.*

PROOF. Suppose $\mathbf{A} \in \mathcal{V}$ is a finite subdirectly irreducible algebra with monolith μ, $\text{typ}(0_A, \mu) = \mathbf{3}$, and there exists $\delta \in \mathsf{Con}\, \mathbf{A}$ that covers μ. Let $U \in M_{\mathbf{A}}(0_A, \mu)$ with $N = \{0, 1\}$ a $(0_A, \mu)$-trace in U. Since $\text{typ}(0_A, \mu) = \mathbf{3}$, the polynomial structure of $\mathbf{A}$ restricted to N is that of a Boolean algebra.

Let $k \geqslant 2$ be arbitrary. We choose and fix a set T of size 2^k. Let R be an equivalence relation on T. We will associate with R an algebra $\mathbf{D}_R \in \mathcal{V}$ that is at most $(2|A|^2k)$-generated, and if R_1 and R_2 are nonisomorphic equivalence relations, then $\mathbf{D}_{R_1} \not\cong \mathbf{D}_{R_2}$. Thus $\mathcal{V}$ has many models by Proposition 2.9.

Consider the Boolean algebra $\mathbf{B} = \{0, 1\}^T$. This algebra has 2^k atoms and is the free Boolean algebra on k free generators. Let $f_1, \ldots, f_k$ freely generate $\mathbf{B}$. For each $t \in T$ let $1^t \in \mathbf{B}$ have a 1 in coordinate t and 0's elsewhere. So each 1^t is an atom of $\mathbf{B}$. Let $B_1, \ldots, B_r$ be the equivalence classes of R. Form the subalgebra $\mathbf{S}$ of $\mathbf{B}$ that has as its atoms all elements of the form $\bigvee \{1^t : t \in B_i\}$ for $1 \leqslant i \leqslant r$. Every subalgebra of a k-generated Boolean algebra is also k-generated, so we may choose $g_1, \ldots, g_k \in S$ that generate $\mathbf{S}$. Note that if $(s, t) \in R$, then $g_i(s) = g_i(t)$.

For each $(a, b) \in A^2$ and $1 \leqslant i \leqslant k$ define $f_i^{ab} : T \to \{a, b\}$ by

$$f_i^{ab}(t) = \begin{cases} a, & \text{if } f_i(t) = 0, \\ b, & \text{if } f_i(t) = 1, \end{cases}$$

and $g_i^{ab} : T \to \{a, b\}$ by

$$g_i^{ab}(t) = \begin{cases} a, & \text{if } g_i(t) = 0, \\ b, & \text{if } g_i(t) = 1. \end{cases}$$

Let $\mathbf{D} = \mathbf{D}_R$ be the subalgebra of $\mathbf{A}^T$ generated by

$$G = \{f_i^{ab} : (a, b) \in \mu, 1 \leqslant i \leqslant k\} \cup \{g_i^{ab} : (a, b) \in \delta, 1 \leqslant i \leqslant k\}.$$

We note that $\mathbf{D}$ is a diagonal subalgebra of $\mathbf{A}^T$ generated by fewer than $2|A|^2k$ elements. The polynomials of $\mathbf{A}$ that make $\mathbf{A}|_N$ a Boolean algebra extend to $\mathbf{D}$, and so the $f_i^{01} = f_i$ generate all of $N^T \subseteq D$. Every $d \in D$ is δ-constant since every member of G is. If $(s, t) \in R$ and $d \in D$, then $(d_s, d_t) \in \mu$ for a similar reason. For all $(a, b) \in \mu$ and for all $s \neq t \in T$ there exists $d \in D$ such that $d_s = a$ and $d_t = b$. This is because $\{f_1, \ldots, f_k\}$ generates all of $\{0, 1\}^T \subseteq D$, so there exists an i for which $f_i(s) \neq f_i(t)$. Thus, f_i^{ab} or f_i^{ba} may serve as d. Similarly, for all $(a, b) \in \delta$ and $(s, t) \notin R$ there exists $d \in D$ such that $d_s = a$ and $d_t = b$.

CLAIM 1: For $t \in T$, if $\sigma > \eta_t$, then $\sigma \geqslant \overline{\mu}$ and $\overline{\mu} \vee \eta_t$ covers η_t.

Proof. Let $(c, d) \in \overline{\mu}$. Since $\sigma > \eta_t$ there exists $(u, v) \in \sigma$ with $u_t \neq v_t$. In **A** we have $(c_t, d_t) \in \mu \leqslant \mathrm{Cg}^{\mathbf{A}}(u_t, v_t)$. There exists a Maltsev chain in **A** from c_t to d_t using polynomials $\mathbf{p}_1, \ldots, \mathbf{p}_m$ applied to u_t and v_t. We extend these polynomials to $\overline{\mathbf{p}_i} \in \mathrm{Pol}\,\mathbf{D}$ and write w_i^0 for $\overline{\mathbf{p}_i}(u)$ and w_i^1 for $\overline{\mathbf{p}_i}(v)$. There exist $\epsilon_1, \ldots, \epsilon_m \in \{0, 1\}$ such that $w_1^{\epsilon_1}(t) = c_t, w_m^{1-\epsilon_m}(t) = d_t, (w_i^{\epsilon_i}, w_i^{1-\epsilon_i}) \in \mathrm{Cg}^{\mathbf{D}}(u, v)$ and $(w_i^{1-\epsilon_i}, w_{i+1}^{\epsilon_{i+1}}) \in \eta_t$. Thus $(c, w_1^{\epsilon_1}) \in \eta_t$ as is $(d, w_m^{1-\epsilon_m})$. Since $\eta_t \leqslant \sigma$ we get $(c, d) \in \sigma$ as desired. Moreover, $(\overline{0}, \overline{1}) \in \overline{\mu}$, so $\overline{\mu} \vee \eta_t > \eta_t$ and if $\overline{\mu} \vee \eta_t \geqslant \sigma > \eta_t$, then $\sigma \geqslant \overline{\mu}$, which implies $\overline{\mu} \vee \eta_t \succ \eta_t$.

We know that $\mathbf{D}/\eta_t \simeq \mathbf{A}$, and so η_t has a unique cover in $\mathsf{Con}\,\mathbf{D}$ and the type of this covering pair is **3**. Thus $\mathrm{typ}(\eta_t, \overline{\mu} \vee \eta_t) = \mathbf{3}$ for all $t \in T$.

CLAIM 2: For every atom $\alpha \in \mathsf{Con}\,\mathbf{D}$ there exist $t \in T$ and $u, v \in D$ such that $u_s = v_s$ for all $s \neq t$, $u_t = 0, v_t = 1$, and $\alpha = \mathrm{Cg}^{\mathbf{D}}(u, v)$. In particular, $\alpha \leqslant \overline{\mu}$.

Proof. Suppose $\alpha = \mathrm{Cg}^{\mathbf{D}}(c, d)$ and t is such that $c_t \neq d_t$. We have $(0, 1) \in \mu \leqslant \mathrm{Cg}^{\mathbf{A}}(c_t, d_t)$. There exists a Maltsev chain in **A** from 0 to 1 using polynomials $\mathbf{p}_1, \ldots, \mathbf{p}_m$, and since $0, 1 \in U$, we may assume each $\mathbf{p}_i : A \to U$. Extend each $\mathbf{p}_i$ to the polynomial $\overline{\mathbf{p}_i}$ on **D**. There exists an i for which $\{\overline{\mathbf{p}_i}(c)_t, \overline{\mathbf{p}_i}(d)_t\} = \{0, a\}$ for an element $a \in U$, $a \neq 0$. Without loss of generality, $\overline{\mathbf{p}_i}(c)_t = 0$. Then $(\overline{\mathbf{p}_i}(c) \vee 1^t, \overline{\mathbf{p}_i}(d) \vee 1^t) \in \alpha$ and thus $(\overline{\mathbf{p}_i}(c) \vee 1^t, \overline{\mathbf{p}_i}(d)) \in \alpha$ since $b \vee 0 = b$ for all $b \in U$ and $a \vee 1 = a$. We let $u = \overline{\mathbf{p}_i}(c) \vee 1^t$ and $v = \overline{\mathbf{p}_i}(c)$. Then $0_D < \mathrm{Cg}^{\mathbf{D}}(u, v) \leqslant \alpha$, and since α is an atom, equality holds, and the argument is complete.

CLAIM 3: For every atom $\alpha \in \mathsf{Con}\,\mathbf{D}$ there exists $t \in T$ for which η_t is the pseudo-complement of α.

Proof. Let α be an atom and suppose u, v and t are as in the previous Claim. Then $\alpha \wedge \eta_t = 0_D$. Also, $\eta_t < \alpha \vee \eta_t \leqslant \overline{\mu} \vee \eta_t$ since $\alpha \leqslant \overline{\mu}$. Thus $\alpha \vee \eta_t = \overline{\mu} \vee \eta_t$ since $\eta_t \prec \overline{\mu} \vee \eta_t$. So the prime intervals $(0_D, \alpha)$ and $(\eta_t, \overline{\mu} \vee \eta_t)$ are projective. Thus, $\mathrm{typ}(0_D, \alpha) = \mathbf{3}$. From [**32**, Lemma 5.12], α has a pseudocomplement in $\mathsf{Con}\,\mathbf{D}$, which must be η_t.

We also note that for every $t \in T$ there is an atom in $\mathsf{Con}\,\mathbf{D}$ whose pseudocomplement is η_t. Namely, take any atom below $\mathrm{Cg}^{\mathbf{D}}(\overline{0}, 1^t)$.

It follows that the set of all projection kernels $\{\eta_t : t \in T\}$ is recoverable from the lattice isomorphism type of $\mathsf{Con}\,\mathbf{D}$. They are precisely the set of all pseudo-complements of the atoms of $\mathsf{Con}\,\mathbf{D}$. We use them to recover the isomorphism type of R from $\mathbf{D}_R$.

For $s \neq t \in T$ and $c, d \in D$ we have $(c, d) \in \eta_s \wedge \eta_t$ if and only if $c_s = d_s$ and $c_t = d_t$. So $|\mathbf{D}/(\eta_s \wedge \eta_t)| = |\{(d_s, d_t) : d \in D\}|$. If $(s, t) \in R$, then, as observed when we defined **D**, $\{(d_s, d_t) : d \in D\} = \mu$, and if $(s, t) \notin R$, then $\{(d_s, d_t) : d \in D\} = \delta$. Since $\mu < \delta$ in $\mathsf{Con}\,\mathbf{D}$, we have $|\mu| < |\delta|$. So by an examination of the cardinalities of all $\mathbf{D}/(\eta_s \wedge \eta_t)$, we can determine the sizes of the equivalence classes of R. □

COROLLARY 9.2. *If $\mathcal{V}$ is a locally finite variety that does not have many models, then every type* **3** *minimal set has an empty tail.*

PROOF. In a finite algebra $\mathbf{A} \in \mathcal{V}$ we let U be a minimal set with respect to the type **3** prime quotient $\alpha \prec \beta$. We project this prime quotient up to $\eta \prec \eta^+$, where η is a congruence relation that is maximal with respect to the property of being above α but not above β. In particular we know that $\mathbf{A}/\eta$ is subdirectly irreducible with a type **3** monolith. From Theorem 9.1 we get that $\mathbf{A}/\eta$ is simple. Moreover, U is also an (η, η^+)–minimal set of $\mathbf{A}$, and consequently U/η is a $(0,1)$–minimal set of the type **3** simple algebra $\mathbf{A}/\eta$. Therefore U/η has no tail, which means that U is contained in a single η^+–class. So the (η, η^+)–tail of U is empty, i.e., $|U| = 2$. This shows that U has no (α, β)–tail. □

Recall that a variety $\mathcal{V}$ is meet semi-distributive if for any $\mathbf{A} \in \mathcal{V}$ and $\alpha, \beta, \gamma \in \mathsf{Con}\,\mathbf{A}$ with $\alpha \wedge \beta = \alpha \wedge \gamma$ we have $\alpha \wedge \beta = \alpha \wedge (\beta \vee \gamma)$. Recall also that according to Theorem 2.7, a locally finite variety $\mathcal{V}$ is congruence meet semi-distributive if and only if $\operatorname{typ}\{\mathcal{V}\} \subseteq \{\mathbf{3}, \mathbf{4}, \mathbf{5}\}$.

COROLLARY 9.3. *If $\mathcal{V}$ is a locally finite, congruence meet semi-distributive variety that does not have many models, then $\mathcal{V}$ is congruence distributive and semisimple.*

PROOF. If $\mathcal{V}$ does not have many models, then types **4** and **5** are not in $\operatorname{typ}\{\mathcal{V}\}$. The previous Corollary 9.2 and Theorem 8.6 of [**32**] show $\mathcal{V}$ is congruence distributive. Every subdirectly irreducible algebra in $\mathcal{V}$ has a monolith of type **3**, so $\mathcal{V}$ is semisimple by Theorem 9.1. □

EXAMPLE 9.4. We present an example of a 5-element algebra that generates a congruence meet semi-distributive variety $\mathcal{V}$ for which $\mathrm{G}_{\mathcal{V}}(k)$ is of triply exponential complexity. We exhibit such a variety by modifying the construction in Proposition 5.2. Let $\mathbf{A} = \langle \{1,2\}, \wedge, \mathbf{c} \rangle$ in which $\wedge$ is the meet operation with $1 < 2$ and $\mathbf{c}$ is complementation: $\mathbf{c}(1) = 2, \mathbf{c}(2) = 1$. Extend $\mathbf{A}$ to $\mathbf{A}' = \langle \{0,1,2,3,4\}, \wedge, \mathbf{c}, \mathbf{h} \rangle$ in which $\wedge$ is a meet semilattice with $0 \prec 1, 0 \prec 3, 0 \prec 4$, and $1 \prec 2$. Let $\mathbf{c}(1) = 2, \mathbf{c}(2) = 1$ and $\mathbf{c}(i) = 0$ otherwise. The operation $\mathbf{h}$ is unary, with $\mathbf{h}(1) = 3, \mathbf{h}(2) = 4$ and $\mathbf{h}(i) = 0$ otherwise.

Let $\mathcal{W}$ be the variety generated by $\mathbf{A}'$ and let $\mathbf{F}$ denote $\mathbf{F}_{\mathcal{W}}(k)$. Every k-ary operation on $\{1,2\}$ is the restriction of some member of $\mathbf{F}$ since $\mathbf{A}$ is a Boolean algebra. So there are at least 2^{2^k} distinct members of F of the form $\mathbf{h}(\mathbf{t})$, where $\mathbf{t}$ is a term built using only $\wedge$ and $\mathbf{c}$. Let T be a transversal of these terms. Let γ' be any equivalence relation on T and let γ be the congruence relation on $\mathbf{F}$ generated by γ'. Any nonconstant unary polynomial $\mathbf{p}$ on the original algebra $\mathbf{A}$ has the property that $\mathbf{p}(3) \in \{0,3\}$ and $\mathbf{p}(4) \in \{0,4\}$. From this it follows that $\gamma|_T = \gamma'$ and that $\gamma \in C_s(\mathbf{F}, X)$, where X is the set of free generators of $\mathbf{F}$. Hence there are at least $\mathrm{Bell}(2^{2^k})$ such γ. Since the subalgebra of $\mathbf{A}$ with universe $\{0,3,4\}$ is term equivalent to a semilattice we get that $\mathbf{F}$ is uniquely generated. So Corollary 3.8 can be invoked and we have that $\mathrm{G}_{\mathcal{W}}(k)$ is at least triply exponential. The variety $\mathcal{W}$ is finitely generated so $\mathrm{G}_{\mathcal{W}}(k)$ is at most triply exponential by Proposition 5.1. Thus $\mathcal{W}$ is of triply exponential generative complexity.

If $\mathbf{B}$ is any algebra in $\mathcal{W}$, then $\mathbf{B}$ has a semilattice reduct. Therefore types **1** and **2** are not in $\operatorname{typ}\{\mathbf{B}\}$ since if $(a,b) \in \theta \in \mathsf{Con}\,\mathbf{B}$, with say, $a \not\leq b$ in the semilattice order, then $(a, a \wedge b)$ is a 2-snag and $(a, a \wedge b) \in \theta$. So $\mathcal{W}$ omits types **1** and **2**, and is therefore a congruence meet semi-distributive variety.

CHAPTER 10

Permutable May also Apply

One of our goals is to characterize locally finite varieties omitting type **1** that do not have many models. In Chapter 11 we show that these varieties must be congruence modular. That result, combined with the next theorem, shows that they are, in fact, congruence permutable varieties.

THEOREM 10.1. *Let $\mathcal{V}$ be a locally finite, congruence modular variety. If $\mathcal{V}$ is not congruence permutable, then $\mathcal{V}$ has many models.*

PROOF. From congruence modularity and [**32**, Theorem 8.5] we have $\operatorname{typ}\{\mathcal{V}\} \subseteq \{\mathbf{2}, \mathbf{3}, \mathbf{4}\}$ and that the tails of the minimal sets of prime quotients are empty. From Theorem 8.1 we know $\mathbf{4} \notin \operatorname{typ}\{\mathcal{V}\}$. Any finite, subdirectly irreducible algebra in $\mathcal{V}$ with type **3** monolith is simple by Theorem 9.1. If $\mathcal{V}$ is not congruence permutable, then by [**34**] there is a finite $\mathbf{A} \in \mathcal{V}$ and two atoms α and β in $\mathsf{Con}\,\mathbf{A}$ that do not permute. We have $\{\operatorname{typ}(0_A, \alpha), \operatorname{typ}(0_A, \beta)\} \subseteq \{\mathbf{2}, \mathbf{3}\}$. Any atom of type **2** permutes with any congruence by [**22**, Theorem 6.2]. So we are left with $\operatorname{typ}(0_A, \alpha) = \operatorname{typ}(0_A, \beta) = \mathbf{3}$ and α and β do not permute. Let δ denote $\alpha \vee \beta$. Since $\mathsf{Con}\,\mathbf{A}$ is modular we have $\alpha \prec \delta$ and $\beta \prec \delta$. There are no other atoms below δ by Lemma 6.6 of [**32**].

We argue as in Theorem 9.1. For each $k \geqslant 2$ let T be a set of size 2^k. For each equivalence relation R on T that has every equivalence class of cardinality at least 2 we construct an algebra $\mathbf{D}_R \in \mathcal{V}$ that is $2k|A|^2$-generated. Since the number of partitions of the integer 2^k with all summands of size at least 2 is at least $\Pi(2^{k-1})$, and $\Pi(2^{k-1})$ is doubly exponential in k, we get doubly exponentially many such algebras. We argue that almost all of these algebras are pairwise nonisomorphic and conclude that $\mathcal{V}$ has many models.

Let $\{0,1\}$ be an arbitrary $(0_A, \alpha)$-trace. So this trace is polynomially equivalent to a Boolean algebra. By Corollary 9.2 the tails of the minimal sets of prime quotients of type **3** are empty so $M_{\mathbf{A}}(0_A, \alpha)$ consists of 2-element traces polynomially isomorphic to the Boolean algebra $\langle \{0,1\}; \wedge, \vee, \neg, \rangle$.

As in the proof of Theorem 9.1, the Boolean algebras $\mathbf{B}$ and $\mathbf{S}$ are in $\{0,1\}^T$ and are generated by the f_i and g_i respectively and we let $\mathbf{D} = \mathbf{D}_R$ be the subalgebra of $\mathbf{A}^T$ generated by

$$G = \{f_i^{ab} : (a,b) \in \alpha, 1 \leqslant i \leqslant k\} \cup \{g_i^{ab} : (a,b) \in \delta, 1 \leqslant i \leqslant k\}.$$

Thus the algebra $\mathbf{D}$ is diagonal subalgebra of $\mathbf{A}^T$ generated by $2|A|^2k$ elements with $\{0,1\}^T \subseteq D$. If N is any $(0_A, \alpha)$-trace, then $N^T \subseteq D$ as well. If $\{a,b\}$ is any $(0_A, \beta)$-trace, then by means of the g_i^{ab} and g_i^{ba} we see that D contains every $c \in \{a,b\}^T$ that is constant on every equivalence class of R. Every $d \in D$ is δ-constant and is α-constant on the equivalence classes of R.

Let $N = \{a, b\}$ be any $(0_A, \beta)$-trace, with a being the zero for the pseudo-meet operation on N. Consider any $(c, d) \in D^2$ and $t \in T$ such that $c_t \neq d_t$ and for which $\mathrm{Cg}^{\mathbf{A}}(c_t, d_t) \geqslant \beta$. Let Z be the equivalence class of R that contains t. There exists $\mathbf{p} \in \mathrm{Pol}_1\mathbf{A}$ with $\mathbf{p}(A) = N$, $\mathbf{p}(c_t) = a$, and $\mathbf{p}(d_t) = b$. Such a $\mathbf{p}$ exists because $N \in M_{\mathbf{A}}(0_A, \beta)$ and $\mathrm{typ}(0_A, \beta) = \mathbf{3}$. Let $\overline{\mathbf{p}}$ be the coordinatewise extension of $\mathbf{p}$ to $\mathbf{D}$. Since $\overline{\mathbf{p}}(c) \in N^T$ and elements of D are α-constant on the coordinates in Z and α and β are disjoint, it follows that $\overline{\mathbf{p}}(c)$ is constant on Z, as is $\overline{\mathbf{p}}(d)$. There exists $w \in N^T$ such that $w_s = a$ for all $s \notin Z$ and $w_s = b$ otherwise. We have

$$(\overline{\mathbf{p}}(c) \wedge w, \overline{\mathbf{p}}(d) \wedge w) = (\overline{a}, w) \in \mathrm{Cg}^{\mathbf{D}}(c, d).$$

Note that $\overline{a}$ and w do not depend on c, d, or t and that $(\overline{a}, w)$ is a 2-snag.

We first characterize the atoms of $\mathsf{Con}\,\mathbf{D}$. To this end, for each $Z \subseteq T$ that is an equivalence class of R define

$$\beta_Z = \big\{(c, d) \in D^2 : (c_t, d_t) \in \beta \text{ for all } t \in Z, \text{ and } c_t = d_t \text{ for all } t \notin Z\big\}.$$

It is immediate that $\beta_Z \in \mathsf{Con}\,\mathbf{D}$. If $(u, v) \in \beta_Z$ and there exists a t for which $u_t \neq v_t$, then for all $s \in Z$ we have $u_s \neq v_s$, since otherwise $u_t \overset{\alpha}{\equiv} u_s = v_s \overset{\alpha}{\equiv} v_t$ would imply $(u_t, v_t) \in \alpha \wedge \beta = 0_A$, a contradiction.

Suppose $0_D \prec \sigma \leqslant \beta_Z$. Let $(c, d) \in \sigma$ and $t \in Z$ be such that $c_t \neq d_t$ and let $N = \{a, b\}$ be any $(0_A, \beta)$-trace. For $w \in D$ as described above, we have $(\overline{a}, w) \in \sigma$ and since $(\overline{a}, w)$ is a 2-snag, we conclude $\mathrm{typ}(0_D, \sigma) = \mathbf{3}$. Let σ^* be the pseudocomplement of σ. As is the case for any pseudocomplement of an atom, σ^* is meet-irreducible in $\mathsf{Con}\,\mathbf{D}$. The two prime intervals $(0_D, \sigma)$ and $(\sigma^*, \sigma \vee \sigma^*)$ are projective, so $\mathrm{typ}(\sigma^*, \sigma \vee \sigma^*) = \mathbf{3}$. All finite, subdirectly irreducible algebras in $\mathcal{V}$ with type $\mathbf{3}$ monolith are simple. This implies $\sigma \vee \sigma^* = 1_D$. We must have $\sigma^* \wedge \beta_Z = 0_D$ for otherwise $\sigma^* \wedge \beta_Z$ contains an atom, say τ with $\tau \neq \sigma$. Now if $\tau = \mathrm{Cg}^{\mathbf{D}}(r, s)$, then as observed above, $(\overline{a}, w) \in \mathrm{Cg}^{\mathbf{D}}(r, s)$ and thus $(\overline{a}, w) \in \sigma^*$. But then $(\overline{a}, w) \in \sigma \wedge \sigma^*$, which is impossible, so $\sigma^* \wedge \beta_Z = 0_D$. If $\sigma < \beta_Z$, then $\{0_D, \sigma, \sigma^*, \beta_Z, 1_D\}$ forms a sublattice that is a pentagon. So we conclude that $\beta_Z = \sigma = \mathrm{Cg}^{\mathbf{A}}(\overline{a}, w)$ is an atom of $\mathsf{Con}\,\mathbf{D}$.

For every $t \in T$ define

$$\alpha_t = \big\{(c, d) \in D^2 : (a_t, b_t) \in \alpha \text{ and } a_s = b_s \text{ for all } s \neq t\big\}.$$

Each α_t is a congruence and the argument given for the β_Z, with the role of α and β interchanged and $Z = \{t\}$, shows that α_t is an atom in $\mathsf{Con}\,\mathbf{D}$.

We next argue that the only atoms in $\mathsf{Con}\,\mathbf{D}$ that are below $\overline{\delta}$ are of the form β_Z and α_t. Let $0_D \prec \sigma \leqslant \overline{\delta}$ and $(c, d) \in \sigma$ with $c_t \neq d_t$. In $\mathsf{Con}\,\mathbf{A}$ we have $\mathrm{Cg}^{\mathbf{A}}(c_t, d_t) \leqslant \delta$ so either $\mathrm{Cg}^{\mathbf{A}}(c_t, d_t) \geqslant \alpha$ or $\mathrm{Cg}^{\mathbf{A}}(c_t, d_t) \geqslant \beta$. We suppose the latter. Let Z be the equivalence class of R that contains t and let $N = \{a, b\}$ be any $(0_A, \beta)$-trace. As above, we have $(\overline{a}, w) \in \sigma$, but since $\beta_Z = \mathrm{Cg}^{\mathbf{D}}(\overline{a}, w)$, we have $\beta_Z = \sigma$. A similar argument applies if $\mathrm{Cg}^{\mathbf{A}}(c_t, d_t) \geqslant \alpha$.

Thus, in $\mathsf{Con}\,\mathbf{D}$ there are precisely $2^k + r$ atoms below $\overline{\delta}$, where r is the number of equivalence classes of R. Using only the isomorphism type of $\mathbf{D}$ we wish to distinguish the α_t from the β_Z, and to determine the pairs (α_t, β_Z) for $t \in Z$.

If $0_D \prec \sigma \leqslant \overline{\delta}$, then we have observed that the pseudocomplement σ^* of σ is covered by 1_D. Thus, by congruence modularity, $\sigma^* \wedge \overline{\delta} \prec \overline{\delta}$. We work with the $\sigma^* \wedge \overline{\delta}$ rather than the atoms themselves.

CLAIM 1: For every $t \in T$ and every equivalence class Z of R

$$\begin{aligned} \alpha_t^* \wedge \overline{\delta} &= \{(c,d) \in D^2 : (c_t, d_t) \in \beta\}, \\ \beta_Z^* \wedge \overline{\delta} &= \{(c,d) \in D^2 : (c_s, d_s) \in \alpha \text{ for all } s \in Z\}. \end{aligned}$$

Proof. We present the β_Z^* case; the α_t^* case is similar. Let γ denote

$$\{(c,d) \in D^2 : (c_s, d_s) \in \alpha \text{ for all } s \in Z\}.$$

Clearly γ is a congruence. Since elements of D are δ-constant, $\gamma \leqslant \overline{\delta}$. Also, $\gamma \wedge \beta_Z = 0_D$ since $\alpha \wedge \beta = 0_A$. Thus, $\beta_Z^* \wedge \overline{\delta} \geqslant \gamma$. For the reverse inclusion, let $(c,d) \in \beta_Z^* \wedge \overline{\delta}$. If $c_s = d_s$ for all $s \in Z$, then $(c,d) \in \gamma$. So suppose there is an $s \in Z$ for which $c_s \neq d_s$. Then either $\mathrm{Cg}^{\mathbf{A}}(c_s, d_s) \geqslant \alpha$ or $\mathrm{Cg}^{\mathbf{A}}(c_s, d_s) \geqslant \beta$. If the latter holds, then we can argue as above that there exist $\overline{a}, w \in D$ with $(\overline{a}, w) \in \mathrm{Cg}^{\mathbf{D}}(c,d) \leqslant \beta_Z^*$ and $(\overline{a}, w) \in \beta_Z$, which is a contradiction. Therefore $\mathrm{Cg}^{\mathbf{A}}(c_s, d_s) = \alpha$. Elements of D are α-constant on Z, so $(c,d) \in \gamma$.

Let σ and τ be binary relations and n a positive integer. By $\sigma \circ^n \tau$ we denote the binary relation $\sigma \circ \tau \circ \sigma \circ \cdots$, in which there are n factors appearing. The basic properties of $\circ^n$ are given in [**55**, p. 196]. If $\sigma \circ^n \tau = \tau \circ^n \sigma$, then σ and τ are said to n-permute. Our algebra $\mathbf{A}$ is finite so there exists an n that is maximal with the property that at least one of $\alpha \circ^n \beta \neq \delta$ or $\beta \circ^n \alpha \neq \delta$ holds. We have $n \geqslant 2$ from the assumption that α and β do not permute.

CLAIM 2:

(1) $\alpha_t^* \wedge \overline{\delta}$ and $\beta_Z^* \wedge \overline{\delta}$ permute for all $t \notin Z$;
(2) $\alpha_t^* \wedge \overline{\delta}$ and $\alpha_s^* \wedge \overline{\delta}$ n-permute for all $s, t \in T$;
(3) $\alpha_t^* \wedge \overline{\delta}$ and $\beta_Z^* \wedge \overline{\delta}$ do not n-permute for all $t \in Z$.

Proof. For (1), let $t \notin Z$. We show $(\alpha_t^* \wedge \overline{\delta}) \circ (\beta_Z^* \wedge \overline{\delta}) = \overline{\delta}$. Let $(c,d) \in \overline{\delta}$ and $s \in Z$ be arbitrary. We have $(c_t, d_s) \in \delta$ since elements of D are δ-constant. By selecting an appropriate g_i^{ab} there is an $h \in D$ for which $h_t = c_t$ and $h_s = d_s$. Then

$$c \overset{\alpha_t^* \wedge \overline{\delta}}{\equiv} h \overset{\beta_Z^* \wedge \overline{\delta}}{\equiv} d$$

since d and h are α-constant on Z.

To prove (2), since s and t are interchangeable, it suffices to show

$$(\alpha_t^* \wedge \overline{\delta}) \circ^n (\alpha_s^* \wedge \overline{\delta}) = \overline{\delta}.$$

Clearly, $\subseteq$ holds. By hypothesis $\alpha \circ^{n+1} \beta = \beta \circ^{n+1} \alpha = \delta$. Let $(c,d) \in \overline{\delta}$ be arbitrary. We have $(c_t, d_s) \in \delta$ since elements of D are δ-constant. Set $m = \lfloor \frac{n+1}{2} \rfloor$. From $\beta \circ^{n+1} \alpha = \delta$ there exist elements $a_0, a_1, \ldots, a_{2m} \in A$ for which

$$c_t = a_0 \overset{\beta}{\equiv} a_1 \overset{\alpha}{\equiv} a_2 \overset{\beta}{\equiv} a_3 \ldots a_{2m-1} \overset{\alpha}{\equiv} a_{2m} \overset{\beta}{\equiv} d_s,$$

where an extra β is padded at the end if n is odd. Thus, m is the number of instances of α in this expression. Use the appropriate f_i^{ab} to give $h_1, \ldots, h_m \in D$

with

$$\begin{array}{ll} h_1(t) = a_1, & h_1(s) = a_2; \\ h_2(s) = a_3, & h_2(t) = a_4; \\ h_3(t) = a_5, & h_3(s) = a_6; \\ \cdot & \cdot \\ \cdot & \cdot \\ \cdot & \cdot \\ h_m(t) = a_{2m-1}, & h_m(s) = a_{2m} \end{array}$$

if m is odd; and $h_m(s) = a_{2m-1}$, $h_m(t) = a_{2m}$ if m is even. Then

$$c \overset{\alpha_t^* \wedge \overline{\delta}}{\equiv} h_1 \overset{\alpha_s^* \wedge \overline{\delta}}{\equiv} h_2 \overset{\alpha_t^* \wedge \overline{\delta}}{\equiv} h_3 \cdots \equiv h_m \overset{\gamma}{\equiv} d,$$

with $\gamma = \alpha_t^* \wedge \overline{\delta}$ if m is even and $\gamma = \alpha_s^* \wedge \overline{\delta}$ if m is odd. So $(c,d) \in (\alpha_t^* \wedge \overline{\delta}) \circ^{m+1} (\alpha_s^* \wedge \overline{\delta})$. Since $m+1 = 1 + \lfloor \frac{n+1}{2} \rfloor \leqslant n$ for all $n \geqslant 2$, we are done with (2).

For (3), we know that α and β do not n-permute so either $\alpha \circ^n \beta \neq \delta$ or $\beta \circ^n \alpha \neq \delta$. Without loss of generality, we assume the former. Let $(a,b) \in \delta \backslash (\alpha \circ^n \beta)$. In $\mathbf{D}$ we have $(\overline{a}, \overline{b}) \in \overline{\delta}$. Thus, if $t \in Z$, then $(\overline{a}, \overline{b}) \notin (\beta_Z^* \wedge \overline{\delta}) \circ^n (\alpha_t^* \wedge \overline{\delta})$ for otherwise the projection on coordinate t would give $(a,b) \in \alpha \circ^n \beta$, as required.

For a fixed $\theta \in \mathsf{Con}\,\mathbf{D}$ a set $Q \subseteq \mathsf{Con}\,\mathbf{D}$ is called θ-*good* if

(1) $|Q| = |T|$;
(2) each $\sigma \in Q$ is an atom of $\mathsf{Con}\,\mathbf{D}$ with $\sigma \leqslant \theta$;
(3) each $\sigma \in Q$ has a pseudocomplement σ^*;
(4) if $\sigma, \tau \in Q$, then $\sigma^* \wedge \theta$ and $\tau^* \wedge \theta$ are n-permutable.

CLAIM 3: There is a unique $\overline{\delta}$-good subset of $\mathsf{Con}\,\mathbf{D}$.

Proof. One such set is $Q = \{\alpha_t : t \in T\}$. If any other $\overline{\delta}$-good set exists, then it must be obtained from Q by replacing some of the α_t with an equal number of β_Z. (This is because every atom below $\overline{\delta}$ is an α_t or a β_Z.) But because of our choice of R, every Z has cardinality at least 2. So adjoining a β_Z forces the removal of all α_t for $t \in Z$, by virtue of (3) of the previous Claim. The different Z are pairwise disjoint, so the net effect of adjoining a β_Z to the proposed set of atoms would be to decrease its cardinality by at least one. So Q is the unique $\overline{\delta}$-good subset.

Define Γ to be the set of all $\gamma \in \mathsf{Con}\,\mathbf{D}$ for which $\mathbf{D}/\gamma \simeq \mathbf{A}/\delta$ and there is a unique γ-good subset $Q_\gamma \subseteq \mathsf{Con}\,\mathbf{D}$. Then $\overline{\delta} \in \Gamma$ by the previous Claim. We have

$$|\Gamma| < |A|^{2k|A|^2}$$

since every $\gamma \in \Gamma$ corresponds to a map of the generators of $\mathbf{D}$ to $\mathbf{A}/\delta$.

For each $\gamma \in \Gamma$ define a relation $\equiv_\gamma$ on Q_γ by $\sigma \equiv_\gamma \tau$ if and only if there is an atom $\nu < \gamma$, $\nu \notin Q_\gamma$, such that

(1) ν has a pseudocomplement ν^*,
(2) $\sigma^* \wedge \gamma$ and $\nu^* \wedge \gamma$ do not permute,
(3) $\tau^* \wedge \gamma$ and $\nu^* \wedge \gamma$ do not permute.

Note that $\alpha_s \equiv_{\overline{\delta}} \alpha_t$ if and only if s and t are in the same equivalence class of R. If γ is such that $\equiv_\gamma$ is an equivalence relation on Q_γ, then we choose R_γ that is an equivalence relation on T isomorphic to $\equiv_\gamma$. In particular, $R_{\overline{\delta}} \simeq R$. Let $\pi(R_\gamma)$ be the partition of the integer $2^k = |T|$ determined by the cardinalities of the equivalence classes of R_γ. The partition $\pi(R_\gamma)$ determines the isomorphism class of R_γ. Define

$$C_R = \{\pi(R_\gamma) : \gamma \in \Gamma \text{ and } \equiv_\gamma \text{ is an equivalence relation on } Q_\gamma\}.$$

Suppose that R and R' are both equivalence relations on T with all equivalence classes having at least two elements and $\mathbf{D}_R$ and $\mathbf{D}_{R'}$ are the corresponding algebras in $\mathcal{V}$. $C_R \neq C_{R'}$ implies $\mathbf{D}_R \not\cong \mathbf{D}_{R'}$ because C_R is determined by algebraic properties of $\mathbf{D}_R$ and $\mathsf{Con}\,\mathbf{D}_R$. We know $|C_R|$ is at most singly exponential in k since $|C_R| \leqslant |A|^{2k|A|^2}$. The number of partitions $\pi(R')$ with $C_R = C_{R'}$ is at most $|C_R|$ because $\pi(R) \in C_R$. So the number of pairwise nonisomorphic $\mathbf{D}_R$ that can be constructed is at least $\Pi(2^{k-1})/|A|^{2k|A|^2}$, which is at least doubly exponential in k. So $\mathcal{V}$ has many models. □

COROLLARY 10.2. *Let $\mathcal{V}$ be a locally finite variety that is congruence meet semi-distributive. If $\mathcal{V}$ does not have many models, then $\mathcal{V}$ is semisimple and arithmetical.*

PROOF. By Theorem 2.7, the congruence meet semi-distributivity of $\mathcal{V}$ gives $\mathrm{typ}\{\mathcal{V}\} \subseteq \{\mathbf{3}, \mathbf{4}, \mathbf{5}\}$. $\mathcal{V}$ is congruence distributive and semisimple by Corollary 9.3. Congruence permutability follows from Theorem 10.1. □

The converse of Corollary 10.2 fails in a very strong sense as witnessed by the locally finite discriminator varieties described in Example 5.12. They show that the G-spectrum in this case can be arbitrarily large. If, however, we restrict to finitely generated varieties, we have the following.

COROLLARY 10.3. *The following conditions are equivalent for any finitely generated, congruence meet semi-distributive, nontrivial variety $\mathcal{V}$.*

(1) *$\mathcal{V}$ has few models.*
(2) *$\mathcal{V}$ does not have many models.*
(3) *$\mathcal{V}$ is semisimple arithmetical.*
(4) *There exist positive constants b and c such that $2^{bk} \leqslant \mathrm{G}_{\mathcal{V}}(k) \leqslant 2^{ck}$ for all k.*

PROOF. Corollary 10.2 shows (2) implies (3) and Corollary 6.6 gives (3) implies (4). The remaining implications are trivial. □

EXAMPLE 10.4. If we apply Corollary 10.3 to varieties of Heyting algebras, then we see that the only nontrivial variety of Heyting algebras with few models is the variety of Boolean algebras since all others contain a finite subdirectly irreducible algebra that is not simple. On the other hand, the variety of implication algebras generated by a 2-element algebra is semisimple and congruence distributive, but does not have few models since it is not arithmetical. In [**6**] this variety of implication algebras is shown to have many models and an explicit doubly exponential lower bound for its G-spectrum is presented.

CHAPTER 11

Forcing Modular Behavior

In this Chapter we consider a finite subdirectly irreducible algebra $\mathbf{A}$ having a type $\mathbf{2}$ monolith and show that if the congruence lattice of $\mathbf{A}$ has certain other properties, then the variety generated by $\mathbf{A}$ has many models. In particular we show that any locally finite variety omitting type $\mathbf{1}$ is either congruence permutable or has many models.

To this end, for an algebra $\mathbf{A}$ we say the triple $(\beta, \gamma, \delta) \in (\mathsf{Con}\ \mathbf{A})^3$ is *skew* if

(1) $0_A \prec \beta \leqslant \gamma \prec \delta$ in $\mathsf{Con}\ \mathbf{A}$ with $\mathrm{typ}(0_A, \beta) = \mathbf{2}$ and $\mathrm{typ}(\gamma, \delta) = \mathbf{3}$;
(2) there exists $\{0, 1\} \in M_{\mathbf{A}}(\gamma, \delta)$ with
(3) $C(\beta, \{0,1\}^2; 0_A)$ and
(4) $\delta = \mathrm{Cg}^{\mathbf{A}}(0, 1)$.

We wish to show that for such an algebra $\mathbf{A}$, the variety generated by $\mathbf{A}$ has many models. The proof splits into two cases depending on whether or not there is a unary polynomial $\mathbf{p}$ for $\mathbf{A}$ such that $\mathbf{p}(0) \neq \mathbf{p}(1)$ with $\{\mathbf{p}(0), \mathbf{p}(1)\}$ contained in a $(0_A, \beta)$-trace. The next two lemmas deal with these two cases. Some ideas in our proof and some of our notation come from the proof of Lemma 10.2 in [**32**].

LEMMA 11.1. *Let $\mathbf{A}$ be a finite algebra with $(\beta, \gamma, \delta) \in (\mathsf{Con}\ \mathbf{A})^3$ a skew triple. If there is unary polynomial $\mathbf{p} \in \mathrm{Pol}_1 \mathbf{A}$ and a body B of a $(0_A, \beta)$-minimal set such that $(\mathbf{p}(0), \mathbf{p}(1)) \in \beta|_B - 0|_B$, then the variety generated by $\mathbf{A}$ has many models.*

PROOF. Let k be an arbitrary positive integer and X a set of size 2^k. In our proof, we first build a $(k + |A|)$-generated diagonal subalgebra $\mathbf{D}$ of $\mathbf{A}^X$. We show that the congruence $\overline{\delta}$ of $\mathbf{D}$ has $|X|$ subcovers and $\mathbf{D}/\overline{\delta}$ is isomorphic to $\mathbf{A}/\delta$. Let R be an arbitrary equivalence relation on X. For each equivalence class Z of R we find a congruence $\theta'_Z \in \mathsf{Con}\ \mathbf{D}$ such that the set of subcovers of $\overline{\delta}$ above θ'_Z is a block of a partition of these subcovers, and the equivalence relation determined by this partition is isomorphic to the relation R. We define the congruence θ_R to be the meet of all the θ'_Z, and we work in the algebra $\mathbf{D}/\theta_R$, which we call $\mathbf{D}_R$. Each $\mathbf{D}_R$ is $(k+|A|)$-generated. We show that there is a bijection from X to the set of all subcovers of the congruence $\overline{\delta}/\theta_R$ of $\mathbf{D}_R$. We let $\Delta = \{\sigma \in \mathsf{Con}\ \mathbf{D}_R : \mathbf{D}_R/\sigma \simeq \mathbf{A}/\delta\}$ and show that

(1) $|\Delta| \leqslant |A|^{k+|A|}$,
(2) $\overline{\delta}/\theta_R \in \Delta$,
(3) for each $\sigma \in \Delta$ we define a particular equivalence relation E_σ on the set of all subcovers of σ, and E_σ can be defined using only algebraic properties of $\mathsf{Con}\ \mathbf{D}_R$,
(4) the equivalence relation $E_{\overline{\delta}/\theta_R}$ is isomorphic to R.

Having made these constructions we argue as follows. Let R' be another equivalence relation on X and form the algebra $\mathbf{D}_{R'}$, the set of congruences Δ' and the congruence $\overline{\delta}/\theta_{R'} \in \Delta'$ for this relation. If there exists a $\sigma \in \Delta$ for which E_σ is not isomorphic to any $E_{\sigma'} \in \Delta'$, then we can conclude by property (3) that the algebras $\mathbf{D}_R$ and $\mathbf{D}_{R'}$ are not isomorphic. Since $\overline{\delta}/R' \in \Delta'$ and (4) holds, we know there are at most $|\Delta|$ pairwise nonisomorphic R' for which $\mathbf{D}_R \simeq \mathbf{D}_{R'}$. From Proposition 2.9 we know that there are, as a function of k, at least doubly exponentially many pairwise nonisomorphic equivalence relations on X. Since $|\Delta|$ is at most singly exponential, we conclude that there are at least doubly exponentially many $(k + |A|)$-generated pairwise nonisomorphic algebras $\mathbf{D}_R$ in the variety generated by $\mathbf{A}$. Thus, this variety has many models.

We now prove a series of claims to flesh out this argument.

Let $\mathbf{D}$ be the diagonal subalgebra of $\mathbf{A}^X$ generated by $\{0,1\}^X \cup \{\overline{c} : c \in A\}$. As in the proof of Theorem 9.1 the set $\{0,1\}^X$ may be viewed as a Boolean algebra with 2^k atoms that is freely generated by elements $f_1, \ldots, f_k$. Since $\{0,1\}$ is a trace of type $\mathbf{3}$, the algebra $\mathbf{D}$ is $(k + |A|)$-generated by $\{f_1, \ldots, f_k\} \cup \{\overline{c} : c \in A\}$.

Because all of the f_i and $\overline{c}$ that generate $\mathbf{D}$ are δ-constant, Proposition 2.1 gives the following.

CLAIM 1: $\mathbf{D}/\overline{\delta} \simeq \mathbf{A}/\delta$.

We next wish to code X using the subcovers in $\mathsf{Con}\,\mathbf{D}$ of $\overline{\delta}$. For each $t \in X$ let $1^t \in D$ be given by

$$1^t(s) = \begin{cases} 1, & \text{if } s = t, \\ 0, & \text{if } s \neq t. \end{cases}$$

CLAIM 2: If $\theta \prec \overline{\delta}$, then $\mathrm{typ}(\theta, \overline{\delta}) = \mathbf{3}$ and there exists a unique $t \in X$ with $(1^t, \overline{0}) \notin \theta$.

Proof. First suppose that $(1^t, \overline{0}) \in \theta$ for all $t \in X$. By extending the pseudo-join on $\{0,1\}$ to $\mathbf{D}$, we have $(\overline{1}, \overline{0}) \in \theta$, and in fact, $(f, g) \in \theta$ for all $f, g \in \mathbf{D}(0,1)$. Therefore each of the $k + |A|$ generators of $\mathbf{D}$ is congruent modulo θ to $\overline{c}$ for some $c \in A$. Consequently, the map $h : \mathbf{A} \to \mathbf{D}/\theta$ given by $h(c) = \overline{c}/\theta$ is a surjective homomorphism with $h(0) = h(1)$. The skew triple condition (4) gives $\delta = \mathrm{Cg}^{\mathbf{A}}(0,1) \subseteq \ker(h)$. Thus, there is a surjective homomorphism from $\mathbf{A}/\delta$ onto $\mathbf{D}/\theta$. But $|D/\overline{\delta}| = |A/\delta|$ by the previous Claim, yet $|A/\delta| \geqslant |D/\theta| > |D/\overline{\delta}|$, which gives a contradiction. So there is at least one t for which $(1^t, \overline{0}) \notin \theta$.

Since $\{0,1\}$ is a (γ, δ)-minimal set by the skew triple condition (2), there exists an idempotent unary polynomial on $\mathbf{A}$ with range $\{0,1\}$. Let $\mathbf{e}$ be the extension of this polynomial to $\mathbf{D}$. Then the polynomial $1^t \wedge \mathbf{e}$ has range $\{1^t, \overline{0}\}$ and does not collapse $\overline{\delta}$ to θ. Consequently $\{1^t, \overline{0}\} \in M_{\mathbf{D}}(\theta, \overline{\delta})$, which shows $\mathrm{typ}(\theta, \overline{\delta}) = \mathbf{3}$.

Finally, note that the map $\psi \mapsto \psi|_{\mathbf{D}(0,1)}$ of $\mathsf{Con}\,\mathbf{D}$ to the congruence lattice of the Boolean algebra $\mathbf{D}(0,1)$ is an onto lattice homomorphism that sends $\overline{\delta}$ to the top element of $\mathsf{Con}\,\mathbf{D}(0,1)$. Since $\theta \prec \overline{\delta}$ and $(1^t, \overline{0}) \notin \theta$ it follows that $\theta|_{\mathbf{D}(0,1)}$ is a coatom in $\mathsf{Con}\,\mathbf{D}(0,1)$. So t is the unique member of X with this property.

CLAIM 3: The map from the set of subcovers of $\overline{\delta}$ to X given by sending θ to the unique $t \in X$ for which $(1^t, \overline{0}) \notin \theta$ is a bijection.

Proof. By the previous Claim this mapping is well defined. We call it h.

To see that h is injective, we note that by [**32**, Lemma 5.15] for $\theta \prec \overline{\delta}$ there is a smallest congruence θ^* with $\theta \vee \theta^* = \overline{\delta}$ since $\mathrm{typ}(\theta, \overline{\delta}) = \mathbf{3}$. Consequently, for any $\alpha \leqslant \overline{\delta}$ we have either $\alpha \leqslant \theta$ or $\theta^* \leqslant \alpha$. If $h(\theta) = t$, then $(1^t, \overline{0}) \notin \theta$, and so $\mathrm{Cg}^{\mathbf{D}}(1^t, \overline{0}) \geqslant \theta^*$. On the other hand, $(1^t, \overline{0}) \in \overline{\delta} = \theta^* \vee \theta$ so that $(1^t, \overline{0}) \in \theta^*|_{\mathbf{D}(0,1)} \vee \theta|_{\mathbf{D}(0,1)}$. But $\mathrm{Cg}^{\mathbf{D}(0,1)}(1^t, \overline{0})$ is an atom in the distributive lattice $\mathsf{Con}\, \mathbf{D}(0,1)$ and $(1^t, \overline{0}) \notin \theta$, so $(1^t, \overline{0}) \in \theta^*|_{\mathbf{D}(0,1)}$. Thus, $\theta^*|_{\mathbf{D}(0,1)} = \mathrm{Cg}^{\mathbf{D}(0,1)}(1^t, \overline{0})$. Now, let θ_1 and θ_2 be two subcovers of $\overline{\delta}$ with $h(\theta_1) = h(\theta_2) = t$. If $\theta_1 \neq \theta_2$, then $\theta_1 \nleqslant \theta_2$. Thus, $\theta_2^* \leqslant \theta_1$. This implies that $(1^t, \overline{0}) \in \theta_2^* \leqslant \theta_1$, which contradicts $h(\theta_1) = t$.

Finally, to show that h is surjective, for $t \in X$ define

$$\theta_t = (\delta \times \cdots \times \delta \times \gamma \times \delta \times \cdots \times \delta)|_D$$

where the γ appears in coordinate t. Then $\theta_t < \overline{\delta}$ since $(1^t, \overline{0}) \notin \theta_t$. On the other hand, if θ is any subcover of $\overline{\delta}$ that contains θ_t, then θ must contain $(1^s, \overline{0})$ for every $t \neq s \in X$. So $h(\theta) = t$.

So Claim 3 gives a correspondence between X and the subcovers of $\overline{\delta}$. For $t \in X$ we define

$$\gamma_t = \text{ the unique subcover of } \overline{\delta} \text{ with } (1^t, \overline{0}) \notin \gamma_t.$$

Moreover, let γ_t^* be the smallest congruence of $\mathbf{D}$ with $\gamma_t \vee \gamma_t^* = \overline{\delta}$. We have seen in the proof of Claim 3 that the following holds:

FACT 4: For every $\alpha \leqslant \overline{\delta}$ and all $t \in X$ either $\alpha \leqslant \gamma_t$ or $\gamma_t^* \leqslant \alpha$.

We also see that $(1^t, \overline{0}) \in \gamma_t^*$ and $(1^t, \overline{0}) \notin \gamma_t$. So, Fact 4 applied to $\mathrm{Cg}^{\mathbf{D}}(1^t, \overline{0})$ yields our next observation.

FACT 5: $\gamma_t^* = \mathrm{Cg}^{\mathbf{D}}(1^t, \overline{0})$.

Let U be a $(0_A, \beta)$-minimal set that has B as its body, and suppose $\mathbf{p}(1) = a$ and $\mathbf{p}(0) = e$, with $(a, e) \in \beta|_B - 0_A$. Let $N = U \cap e/\beta$ be a $(0_A, \beta)$-trace. Since $\mathrm{typ}(0_A, \beta) = \mathbf{2}$, the algebra $\mathbf{A}|_N$ induced on the trace N is polynomially equivalent to a vector space over some finite field. Without loss of generality we may assume that e is the zero element of this vector space.

We introduce some notation. Define $a^t \in A^X$ by

$$a^t(s) = \begin{cases} a, & \text{if } s = t, \\ e, & \text{if } s \neq t. \end{cases}$$

Note that $a^t \in D$ since the extension of $\mathbf{p}$ to $\mathbf{D}$ sends 1^t to a^t. For $Z \subseteq X$ put

$$\begin{aligned} \Sigma_Z &:= \left\{ f \in D : (\forall t \in Z)(f(t) \in N) \text{ and } \sum_{t \in Z} f(t) = e \right\}, \\ \theta_Z &:= \mathrm{Cg}^{\mathbf{D}}(\Sigma_Z \times \Sigma_Z), \\ \eta_Z &:= \{(f, g) \in D^2 : (\forall t \in Z) f(t) = g(t)\}, \\ \theta'_Z &:= \eta_Z \vee \theta_Z. \end{aligned}$$

Initially in our proof Z is an arbitrary subset of X but later each Z will be an equivalence class of the equivalence relation R on X.

CLAIM 6: $\overline{e}/\theta_Z \cap \mathbf{D}(U) = \Sigma_Z \cap \mathbf{D}(U)$.

Proof. The inclusion $\supseteq$ is immediate since $\overline{e} \in \Sigma_Z$. In order to obtain the reverse inclusion we define

$$\theta^* = \{(f,g) \in D^2 : \text{ for all } \mathbf{h} \in \operatorname{Pol}_1\mathbf{D}, \text{ if } \mathbf{h}(f), \mathbf{h}(g) \in \mathbf{D}(U), \\ \text{then } \mathbf{h}(f) \in \Sigma_Z \text{ iff } \mathbf{h}(g) \in \Sigma_Z\}.$$

It is easily checked that θ^* is a congruence relation of $\mathbf{D}$. The set U is the image of an idempotent polynomial of $\mathbf{A}$ and so $D(U)$ is the image of an idempotent polynomial of $\mathbf{D}$. Thus, since $\overline{e} \in \Sigma_Z$, it follows that $\overline{e}/\theta^* \cap \mathbf{D}(U) \subseteq \Sigma_Z$. So in order to show the $\subseteq$ inclusion of the Claim it suffices to show $\theta_Z \leqslant \theta^*$ in $\mathsf{Con}\,\mathbf{D}$. We do this by proving

$$\Sigma_Z \times \Sigma_Z \subseteq \theta^*.$$

To establish this, let $f^0, f^1 \in \Sigma_Z, \mathbf{h} \in \operatorname{Pol}_1\mathbf{D}$, with $\mathbf{h}(f^0), \mathbf{h}(f^1) \in \mathbf{D}(U)$, and $\mathbf{h}(f^0) \in \Sigma_Z$. We wish to show $\mathbf{h}(f^1) \in \Sigma_Z$. For the unary polynomial $\mathbf{h}$ we can find an integer m, an $\mathbf{h}' \in \operatorname{Pol}_{m+1}\mathbf{A}$, and an m-tuple $\overline{g}$ of elements of $\mathbf{D}(0,1)$ for which $\mathbf{h}(x)(t) = \mathbf{h}'(\overline{g}(t), x(t))$ for all $t \in X$. By prefacing $\mathbf{h}'$ with the appropriate idempotent polynomial, we may, without loss of generality, assume that the range of $\mathbf{h}'$ is contained in U. For all $t \in Z$, our hypotheses give $\mathbf{h}'(\overline{g}(t), f^0(t)) = \mathbf{h}(f^0(t)) \in N$, and therefore for all $t \in Z$ and all $u \in N$, we have $\mathbf{h}'(\overline{g}(t), u) \in N$.

Choose and fix $t_0 \in Z$. We invoke the assumption $C(\beta, \{0,1\}^2; 0_A)$ to

$$\mathbf{h}'(\overline{g}(t), \underline{e}) - \mathbf{h}'(\overline{g}(t), e) = e = \mathbf{h}'(\overline{g}(t_0), \underline{e}) - \mathbf{h}'(\overline{g}(t_0), e)$$

and replace the underlined elements to obtain

$$\mathbf{h}'(\overline{g}(t), u) - \mathbf{h}'(\overline{g}(t), e) = \mathbf{h}'(\overline{g}(t_0), u) - \mathbf{h}'(\overline{g}(t_0), e)$$

for all $t \in Z$ and all $u \in N$. Thus,

$$\mathbf{h}'(\overline{g}(t), u) = \mathbf{h}'(\overline{g}(t_0), u) - \mathbf{h}'(\overline{g}(t_0), e) + \mathbf{h}'(\overline{g}(t), e).$$

But the map $u \mapsto \mathbf{h}'(\overline{g}(t_0), u) - \mathbf{h}'(\overline{g}(t_0), e)$ is a polynomial of the algebra $\mathbf{A}|_N$ sending e to e. So it must be of the form $u \mapsto \lambda \cdot u$ for some element λ in the field. So, $\mathbf{h}'(\overline{g}(t), u) = \lambda \cdot u + \mathbf{h}'(\overline{g}(t), e)$ for all $t \in Z$ and all $u \in N$. In particular, for all $t \in Z$ and $i = 0$ or 1,

$$\mathbf{h}(f^i)(t) = \mathbf{h}'(\overline{g}(t), f^i(t)) = \lambda \cdot f^i(t) + \mathbf{h}'(\overline{g}(t), \overline{e}(t)) = \lambda \cdot f^i(t) + \mathbf{h}(\overline{e})(t).$$

Thus, for all $t \in Z$,

$$\mathbf{h}(f^1)(t) - \mathbf{h}(f^0)(t) = \lambda(f^1(t) - f^0(t)),$$

which gives

$$\sum_{t\in Z}(\mathbf{h}(f^1)(t) - \mathbf{h}(f^0)(t)) = \lambda(\sum_{t\in Z} f^1(t) - \sum_{t\in Z} f^0(t)) = \lambda \cdot (e - e) = e.$$

Therefore,

$$\sum_{t\in Z}\mathbf{h}(f^1)(t) = \sum_{t\in Z}\mathbf{h}(f^0)(t), \text{ and by hypothesis } \sum_{t\in Z}\mathbf{h}(f^0)(t) = e,$$

and so $\mathbf{h}(f^1) \in \Sigma_Z$.

We give some immediate consequences of this result.

CLAIM 7: For every $s, t \in Z \subseteq X$ we have

(1) $(a^t, \overline{e}) \notin \theta_Z$,
(2) $(a^t, a^s) \in \theta_Z$,

(3) $(a^t, \overline{e}) \notin \theta'_Z$.

Proof. The first is because $a^t \notin \Sigma_Z$. The second holds because $a^t - a^s \in \Sigma_Z$ so $(a^t - a^s, \overline{e}) \in \theta_Z$, which gives $(a^t - a^s + a^s, \overline{e} + a^s) = (a^t, a^s) \in \theta_Z$. For the third part, note that for all $f \in \Sigma_Z$ we have $f/\eta_Z \subseteq \Sigma_Z$. Thus, by Claim 6, $\overline{e}/\theta'_Z \cap \mathbf{D}(U) = \overline{e}/\theta_Z \cap \mathbf{D}(U)$. This shows the third part follows from the first.

The next Claim will allow us to partition the subcovers of $\overline{\delta}$ using the congruences θ'_Z.

CLAIM 8: For $Z \subseteq X$ and $t \in X$ we have $\theta'_Z \leqslant \gamma_t$ if and only if $t \in Z$.

Proof. If $t \notin Z$, then $(1^t, \overline{0}) \in \eta_Z \leqslant \theta'_Z$. So $\theta'_Z \leqslant \gamma_t$ would give $(1^t, \overline{0}) \in \gamma_t$, contradicting the definition of γ_t. For the other direction, suppose $\theta'_Z \nleqslant \gamma_t$. Then $\gamma_t^* \leqslant \theta'_Z$ and thus $(1^t, \overline{0}) \in \theta'_Z$ by Facts 4 and 5. This means $(a^t, \overline{e}) = (\mathbf{p}(1^t), \mathbf{p}(\overline{0})) \in \theta'_Z$, which gives $t \notin Z$ by Claim 7.

For the remainder of the proof of the Lemma, $Z \subseteq X$ denotes an equivalence class of the equivalence relation R. We write X/R for the set of equivalence classes of R. Our coding of R is based on the following definitions.

$$\theta_R = \bigcap_{t \in X} \theta'_{t/R} \quad \text{and} \quad \mathbf{D}_R = \mathbf{D}/\theta_R.$$

Thus, θ_R is the intersection of all θ'_Z where Z ranges over all the equivalence classes of R.

Our next Claim allows us to work in $\mathsf{Con}\, \mathbf{D}$ instead of $\mathsf{Con}\, \mathbf{D}_R$.

CLAIM 9: $\theta_R \leqslant \gamma_t$ for every $t \in X$.

Proof. Suppose instead there is a t for which $\theta_R \nleqslant \gamma_t$. From Facts 4 and 5 we have $(1^t, \overline{0}) \in \gamma_t^* \leqslant \theta_R \leqslant \theta'_{t/R}$. But from the previous Claim we have $\theta'_{t/R} \leqslant \gamma_t$. This gives $(1^t, \overline{0}) \in \gamma_t$, a contradiction.

We define

$$\Delta = \{\sigma \in \mathsf{Con}\, \mathbf{D}_R : \mathbf{D}_R/\sigma \simeq \mathbf{A}/\delta\}.$$

The cardinality of Δ is bounded above by the number of maps of a generating set of $\mathbf{D}_R$ to A/δ. So $|\Delta| \leqslant |A|^{k+|A|}$. This establishes the first of the four properties of Δ in our outline of the proof, and the next Claim gives the second.

CLAIM 10: $\overline{\delta}/\theta_R \in \Delta$.

Proof. The previous Claim shows $\theta_R \leqslant \overline{\delta}$, so $\overline{\delta}/\theta_R \in \mathsf{Con}\, \mathbf{D}_R$. We have seen that $\mathbf{D}/\overline{\delta} \simeq \mathbf{A}/\delta$. But $\mathbf{D}_R$ is $\mathbf{D}/\theta_R$, so $\mathbf{D}_R/(\overline{\delta}/\theta_R) \simeq \mathbf{D}/\overline{\delta}$ by a basic isomorphism theorem.

We say a set $\Psi \subseteq \mathsf{Con}\, \mathbf{D}_R$ *determines a partition* of $\sigma \in \Delta$ if the following three conditions hold:

(1) $\psi \leqslant \sigma$ for all $\psi \in \Psi$,
(2) $\bigwedge \Psi = 0_{D_R}$,
(3) For all $\theta \prec \sigma$ there exists a unique $\psi \in \Psi$ with $\psi \leqslant \theta$.

If Ψ determines a partition of σ, then we let $E(\sigma, \Psi)$ denote the equivalence relation defined on the subcovers of σ by

$$E(\sigma, \Psi) = \{(\theta_1, \theta_2) : \theta_1, \theta_2 \prec \sigma \text{ and for all } \psi \in \Psi\ (\psi \leqslant \theta_1 \text{ iff } \psi \leqslant \theta_2)\}.$$

CLAIM 11: The set $\Psi_R = \{\theta'_Z/\theta_R : Z \in X/R\}$ determines a partition of $\overline{\delta}/\theta_R$ and the equivalence relation $E(\overline{\delta}/\theta_R, \Psi_R)$ on the subcovers of $\overline{\delta}/\theta_R$ is isomorphic to the relation R on X.

Proof. Of the three conditions that show Ψ_R determines a partition of $\overline{\delta}/\theta_R$, the first is immediate, the second follows from the definition of θ_R, and the third holds because the subcovers of $\overline{\delta}/\theta_R$ are the congruences γ_t/θ_R for all $t \in X$. The map $\gamma_t \mapsto \gamma_t/\theta_R$ is a bijection from the subcovers of $\overline{\delta}$ to the subcovers of $\overline{\delta}/\theta_R$ since $\theta_R \leqslant \gamma_t$ for all $t \in X$ by Claim 9. The desired isomorphism is given by Claim 8.

In the next Claim we essentially show that the initial equivalence relation R is the smallest one of the form $E(\overline{\delta}/\theta_R, \Psi)$.

CLAIM 12: If $\Psi \subseteq \mathsf{Con}\,\mathbf{D}$ is such that

(1) $\psi \leqslant \overline{\delta}$ for all $\psi \in \Psi$,
(2) $\theta_R = \bigcap \Psi$, and
(3) for all $t \in X$ there exists a unique $\psi \in \Psi$ such that $\psi \leqslant \gamma_t$,

then for all $(t, s) \in R$ and $\psi \in \Psi$ we have $\psi \leqslant \gamma_t$ if and only if $\psi \leqslant \gamma_s$.

Proof. Suppose, contrary to the Claim, that there exist $\psi \in \Psi$ and $(t, s) \in R$ with $\psi \leqslant \gamma_t$ and $\psi \nleqslant \gamma_s$. Let $Z = t/R = s/R$.

We first show that $(a^t, \overline{e}) \in \xi$ for all $\xi \in \Psi - \{\psi\}$. Indeed, if $\xi \in \Psi$ and $\xi \neq \psi$, then $\xi \nleqslant \gamma_t$ by condition (3) of the Claim. Consequently, $\gamma_t^* \leqslant \xi$. So $(a^t, \overline{e}) \in \mathrm{Cg}^{\mathbf{D}}(1^t, \overline{0}) = \gamma_t^* \leqslant \xi$.

We next show that $(a^t, \overline{e}) \in \psi$. If $W \in X/R$ and $W \neq Z$, then we certainly have $(a^t, a^s) \in \theta'_W$. Moreover, $(a^t, a^s) \in \theta_Z$ by Claim 7 and so $(a^t, a^s) \in \theta'_Z$. Thus, $(a^t, s^s) \in \theta_R$. By (2) of the Claim we therefore have $(a^t, a^s) \in \bigcap \Psi \subseteq \psi$. On the other hand, $\psi \nleqslant \gamma_s$ so that $\gamma_s^* \leqslant \psi$. This gives $(a^s, \overline{e}) \in \psi$. So (a^t, a^s) and $(a^s, \overline{e})$ in ψ give $(a^t, \overline{e}) \in \psi$.

By combining these two observations we have $(a^t, \overline{e}) \in \bigcap \Psi = \theta_R \leqslant \theta'_Z$, which contradicts Claim 7, and the proof of the Claim is complete.

For any $\sigma \in \Delta$ we define

$$E_\sigma = \bigcap \{E(\sigma, \Psi) : \Psi \text{ determines a partition of } \sigma\}.$$

Clearly, E_σ is an equivalence relation on the set of subcovers of σ. Thus we have established the third condition on Δ.

By Claim 10 we know $\overline{\delta}/\theta_R \in \Delta$ and so we may consider $E_{\overline{\delta}/\theta_R}$. From Claims 11 and 12 we know that $E_{\overline{\delta}/\theta_R}$ as a relation on the subcovers of $\overline{\delta}/\theta_R$ is isomorphic to the relation R on the set X. This verifies the fourth and final item in the list of properties of Δ given in our initial outline of the proof, and thereby completes our argument. □

We next handle the case that there is no $\mathbf{p} \in \mathrm{Pol}_1 \mathbf{A}$ that satisfies the hypotheses of Lemma 11.1. Our proof follows the same outline as the proof of that Lemma.

LEMMA 11.2. *Let* $\mathbf{A}$ *be a finite algebra with* $(\beta, \gamma, \delta) \in (\mathsf{Con}\,\mathbf{A})^3$ *skew. Suppose* $\{0, 1\}$ *is contained in a* $(0_A, \beta)$*-minimal set with* 0 *in the body* B *and* 1 *in the tail. If there is no unary polynomial* $\mathbf{p} \in \mathrm{Pol}_1 \mathbf{A}$ *with* $(\mathbf{p}(0), \mathbf{p}(1)) \in \beta|_B - 0_B$*, then the variety generated by* $\mathbf{A}$ *has many models.*

PROOF. Let $X, \mathbf{D}, f_1, \ldots, f_k, 1^t, \gamma_t, \gamma_t^*$, and η_Z be as in the proof of the previous Lemma and let R be an arbitrary equivalence relation on X. The arguments for Claims 1 to 3 and Facts 4 and 5 given there use only the skew triple conditions and so they hold again here. In particular $\gamma_t^* = \mathrm{Cg}^{\mathbf{D}}(\overline{0}, 1^t)$ and $\gamma_t \prec \overline{\delta}$ for all $t \in X$.

Let $U \in M_{\mathbf{A}}(0_A, \beta)$ be the minimal set containing $\{0,1\}$ and let B denote the body of U. Since $0 \in B$ we can choose $a \in U$ with $(0,a) \in \beta|_U - 0_U$. From [**32**, Lemma 4.22] there is a pseudo-Maltsev operation $\mathbf{d}$ for U. Among the properties of $\mathbf{d}$ are that $\mathbf{d}(x,x,y) = y = \mathbf{d}(y,x,x)$ for all $x \in B$ and all $y \in U$ and that the three unary polynomials $\mathbf{d}(b,c,x), \mathbf{d}(b,x,c), \mathbf{d}(x,b,c)$ are permutations of U for every $b, c \in B$. Additionally we may assume that the range of $\mathbf{d}$ is contained in U.

Since $(0,a) \in \beta$ we get $(\mathbf{d}(1,0,0), \mathbf{d}(1,0,a)) \in \beta|_U$. But $\mathbf{d}(1,0,0) = 1 \notin B$, so $\mathbf{d}(1,0,a) = 1$.

For $Y \subseteq X$ and $c \in A$ we define $c^Y \in A^X$ by

$$c^Y(t) = \begin{cases} c, & \text{if } t \in Y, \\ 1, & \text{if } t \in X - Y. \end{cases}$$

Since $\mathbf{d}(1,0,a) = 1$ the element $a^Y = \mathbf{d}(0^Y, \overline{0}, \overline{a})$ belongs to $\mathbf{D}$. We write $Y - t$ for $Y - \{t\}$.

CLAIM 6′: If $t \in Y \subseteq X$, then $(0^Y, 0^{Y-t})$ and (a^Y, a^{Y-t}) are both in γ_t^*.

Proof. Indeed,

$$0^{Y-t} = \mathbf{d}(1^t, \overline{0}, 0^Y) \overset{\gamma_t^*}{\equiv} \mathbf{d}(\overline{0}, \overline{0}, 0^Y) = 0^Y$$

and

$$a^{Y-t} = \mathbf{d}(1^t, \overline{0}, a^Y) \overset{\gamma_t^*}{\equiv} \mathbf{d}(\overline{0}, \overline{0}, a^Y) = a^Y,$$

as required.

For $Z \subseteq X$ we let

$$\theta'_Z := \bigvee_{Y \subset Z} \mathrm{Cg}^{\mathbf{D}}(0^Y, a^Y) \vee \eta_Z.$$

The following is the analog of Claim 7 in the proof of Lemma 11.1.

CLAIM 7′: $(0^Z, a^Z) \notin \theta'_Z$ for every nonvoid $Z \subseteq X$.

Proof. We first show

(9) if $g, h \in D(U), h|_Z = 0, (g,h) \in \mathrm{Cg}^{\mathbf{D}}(0^Y, a^Y)$, and $Y \subset Z$, then $g = h$.

To verify this, if $Y = \emptyset$, then $\mathrm{Cg}^{\mathbf{D}}(0^Y, a^Y) = 0_D$, so (9) holds trivially. So let $\emptyset \neq Y \subset Z$. Since $(0^Y, a^Y) \in \eta_{X-Y}$ we have $g|_{X-Y} = h|_{X-Y}$. Choose and fix an arbitrary $s \in Z - Y$. We need to show that $g(t) = 0$ for all $t \in Y$.

From $(g,h) \in \mathrm{Cg}^{\mathbf{D}}(0^Y, a^Y)$ there is a Maltsev chain $h = h_0, h_1, \ldots, h_m = g$ obtained by projecting $(0^Y, a^Y)$ using unary polynomials of $\mathbf{D}$ whose ranges are contained in $D(U)$. Assuming that this is the shortest such chain, we get $h_0 \neq h_1$ and $\{h_0, h_1\} = \{\mathbf{h}(0^Y), \mathbf{h}(a^Y)\}$ for some $\mathbf{h} \in \mathrm{Pol}_1\mathbf{D}$. Since the unary polynomial $\mathbf{d}(0^Y, x, a^Y)$ switches 0^Y and a^Y, we assume that $h_0 = \mathbf{h}(0^Y)$ and $h_1 = \mathbf{h}(a^Y)$. As $f_1, \ldots, f_k$ together with the diagonal generate $\mathbf{D}$, we may represent $\mathbf{h}(x)$ as $\mathbf{h}'(x, f_1, \ldots, f_k)$ for some polynomial $\mathbf{h}'(x, y_1, \ldots, y_k)$ of $\mathbf{A}$. Note that

$$\begin{aligned} h_1(s) = \mathbf{h}'(1, \overline{f}(s)) = h_0(s) = h(s) = 0 = h_0(t) = \\ \mathbf{h}'(0, \overline{f}(t)) \overset{\beta}{\equiv} \mathbf{h}'(a, \overline{f}(t)) = h_1(t). \end{aligned} \tag{10}$$

Let $\mathbf{e}_{0,1}(x)$ be a unary idempotent polynomial of $\mathbf{A}$ with range $\{0,1\}$ and let $\neg x$ be a Boolean complementation polynomial of $\mathbf{A}|_{\{0,1\}}$. We substitute $\tau(y_i)$ for each y_i in the polynomial $\mathbf{h}'(x,\overline{y})$ where

$$\tau(y_i) = \begin{cases} 0, & \text{if } f_i(t)=0 \ \& \ f_i(s)=0, \\ \mathbf{e}_{0,1}(x), & \text{if } f_i(t)=0 \ \& \ f_i(s)=1, \\ \neg\mathbf{e}_{0,1}(x), & \text{if } f_i(t)=1 \ \& \ f_i(s)=0, \\ 1, & \text{if } f_i(t)=1 \ \& \ f_i(s)=1, \end{cases}$$

to create the unary polynomial $\mathbf{p}(x) = \mathbf{h}'(\mathbf{d}(a,0,x),\tau(y_1),\ldots,\tau(y_m))$ on $\mathbf{A}$. We have

$$\begin{array}{rclclcl} \mathbf{p}(0) & = & \mathbf{h}'(\mathbf{d}(a,0,0),\overline{f}(t)) & = & \mathbf{h}'(a,\overline{f}(t)) & = & h_1(t) \\ \mathbf{p}(1) & = & \mathbf{h}'(\mathbf{d}(a,0,1),\overline{f}(s)) & = & \mathbf{h}'(1,\overline{f}(s)) & = & h_1(s) \end{array}$$

so that (10) gives $(\mathbf{p}(0),\mathbf{p}(1)) \in \beta$. But our original hypothesis in the Lemma is that no such $\mathbf{p}$ with $\mathbf{p}(0) \neq \mathbf{p}(1)$ exists for $\mathbf{A}$. So $\mathbf{p}(0) = \mathbf{p}(1)$, and thus $h_1(s) = h_1(t) = 0$. This proves that $h_1|_Y = 0$, which together with $h_1|_{X-Y} = h_0|_{X-Y}$, gives $h_1 = h_0$ and shows there is no Maltsev chain starting at $h = h_0$. Consequently, the only element that is $\mathrm{Cg}^{\mathbf{D}}(0^Y,a^Y)$-congruent to h is h itself, so that (9) is proved.

We now return to the proof of the Claim. Assuming that $(0^Z,a^Z) \in \theta'_Z$ we can find a chain

$$0^Z = h_0 \equiv h_1 \equiv \cdots \equiv h_n = a^Z$$

with $(h_i,h_{i+1}) \in \mathrm{Cg}^{\mathbf{D}}(0^Y,a^Y)$ for some $Y \subset Z$ or $(h_i,h_{i+1}) \in \eta_Z$. But then, inducting on i and using (9) one can show that $h_i|_Z = 0$ for all $i = 0,1,\ldots,n$. This contradicts $h_n = a^Z$.

CLAIM 8′: For $\emptyset \neq Z \subseteq X$ we have $\theta'_Z \leqslant \gamma_t$ if and only if $t \in Z$.

Proof. If $t \notin Z$, then the argument in the proof of Claim 8 in the previous Lemma shows $\theta'_Z \not\leqslant \gamma_t$. For the other direction, let $t \in Z$. From Claim 6′ and the definition of θ'_Z we have

$$0^Z \stackrel{\gamma_t^*}{\equiv} 0^{Z-t} \stackrel{\theta'_Z}{\equiv} a^{Z-t} \stackrel{\gamma_t^*}{\equiv} a^Z.$$

If $\theta'_Z \not\leqslant \gamma_t$, then $\gamma_t^* \leqslant \theta'_Z$, so $(0^Z,a^Z) \in \theta'_Z$, contradicting Claim 7′.

For the remainder of the proof we let $Z \in X/R$, that is, Z denotes an equivalence class of R. We define

$$\theta_R = \bigcap_{t\in X} \theta'_{t/R} \text{ and } \mathbf{D}_R = \mathbf{D}/\theta_R.$$

and observe that Claims 9 to 11 hold exactly as in the previous Lemma.

For the final Claim 12, we argue that if $(t,s) \in R$ and there exists $\psi \in \Psi$ with $\psi \leqslant \gamma_t$ and $\psi \not\leqslant \gamma_s$, then for $Z := t/R = s/R$ we have $(0^Z,a^Z) \in \bigcap \Psi = \theta_R \leqslant \theta'_Z$, contradicting Claim 7'.

In order to prove this we first observe that if $Y \subset Z$, then $(0^Y,a^Y) \in \theta_R$. This is because for $T \in X/R$, if $T \neq Z$, then $(0^Y,a^Y) \in \eta_T \leqslant \theta'_T$ while if $T = Z$, then $(0^Y,a^Y) \in \theta'_T$ by definition. So

$$(0^Y,a^Y) \in \bigcap_{T\in X/R} \theta'_T = \theta_R.$$

Now, if $\xi \in \Psi - \{\psi\}$, then $\xi \not\leqslant \gamma_t$ by condition (3) of Claim 12, so $\gamma_t^* \leqslant \xi$. By the previous observation we have $(0^{Z-t}, a^{Z-t}) \in \theta_R \leqslant \xi$. But then

$$0^Z \overset{\gamma_t^*}{\equiv} 0^{Z-t} \overset{\xi}{\equiv} a^{Z-t} \overset{\gamma_t^*}{\equiv} a^Z,$$

so $(0^Z, a^Z) \in \xi$.

It remains to show $(0^Z, a^Z) \in \psi$. We have assumed $\psi \not\leqslant \gamma_s$ and so $\gamma_s^* \leqslant \psi$. We have $(0^{Z-s}, a^{Z-s}) \in \theta_R \leqslant \psi$ and

$$0^Z \overset{\gamma_s^*}{\equiv} 0^{Z-s} \overset{\psi}{\equiv} a^{Z-s} \overset{\gamma_s^*}{\equiv} a^Z,$$

which gives $(0^Z, a^Z) \in \psi$ as desired and thereby completes the proof of the Lemma. □

THEOREM 11.3. *If a locally finite variety $\mathcal{V}$ does not have many models, then the type* **2** *minimal sets of finite algebras in $\mathcal{V}$ have empty tails.*

PROOF. Recall that a lattice $\mathbf{L}$ is an $[\alpha, \beta, \gamma]$-pentagon if $\mathbf{L}$ is the 5-element nonmodular lattice and α, β, and γ are in L with $\alpha < \beta$, $\alpha \vee \gamma = \beta \vee \gamma$ and $\alpha \wedge \gamma = \beta \wedge \gamma$. In the event that $\mathbf{L}$ is a $[\alpha, \beta, \gamma]$-pentagon contained in a congruence lattice of a finite algebra $\mathbf{A}$ and $\alpha \prec \beta$ with $\mathrm{typ}(\alpha, \beta) = \mathbf{i}$, then we say $\mathbf{L}$ is a *pentagon of type* $\mathbf{i}$. Suppose that there is a finite algebra in $\mathcal{V}$ that has a type **2** minimal set with a nonempty tail. By [**44**] there is a finite algebra $\mathbf{A} \in \mathcal{V}$ whose congruence lattice contains an $[\alpha, \beta, \theta]$-pentagon $\mathbf{L}$ of type **2**. By taking homomorphic images if necessary, we may choose $\mathbf{A}$ so that the following minimality conditions hold:

(1) The least element of $\mathbf{L}$ is 0_A.
(2) $\mathbf{A}$ is of minimal cardinality among all finite algebras in $\mathcal{V}$ that contain a pentagon of type **2**.
(3) If δ denotes the top element of $\mathbf{L}$, then δ is minimal among the congruences of $\mathbf{A}$ such that the interval $I[0_A, \delta]$ contains a pentagon of type **2**.
(4) θ is minimal such that $\alpha \vee \theta = \delta$.

We prove a series of facts about this pentagon that will allow us to invoke Lemmas 11.1 and 11.2.

Let $\theta' \prec \theta$. We argue that $\theta' = 0_A$, that is, θ is an atom in $\mathsf{Con}\,\mathbf{A}$. To this end we first show that $\beta \vee \theta' < \delta$. Suppose otherwise. We construct a sublattice of $\mathsf{Con}\,\mathbf{A}$ isomorphic to the 7-element lattice $\mathbf{D}_2$ of Figure 11.1 and invoke [**32**, Lemma 6.3] to obtain a contradiction.

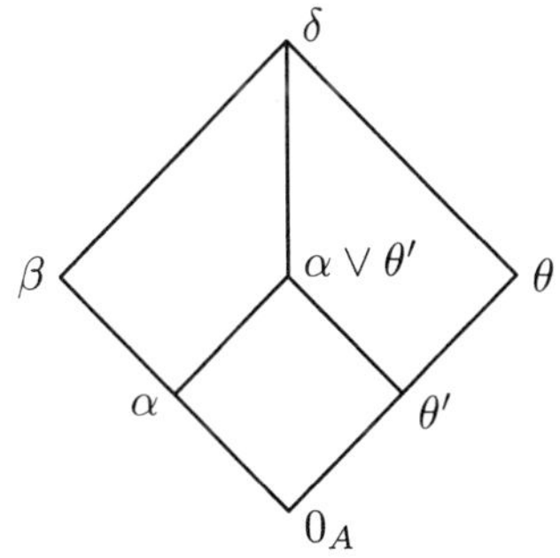

FIGURE 11.1. The lattice $\mathbf{D}_2$.

We know that $\beta \wedge \theta' = \alpha \wedge \theta' = 0_A$ and by our minimality conditions, $\alpha \vee \theta' < \delta$. We claim that the seven congruences $0_A, \alpha, \theta', \beta, \theta, \alpha \vee \theta'$, and δ form a sublattice isomorphic to $\mathbf{D}_2$. The only relations that need checking are $\beta \wedge (\alpha \vee \theta') = \alpha$ and $\theta \wedge (\alpha \vee \theta') = \theta'$. If $\beta \wedge (\alpha \vee \theta') = \beta$, then $\beta \leqslant \alpha \vee \theta'$. This gives $\delta = \beta \vee \theta' \leqslant \alpha \vee \theta' < \delta$, which is absurd. Since $\alpha \prec \beta$ we conclude that $\beta \wedge (\alpha \vee \theta') = \alpha$. The argument for the other covering pair $\theta' \prec \theta$ is similar, and thus we have a copy of $\mathbf{D}_2$ in $\mathsf{Con}\,\mathbf{A}$. Let $\alpha \vee \theta' \leqslant \tau \prec \delta$ in $\mathsf{Con}\,\mathbf{A}$. Lemma 6.3 in [**32**] gives $\mathrm{typ}(\tau, \delta) \in \{\mathbf{1}, \mathbf{5}\}$. But $I[\tau, \delta]$ is projective to $I[\alpha, \beta]$, which is of type $\mathbf{2}$. This is impossible, so $\beta \vee \theta' < \delta$.

If $\alpha \vee \theta' = \beta \vee \theta'$, then we have an $[\alpha, \beta, \theta']$-pentagon of type $\mathbf{2}$, which would violate the minimality condition (3). So we assume $\alpha \vee \theta' < \beta \vee \theta'$ and we next argue that there is a $[\alpha \vee \theta', \beta \vee \theta', \theta]$-pentagon in $\mathsf{Con}\,\mathbf{A}$. Indeed, we have $\theta \vee (\alpha \vee \theta') = \delta$ since $\theta \vee \alpha = \delta$. Also, $\theta \wedge (\beta \vee \theta') = \theta'$, as otherwise this meet would be θ, which would imply $\theta \leqslant \beta \vee \theta'$ and thus $\delta = \beta \vee \theta \leqslant \beta \vee \theta'$, contrary to the result in the previous paragraph. Hence there is a $[\alpha \vee \theta', \beta \vee \theta', \theta]$-pentagon with least element θ' and largest element δ. If τ is any congruence with $\alpha \vee \theta' \leqslant \tau \prec \beta \vee \theta'$, then the prime intervals $I[\alpha, \beta]$ and $I[\tau, \beta \vee \theta']$ are projective, and this gives a $[\tau, \beta \vee \theta', \theta]$-pentagon with largest element δ that is of type $\mathbf{2}$. Our minimality conditions give $\theta' = 0_A$. So we have shown that θ is an atom in $\mathsf{Con}\,\mathbf{A}$.

If θ is an atom, then we may consider $\mathrm{typ}(0_A, \theta)$. By hypothesis, $\mathcal{V}$ does not have many models so $\mathbf{4}, \mathbf{5} \notin \mathrm{typ}\{\mathcal{V}\}$. If $\mathrm{typ}(0_A, \theta) \in \{\mathbf{1}, \mathbf{2}\}$, then [**32**, Lemma 6.5] would give $\mathrm{typ}(\alpha, \beta) = \mathbf{1}$, contrary to our assumptions. We conclude that $\mathrm{typ}(0_A, \theta) = \mathbf{3}$.

Let $U \in M_{\mathbf{A}}(\alpha, \beta)$ have body B and tail T and let $\mathbf{e} \in \mathrm{Pol}_1\mathbf{A}$ be an idempotent polynomial with $\mathbf{e}(A) = U$. The proof of [**44**, Theorem 2.1] shows $\theta \cap (B \times T) \neq \emptyset$. If, say, $(c, d) \in \theta \cap (B \times T)$, then c and d can be connected by a series of overlapping $(0_A, \theta)$-traces. We apply $\mathbf{e}$ to this chain. The image of each trace is either a singleton or is still a trace. Since c is in the body of U and d is in the tail, then one of these traces must contain elements from both B and T. So there exist $0 \in B$ and $1 \in T$ with $\{0, 1\} \in M_{\mathbf{A}}(0_A, \theta)$. Since θ is an atom, $\theta = \mathrm{Cg}^{\mathbf{A}}(0, 1)$.

We note that $C(\beta, \theta; \alpha)$ holds in $\mathbf{A}$ by virtue of [**32**, Proposition 3.4(4)].

Choose γ to be any congruence such that $\beta \leqslant \gamma \prec \delta$. The prime intervals $I[0_A, \theta]$ and $I[\gamma, \delta]$ are projective, so $\mathrm{typ}(\gamma, \delta) = \mathbf{3}$. Note that $(0, 1) \notin \gamma$ since $\gamma \not\geqslant \theta$.

We now replace $\mathbf{A}$ by $\mathbf{A}/\alpha$, and we change notation so that β, γ, δ are the congruences $\beta/\alpha, \gamma/\alpha, \delta/\alpha$ and 0 and 1 are the elements $0/\alpha$ and $1/\alpha$. We have:

(1) $0_A \prec \beta \leqslant \gamma \prec \delta$ with $\mathrm{typ}(0_A, \beta) = \mathbf{2}$ and $\mathrm{typ}(\gamma, \delta) = \mathbf{3}$ since type is preserved by homomorphism.
(2) $\{0, 1\} \in M_{\mathbf{A}}(\gamma, \delta)$ since $(0, 1) \in \delta - \gamma$, and the polynomial in the original algebra that establishes $\{0, 1\} \in M_{\mathbf{A}}(0_A, \theta)$, when suitably modified, can be used here.
(3) $C(\beta, \{0, 1\}; 0_A)$ holds since in the original algebra $\mathbf{A}$ we have $C(\beta, \theta; \alpha)$ and $(0, 1) \in \theta$.
(4) $\delta = \mathrm{Cg}^{\mathbf{A}}(0, 1)$ since in the original algebra $\alpha \vee \mathrm{Cg}^{\mathbf{A}}(0, 1) = \delta$.

Hence (β, γ, δ) is skew. Moreover, $\{0, 1\}$ is contained in a $\langle 0_A, \beta \rangle$-minimal set with 0 in the body B and 1 in the tail. Now, depending on whether or not there is a polynomial $\mathbf{p} \in \mathrm{Pol}_1\mathbf{A}$ with $(\mathbf{p}(0), \mathbf{p}(1)) \in \beta|_B - 0|_B$ we may apply Lemma 11.1 or 11.2 to show that the variety generated by $\mathbf{A}$ has many models. □

COROLLARY 11.4. *Let $\mathcal{V}$ be a locally finite variety with $\mathbf{1} \notin \operatorname{typ}\{\mathcal{V}\}$. If $\mathcal{V}$ does not have many models, then $\mathcal{V}$ is congruence permutable.*

PROOF. By Theorem 8.1 and the hypothesis we know that $\operatorname{typ}\{\mathcal{V}\} \subseteq \{\mathbf{2}, \mathbf{3}\}$. From Corollary 9.2 and Theorem 11.3 every minimal set of a prime quotient of congruences of a finite algebra in $\mathcal{V}$ has an empty tail. Thus, $\mathcal{V}$ is congruence modular by [**32**, Lemma 8.5]. An appeal to Theorem 10.1 completes the proof. □

CHAPTER 12

Restricting Solvable Behavior

We are working towards a characterization of locally finite varieties that do not have many models. We do this by analyzing the finite subdirectly irreducible algebras in such varieties. If $\mathbf{A}$ is a finite subdirectly irreducible algebra with monolith μ, then a result of Chapter 8 tells us that $\operatorname{typ}(0,\mu) \neq \mathbf{4}, \mathbf{5}$ and from Chapter 9 we fully understand the structure of $\mathbf{A}$ when $\operatorname{typ}(0,\mu) = \mathbf{3}$. We now consider the situation that $\operatorname{typ}(0,\mu) = \mathbf{2}$. In this Chapter we prove that the centralizer of such a μ is the largest solvable congruence of $\mathbf{A}$ and is either 1_A or the unique coatom in the congruence lattice of $\mathbf{A}$. If we further restrict to varieties that omit type $\mathbf{1}$, then as a consequence we get that solvable congruences in finite algebras are nilpotent.

We start with the following two technical Lemmas. The first one can be obtained by combining Lemma 1 and Corollary 3 of J. Jeong [**40**].

LEMMA 12.1 (Jeong). *If $\mathbf{A}$ is a finite subdirectly irreducible algebra for which the monolith μ is of type $\mathbf{2}$ and the $(0_A,\mu)$-minimal sets have no tails, then for all congruence relations α, β of $\mathbf{A}$ the condition $C(\alpha,\beta;0_A)$ is equivalent to $C(\beta,\alpha;0_A)$.* □

LEMMA 12.2. *Let $\mathbf{A}$ be a finite subdirectly irreducible algebra with monolith μ of type $\mathbf{2}$ and suppose β is an arbitrary congruence relation on $\mathbf{A}$ for which $C(\beta,\mu;0_A)$ does not hold. If $U \in M_{\mathbf{A}}(0_A,\mu)$ has an empty tail and $(e,a) \in \mu|_U - 0_U$, then there exist $\mathbf{s}(x,y) \in \operatorname{Pol}_2\mathbf{A}$ and $(0,1) \in \beta$ such that $\mathbf{s} : A^2 \to U$ and*

(1) *$\mathbf{s}(x,1) = x$ for all $x \in U$;*
(2) *$\mathbf{s}(x,0) = e$ for all $x \in U \cap e/\mu$;*
(3) *$\mathbf{s}(x,0) = \mathbf{s}(y,0)$ for all $(x,y) \in \mu|_U$;*
(4) *$\mathbf{s}(\mathbf{s}(x,0),0) = \mathbf{s}(x,0)$ for all $x \in U$.*

PROOF. From [**32**, Lemma 4.20] there is a ternary polynomial $\mathbf{d}$ on $\mathbf{A}$ with $\mathbf{d}(U,U,U) \subseteq U$ that is a Maltsev operation on U, and for every $u, v \in U$ the unary polynomial $\mathbf{d}(x,u,v)$ is a permutation of U. The hypotheses on $\mathbf{A}$ allow us to apply Lemma 12.1 and conclude that $C(\mu,\beta;0_A)$ does not hold. The failure of $C(\mu,\beta;0_A)$ means that there exist a polynomial $\mathbf{t}(x,\overline{y})$ of $\mathbf{A}$, a pair $(c,d) \in \mu$, and two sequences $\overline{u} = u_1, u_2, \ldots, u_m$ and $\overline{v} = v_1, v_2, \ldots, v_m$ with $(u_i,v_i) \in \beta$ for all i such that $\mathbf{t}(c,\overline{u}) = \mathbf{t}(c,\overline{v})$, but $\mathbf{t}(d,\overline{u}) \neq \mathbf{t}(d,\overline{v})$. We have $\operatorname{Cg}^{\mathbf{A}}(e,a) = \mu$ and since $(c,d) \in \mu$ there must exist unary polynomials $\mathbf{p}_1, \ldots, \mathbf{p}_n$ with $\mathbf{p}_1(e) = c$, $\mathbf{p}_n(a) = d$, and $\mathbf{p}_i(a) = \mathbf{p}_{i+1}(e)$ for all $1 \leqslant i < n$. We have $\mathbf{t}(\mathbf{p}_1(e),\overline{u}) = \mathbf{t}(\mathbf{p}_1(e),\overline{v})$ and $\mathbf{t}(\mathbf{p}_n(e),\overline{u}) \neq \mathbf{t}(\mathbf{p}_n(e),\overline{v})$. Therefore there must exist an i for which

$$\mathbf{t}(\mathbf{p}_i(e),\overline{u}) = \mathbf{t}(\mathbf{p}_i(e),\overline{v})$$

and

$$\mathbf{t}(\mathbf{p}_i(a), \overline{u}) = \mathbf{t}(\mathbf{p}_{i+1}(e), \overline{u}) \neq \mathbf{t}(\mathbf{p}_{i+1}(e), \overline{v}) = \mathbf{t}(\mathbf{p}_i(a), \overline{v}).$$

We replace $\mathbf{t}(x, \overline{y})$ with $\mathbf{t}(\mathbf{p}_i(x), \overline{y})$ so that we may assume

$$\mathbf{t}(e, \overline{u}) = \mathbf{t}(e, \overline{v})$$

$$\mathbf{t}(a, \overline{u}) \neq \mathbf{t}(a, \overline{v}).$$

Since $(a, e) \in \mu$ we see that $(\mathbf{t}(a, \overline{u}), \mathbf{t}(a, \overline{v})) \in \mu|_U - 0_U$. By prefacing $\mathbf{t}$ with a suitable unary polynomial (e.g., [**32**, Theorem 2.8(4)]), we may assume that the range of $\mathbf{t}$ is a subset of U. Define $\mathbf{t}'(x, \overline{y}) = \mathbf{d}(\mathbf{t}(x, \overline{y}), \mathbf{t}(x, \overline{u}), \mathbf{t}(a, \overline{u}))$. Note that the range of $\mathbf{t}'$ is also contained in U. We see that

$$\mathbf{t}'(e, \overline{u}) = \mathbf{d}(\mathbf{t}(e, \overline{u}), \mathbf{t}(e, \overline{u}), \mathbf{t}(a, \overline{u})) = \mathbf{t}(a, \overline{u})$$

$$\mathbf{t}'(a, \overline{u}) = \mathbf{d}(\mathbf{t}(a, \overline{u}), \mathbf{t}(a, \overline{u}), \mathbf{t}(a, \overline{u})) = \mathbf{t}(a, \overline{u})$$

$$\mathbf{t}'(e, \overline{v}) = \mathbf{d}(\mathbf{t}(e, \overline{v}), \mathbf{t}(e, \overline{u}), \mathbf{t}(a, \overline{u})) = \mathbf{t}(a, \overline{u})$$

$$\mathbf{t}'(a, \overline{v}) = \mathbf{d}(\mathbf{t}(a, \overline{v}), \mathbf{t}(a, \overline{u}), \mathbf{t}(a, \overline{u})) = \mathbf{t}(a, \overline{v})$$

that is,

$$\mathbf{t}'(a, \overline{u}) = \mathbf{t}'(e, \overline{u}) = \mathbf{t}'(e, \overline{v}) \neq \mathbf{t}'(a, \overline{v}).$$

Thus,

$$\mathbf{t}'(e, u_1, u_2, \ldots, u_m) = \mathbf{t}'(a, u_1, u_2, \ldots, u_m)$$

and

$$\mathbf{t}'(e, v_1, v_2, \ldots, v_m) \neq \mathbf{t}'(a, v_1, v_2, \ldots, v_m)$$

which together imply that there exists a minimal index i with $1 \leqslant i \leqslant m$ such that

$$\mathbf{t}'(e, v_1, \ldots, v_i, u_{i+1}, \ldots, u_m) \neq \mathbf{t}'(a, v_1, \ldots, v_i, u_{i+1}, \ldots, u_m).$$

We define $\mathbf{s}'(x, y) = \mathbf{t}'(x, v_1, \ldots, v_{i-1}, y, u_{i+1}, \ldots, u_m)$ and let $0 = u_i$ and $1 = v_i$. With these definitions we see that $\mathbf{s}'(e, 0) = \mathbf{s}'(a, 0)$ and $\mathbf{s}'(e, 1) \neq \mathbf{s}'(a, 1)$. Next we let $\mathbf{s}''(x, y) = \mathbf{d}(\mathbf{s}'(x, y), \mathbf{s}'(e, 0), e)$. Then $\mathbf{s}''(e, 0) = \mathbf{s}''(a, 0) = e$. As observed earlier in the proof, the map $z \mapsto \mathbf{d}(z, \mathbf{s}'(e, 0), e)$ is a permutation of U. Therefore, $\mathbf{s}'(e, 1) \neq \mathbf{s}'(a, 1)$ implies that $\mathbf{s}''(e, 1) \neq \mathbf{s}''(a, 1)$. So $\mathbf{s}''(x, 1)$ is a permutation of U. We iterate $\mathbf{s}''$ in the first variable a sufficient number of times to give a polynomial $\mathbf{s}(x, y)$ for which $\mathbf{s}(\mathbf{s}(x, y), y) = \mathbf{s}(x, y)$. This polynomial has range contained in U and satisfies condition (4) of the Lemma. We have $\mathbf{s}(e, 1) \neq \mathbf{s}(a, 1)$, so $\mathbf{s}(x, 1)$ does not collapse $\mu|_U$ to 0_U. Therefore, $\mathbf{s}(x, 1)$ is an idempotent permutation of the minimal set U, that is,

$$\mathbf{s}(x, 1) = x \quad \text{for all} \quad x \in U,$$

so condition (1) holds. Obviously $\mathbf{s}(e, 0) = e$. We know $\mathbf{s}(e, 0) = \mathbf{s}(a, 0)$, so $\mathbf{s}(x, 0)$ is not a permutation on U. Therefore $\mathbf{s}(x, 0)$ collapses $\mu|_U$ to 0_U. In particular,

$$\mathbf{s}(x, 0) = e \quad \text{for all} \quad x \in U \cap e/\mu$$

and

$$\mathbf{s}(x, 0) = \mathbf{s}(y, 0) \quad \text{for all} \quad (x, y) \in \mu|_U$$

establishing (2) and (3). □

THEOREM 12.3. *If* $\mathbf{A}$ *is a finite subdirectly irreducible algebra with monolith of type* $\mathbf{2}$ *and the variety generated by* $\mathbf{A}$ *does not have many models, then the centralizer of the monolith of* $\mathbf{A}$ *is solvable.*

PROOF. Let β be the monolith of $\mathbf{A}$. Since the variety generated by $\mathbf{A}$ does not have many models, we know that $\text{typ}\{\mathbf{A}\} \subseteq \{\mathbf{1}, \mathbf{2}, \mathbf{3}\}$. If the centralizer of β is not solvable, then there is a covering pair $\gamma \prec \delta$ with $\text{typ}(\gamma, \delta) = \mathbf{3}$ and $C(\delta, \beta; 0_A)$. Let $U \in M_{\mathbf{A}}(0_A, \beta)$ be arbitrary. From Theorem 11.3 we know that U has an empty tail and thus $\mathbf{A}|_U$ has a Maltsev term. Lemma 12.1 applies so $C(\beta, \delta; 0_A)$ holds. Moreover, we may choose δ so that it is minimal with respect to the two properties that $C(\delta, \beta; 0_A)$ and δ is the top element of a covering pair of type $\mathbf{3}$. These two conditions are easily seen to imply that δ is join-irreducible in $\mathsf{Con}\,\mathbf{A}$. We have $\text{typ}(\gamma, \delta) = \mathbf{3}$. Every $\langle \gamma, \delta \rangle$-minimal set has an empty tail because of Corollary 9.2. Let $\{0, 1\} \in M_{\mathbf{A}}(\gamma, \delta)$. Thus, $\delta = \text{Cg}^{\mathbf{A}}(0, 1)$ since δ is join-irreducible. That is, the triple (β, γ, δ) is skew as in Chapter 11.

We claim there exists $\mathbf{p} \in \text{Pol}_1\mathbf{A}$ with $(\mathbf{p}(0), \mathbf{p}(1)) \in \beta|_U - 0_U$. To see this, let $(a, b) \in \beta|_U - 0_U$. Then $(a, b) \in \beta \leqslant \delta = \text{Cg}^{\mathbf{A}}(0, 1)$. So there is a Maltsev chain connecting a and b using links of the form $\{\mathbf{q}(0), \mathbf{q}(1)\}$. We may assume that the range of each $\mathbf{q}$ is contained in U. But $\mathbf{A}|_U$ has a Maltsev term so by Proposition 2.2 it is possible to connect a and b with exactly one link. Let $\mathbf{p}$ be the polynomial that does this.

The existence of such a polynomial $\mathbf{p}$ and the skew condition for the triple (β, γ, δ) allow us to apply Lemma 11.1. This produces a contradiction to the assumption that the variety generated by $\mathbf{A}$ does not have many models. So we conclude that the centralizer of β is solvable. □

LEMMA 12.4. *Let $\mathbf{A}$ be a finite subdirectly irreducible algebra with monolith μ and $\text{typ}(0_A, \mu) = \mathbf{2}$. Suppose there exist $\mu \leqslant \alpha \prec \beta$ with $C(\alpha, \mu; 0_A)$ but not $C(\beta, \mu; 0_A)$. If the variety generated by $\mathbf{A}$ does not have many models, then $\beta = 1_A$.*

PROOF. We assume the contrary and obtain a contradiction. Thus, let β be minimal in $\mathsf{Con}\,\mathbf{A}$ with these hypotheses and suppose $\beta < 1_A$. Let δ be such that $\beta \prec \delta \leqslant 1_A$. So we have in $\mathsf{Con}\,\mathbf{A}$:

$$0_A \prec \mu \leqslant \alpha \prec \beta \prec \delta.$$

We first observe that β is join-irreducible. For if not, then there exists another subcover α' of β. By the minimality of β we have $C(\alpha', \mu; 0_A)$. Therefore we have $C(\alpha \vee \alpha', \mu; 0_A)$, which contradicts the hypothesis that $C(\beta, \mu; 0_A)$ does not hold.

We choose $(0', 1') \in \delta - \beta$ arbitrarily. We also choose $(a, e) \in \mu - 0_A$ with $\{a, e\} \subseteq N \subseteq U$ for $U \in M_{\mathbf{A}}(0_A, \mu)$ and N a $\langle 0_A, \mu \rangle$-trace. From Theorem 11.3 the minimal set U has an empty tail. So we may apply Lemma 12.2 to establish the existence of a binary polynomial $\mathbf{s}$ and elements $(0, 1) \in \beta$ for which

$$\mathbf{s}(e, 0) = \mathbf{s}(e, 1) = \mathbf{s}(a, 0) = e \quad \text{and} \quad \mathbf{s}(a, 1) = a.$$

We have $(0, 1) \notin \alpha$ since $C(\alpha, \mu; 0_A)$ holds. Thus, $\beta = \text{Cg}^{\mathbf{A}}(0, 1)$ since β is join-irreducible.

We now construct an algebra $\mathbf{D}$ in a manner similar to that in the proof of Theorem 9.1. Let $k \geqslant 2$ and T be a set of size $n = 2^k$. Let R be an arbitrary equivalence relation on T. We consider the Boolean algebra $\mathbf{B}$ with universe $\{0, 1\}^T$. The atoms of $\mathbf{B}$ have a 1 in position $t \in T$ and all other entries 0. Let $f_1, \ldots, f_k$ generate $\mathbf{B}$. We form the subalgebra $\mathbf{S}$ of $\mathbf{B}$ whose atoms are of the form $\bigvee C$, where C is a collection of atoms of $\mathbf{B}$ corresponding to an equivalence class of the relation R. For every $1 \leqslant i \leqslant k$ form the elements f_i^{01}, f_i^{10}, and $g_i^{0'1'}$ of A^T as in the proof of Theorem 9.1. Let $\mathbf{D} = \mathbf{D}_R$ be the diagonal subalgebra of $\mathbf{A}^T$ generated by these $3k$ elements and the $|A|$ diagonal elements. The generating elements of D are δ constant, so by Proposition 2.1 every element of D is δ-constant and $\mathbf{D}/\overline{\delta} \simeq \mathbf{A}/\delta$. Elements of D are also β-constant on equivalence classes of R. For every $s \neq t \in T$ there is a $d \in \mathbf{D}(0,1)$ for which $d_s = 0$ and $d_t = 1$ since a suitable choice of one of the f_i^{01} or f_i^{10} will serve.

For each $t \in T$ we have the projection kernel $\eta_t \in \operatorname{Con} \mathbf{D}$. We may argue exactly as in Claim 1 of the proof of Theorem 9.1 that if $\sigma > \eta_t$, then $\sigma \geqslant \overline{\mu}$. For $t \in T$ let

$\mu_t = \{(f,g) \in D^2 : (f_t, g_t) \in \mu \text{ and } f_s = g_s \text{ for all } s \neq t\}$,
$1^t \in A^T$ be given by $1^t(t) = 1$ and $1^t(s) = 0$ for all $s \neq t$,
$a^t \in A^T$ be given by $a^t(t) = a$ and $a^t(s) = e$ for all $s \neq t$.

CLAIM 1: $a^t \in D$ for every $t \in T$.

Proof. We assume $t = 1$, that is, $a^t = (a, e, e, \ldots, e)$. Let $h^2 \in \mathbf{D}(0,1)$ be such that $h_1^2 = 1$ and $h_2^2 = 0$. We have already observed that such an element exists in D. Let $f^2 = \overline{\mathbf{s}}(\overline{a}, h^2)$ where $\overline{\mathbf{s}}$ is the extension to $\mathbf{D}$ of the polynomial $\mathbf{s} \in \operatorname{Pol}_2 \mathbf{A}$. Thus, $f_1^2 = a$, $f_2^2 = e$ and $f^2 \in \{a,e\}^T$. Next, let $h^3 \in \mathbf{D}(0,1)$ with $h_1^3 = 1$ and $h_3^3 = 0$. Then set $f^3 = \overline{\mathbf{s}}(f^2, h^3) \in \{a,e\}^T$. So $f_1^3 = a$, $f_2^3 = e$, and $f_3^3 = e$. Continuing in this way we obtain a^1.

CLAIM 2: If $0_D \prec \sigma \in \operatorname{Con} \mathbf{D}$, then there exists a unique $t \in T$ such that $\sigma = \operatorname{Cg}^{\mathbf{D}}(\overline{e}, a^t)$. Moreover, for every $t \in T$ the congruence $\operatorname{Cg}^{\mathbf{D}}(\overline{e}, a^t)$ is an atom in $\operatorname{Con} \mathbf{D}$.

Proof. Let $f = (f_1, f_2, \ldots, f_n)$ and $g = (g_1, g_2, \ldots, g_n)$ be such that $\sigma = \operatorname{Cg}^{\mathbf{D}}(f,g)$. Suppose $f_1 \neq g_1$. We show that $t = 1$ in the Claim. In $\mathbf{A}$ we have $\operatorname{Cg}^{\mathbf{A}}(f_1, g_1) \geqslant \mu$, so $(a,e) \in \operatorname{Cg}^{\mathbf{A}}(f_1, g_1)$. If the unary polynomials $\mathbf{p}_i$, $1 \leqslant i \leqslant m$, give a Maltsev chain from a to e with links of the form $\{\mathbf{p}_i(f_1), \mathbf{p}_i(g_1)\}$, then by prefacing the $\mathbf{p}_i$ with an idempotent unary polynomial that maps A onto U we obtain a chain whose links are all in U. Since U has a ternary Maltsev polynomial we can find a chain with one link, that is, there is a unary polynomial $\mathbf{p} \in \operatorname{Pol}_1 \mathbf{A}$ whose range is U and for which $\mathbf{p}(f_1) = a$ and $\mathbf{p}(g_1) = e$. If $\overline{\mathbf{p}}$ is the extension of $\mathbf{p}$ to $\mathbf{D}$, then we have $\overline{\mathbf{p}}(f)$, $\overline{\mathbf{p}}(g) \in U^T$ with $\overline{\mathbf{p}}(f)_1 = a$ and $\overline{\mathbf{p}}(g)_1 = e$. Since σ is an atom, $\sigma = \operatorname{Cg}^{\mathbf{D}}(\overline{\mathbf{p}}(f), \overline{\mathbf{p}}(g))$. We rename f and g so as to assume they are of this form at the outset, that is, $f = (a, f_2, \ldots, f_n)$ and $g = (e, g_2, \ldots, g_n)$.

From $\mathbf{s}(a,0) = \mathbf{s}(e,0)$ we see that $\operatorname{Cg}^{\mathbf{D}}(\mathbf{s}(f,\overline{0}), \mathbf{s}(g,\overline{0})) \leqslant \eta_1$ so

$$\sigma > \operatorname{Cg}^{\mathbf{D}}(\mathbf{s}(f,\overline{0}), \mathbf{s}(g,\overline{0})) = 0_D.$$

Therefore, $\mathbf{s}(f_t, 0) = \mathbf{s}(g_t, 0)$ for all $t > 1$. Suppose there is a coordinate $m \geqslant 2$ for which $f_m \neq g_m$. Choose $h \in D$ such that $h_1 = 1$ and $h_m = 0$. Let $f' = \overline{\mathbf{s}}(f,h)$ and $g' = \overline{\mathbf{s}}(g,h)$. Note that $f'_1 = a$, $g'_1 = e$, and $f'_m = g'_m$. We also have $(f', g') \in \sigma$ and $0_D < \operatorname{Cg}^{\mathbf{D}}(f', g') \leqslant \eta_m$, but $\sigma \nleqslant \eta_m$ since $f_m \neq g_m$. This is impossible, so no such coordinate m exists. Thus, $f = (a, f_2, \ldots, f_n)$ and $g = (e, f_2, \ldots, f_n)$ with $f_j \in U$ for all $j \geqslant 2$. In particular, $(f,g) \in \mu_1$, so $\sigma \leqslant \mu_1$.

To complete the proof of the Claim it suffices to show $\sigma|_{\mathbf{D}(U)} = \mu_1|_{\mathbf{D}(U)}$. For if this equality holds, then $(\overline{e}, a^1) \in \mu_1$, so $\sigma = \operatorname{Cg}^{\mathbf{D}}(\overline{e}, a^1)$ and the choice $t = 1$ is unique here.

So suppose instead that $\sigma|_{\mathbf{D}(U)} \neq \mu_1|_{\mathbf{D}(U)}$. We show that in this case we have an $[\sigma|_{\mathbf{D}(U)}, \mu_1|_{\mathbf{D}(U)}, \eta_1|_{\mathbf{D}(U)}]$-pentagon in the congruence lattice of $\mathbf{D}(U)$. But on

$\mathbf{A}|_U$ and hence on $\mathbf{D}(U)$ we have a Maltsev polynomial, so no such pentagon can exist.

To build this pentagon we first note $\eta_1|_{\mathbf{D}(U)} \wedge \mu_1|_{\mathbf{D}(U)} = 0_{\mathbf{D}(U)}$. We also have $0_{\mathbf{D}(U)} < \sigma|_{\mathbf{D}(U)}$. To show that the join operation behaves as desired here we show that $\{(c,d) \in \mathbf{D}(U)^2 : (c_1, d_1) \in \mu\} \subseteq \eta_1|_{\mathbf{D}(U)} \vee \sigma|_{\mathbf{D}(U)}$. If $(c,d) \in \mathbf{D}(U)^2$ satisfies $(c_1, d_1) \in \mu$, then since $\mu = \mathrm{Cg}^{\mathbf{A}}(a,e)$ and there is a Mal'cev polynomial on $\mathbf{A}|_U$, it follows that there exists $\mathbf{p} \in \mathrm{Pol}_1 \mathbf{A}$ such that $\mathbf{p}(a) = c_1$ and $\mathbf{p}(e) = d_1$. We apply $\overline{\mathbf{p}}$ to f and g. Thus, $\overline{\mathbf{p}}(f)_1 = c_1$ and $\overline{\mathbf{p}}(g)_1 = d_1$. So $c \overset{\eta_1}{\equiv} \overline{\mathbf{p}}(f) \overset{\sigma}{\equiv} \overline{\mathbf{p}}(g) \overset{\eta_1}{\equiv} d$, which shows $(c,d) \in \eta_1|_{\mathbf{D}(U)} \vee \sigma|_{\mathbf{D}(U)}$. Therefore,

$$\eta_1|_{\mathbf{D}(U)} \vee \mu_1|_{\mathbf{D}(U)} \subseteq \{(c,d) \in \mathbf{D}(U)^2 : (c_1, d_1) \in \mu\} \subseteq \\ \eta_1|_{\mathbf{D}(U)} \vee \sigma|_{\mathbf{D}(U)} \subseteq \eta_1|_{\mathbf{D}(U)} \vee \mu_1|_{\mathbf{D}(U)},$$

and so $\eta_1|_{\mathbf{D}(U)} \vee \sigma|_{\mathbf{D}(U)} = \eta_1|_{\mathbf{D}(U)} \vee \mu_1|_{\mathbf{D}(U)}$.

To prove the "Moreover", let $r \in T$ and let σ be any atom such that $\sigma \leqslant \mathrm{Cg}^{\mathbf{D}}(\overline{e}, a^r)$. Then for this σ, the t in the proof of the Claim must be r.

Our next claim allows us to use the atoms of $\mathsf{Con}\,\mathbf{D}$ to recover all the η_t and thereby code the elements of T.

CLAIM 3: If $0_D \prec \sigma \in \mathsf{Con}\,\mathbf{D}$, then there exists a unique $t \in T$ such that η_t is the pseudocomplement of σ in $\mathsf{Con}\,\mathbf{D}$.

Proof. Let σ be an atom in $\mathsf{Con}\,\mathbf{D}$. From the previous Claim we may assume that $\sigma = \mathrm{Cg}^{\mathbf{D}}(\overline{e}, a^1)$. We show that η_1 is the pseudocomplement of σ.

Suppose instead that η_1 is not the pseudocomplement of σ. Let θ be minimal among those congruence relations τ of $\mathbf{D}$ for which $\tau \nleqslant \eta_1$ and $\sigma \wedge \tau = 0_D$. This minimality condition forces θ to be join-irreducible in $\mathsf{Con}\,\mathbf{D}$. Let $\theta_- \prec \theta$. Since $\theta_- < \theta$ we have $\theta_- \leqslant \eta_1$. We have $\theta_- \neq 0_D$ for if otherwise, then θ is an atom in $\mathsf{Con}\,\mathbf{D}$ and therefore $\theta = \mathrm{Cg}^{\mathbf{D}}(\overline{e}, a^t)$ for some $t \in T$ by the previous Claim. But since $\theta \nleqslant \eta_1$ we have $t = 1$, so $\theta = \sigma$, which is not the case.

There exist $f, g \in D$ with $\theta = \mathrm{Cg}^{\mathbf{D}}(f,g)$ since θ is join-irreducible. From $\theta \nleqslant \eta_1$ we have $f_1 \neq g_1$. Note that $(a,e) \in \mathrm{Cg}^{\mathbf{A}}(f_1, g_1)$. We may argue as in the previous Claim that there is a unary polynomial $\mathbf{p}$ on $\mathbf{A}$ for which $\mathbf{p}(f_1) = e, \mathbf{p}(g_1) = a$, and $\mathbf{p} : A \to U$. Let $\overline{\mathbf{p}}$ be the extension of $\mathbf{p}$ to $\mathbf{D}$.

For $i = 2, \ldots, n$ we can find $h^i \in \mathbf{D}(0,1)$ such that $h^i_1 = 1$ and $h^i_i = 0$. Put $f^1 = \overline{\mathbf{p}}(f)$ and $f^i = \overline{\mathbf{s}}(f^{i-1}, h^i)$ and analogously for g^i. Then we have $f^n = (e, f^n_2, \ldots, f^n_n)$ and $g^n = (a, g^n_2, \ldots, g^n_n)$, with $f^n, g^n \in \mathbf{D}(U)$. Parts (1) and (4) of Lemma 12.2 give that $\mathbf{s}(f^n_i, 0) = f^n_i$ and $\mathbf{s}(g^n_i, 0) = g^n_i$ for all $i \geqslant 2$.

Clearly, $(f^n, g^n) \nleqslant \eta_1$. Also, $\mathrm{Cg}^{\mathbf{D}}(f^n, g^n) \leqslant \theta$ since $\theta = \mathrm{Cg}^{\mathbf{D}}(f,g)$. We cannot have $\mathrm{Cg}^{\mathbf{D}}(f^n, g^n) < \theta$, so $\mathrm{Cg}^{\mathbf{D}}(f^n, g^n) = \theta$. Let $f' = \overline{\mathbf{s}}(f^n, \overline{0}), g' = \overline{\mathbf{s}}(g^n, \overline{0})$. Note that $f'_1 = g'_1 = e$, so $\mathrm{Cg}^{\mathbf{D}}(f', g') \leqslant \eta_1$. Therefore $\mathrm{Cg}^{\mathbf{D}}(f', g') \leqslant \theta_-$. Also note that $f^n_i = f'_i$ and $g^n_i = g'_i$ for all $i \geqslant 2$.

We obtain our contradiction by exhibiting a $[\theta_-|_{\mathbf{D}(U)}, \theta|_{\mathbf{D}(U)}, \sigma|_{\mathbf{D}(U)}]-$ pentagon in the modular lattice $\mathsf{Con}\,\mathbf{D}(U)$. To this end we first show that the proposed five elements of this pentagon are distinct:

- $(\overline{e}, a^1) \in \sigma$ witnesses $\sigma|_{\mathbf{D}(U)} \neq 0_{\mathbf{D}(U)}$.
- To separate $\theta_-|_{\mathbf{D}(U)}$ and $0_{\mathbf{D}(U)}$, let $(u,w) \in \theta_- - 0_D$ and $t \in T$ be such that $u_t \neq w_t$. There exists $\mathbf{q} \in \mathrm{Pol}_1 \mathbf{A}$ such that $\mathbf{q}(A) = U$ and $\mathbf{q}(u_t) \neq \mathbf{q}(w_t)$. Thus $(\overline{\mathbf{q}}(u), \overline{\mathbf{q}}(w))$ witnesses $\theta_-|_{\mathbf{D}(U)} \neq 0_{\mathbf{D}(U)}$.

- That (f^n, g^n) generates θ shows $\theta|_{\mathbf{D}(U)} \neq \theta_-|_{\mathbf{D}(U)}$.
- Since $\sigma|_{\mathbf{D}(U)}$ and $\theta|_{\mathbf{D}(U)}$ are not comparable, their join is distinct from each of them.

To complete the argument, since $\theta \wedge \sigma = 0_D$ we need only show that $\sigma|_{\mathbf{D}(U)} \vee \theta_-|_{\mathbf{D}(U)} \geqslant \theta|_{\mathbf{D}(U)}$. Let $\overline{\mathbf{d}}$ be a Maltsev polynomial on $\mathbf{D}(U)$. From $(\overline{e}, a^1) \in \sigma$ and $(f', g') \in \theta_-$ we see $(f^n, g^n) = (\overline{\mathbf{d}}(\overline{e}, \overline{e}, f'), \overline{\mathbf{d}}(a^1, \overline{e}, g')) \in \sigma \vee \theta_-$. Since $\theta = \mathrm{Cg}^{\mathbf{D}}(f^n, g^n)$ we are done.

It follows from the previous two Claims that the set of all projection kernels $\{\eta_t : t \in T\}$ is recoverable from the atoms of $\mathsf{Con}\,\mathbf{D}$. An argument identical to that at the end of the proof of Theorem 9.1 shows that the sizes of the equivalence classes of R can be recovered from $\mathsf{Con}\,\mathbf{D}$ and thus the variety generated by $\mathbf{A}$ has many models. This contradicts the hypothesis in the Lemma, so $\beta = 1_A$. □

COROLLARY 12.5. *Let* $\mathbf{A}$ *be a finite subdirectly irreducible algebra with monolith* μ *and* $\mathrm{typ}(0_A, \mu) = \mathbf{2}$. *If the variety generated by* $\mathbf{A}$ *does not have many models, then* $(0_A : \mu)$, *the centralizer of the monolith of* $\mathbf{A}$, *is comparable with all congruences of* $\mathbf{A}$. *Moreover,* $\mathsf{Con}\,\mathbf{A} = I\,[0_A, (0_A : \mu)] \cup \{1_A\}$.

PROOF. Let $\nu = (0_A : \mu)$ be the centralizer of the monolith of $\mathbf{A}$. Suppose that $\mathbf{A}$ has a congruence θ that is not comparable with ν. Then for $\alpha = \nu \cap \theta$ and β such that $\alpha \prec \beta \leqslant \theta$ we have $C(\alpha, \mu; 0_A)$ but not $C(\beta, \mu; 0_A)$. Lemma 12.4 then gives $\beta = 1_A$. Consequently $\theta = 1_A$ contradicting our choice of θ.

This shows that ν is comparable with all congruences. Applying Lemma 12.4 once more we get that if $\nu \neq 1_A$, then any cover of ν is equal to 1_A. □

From Corollary 12.5 we know that the congruence lattice of a finite subdirectly irreducible algebra $\mathbf{A}$ (with monolith of type $\mathbf{2}$) from a variety that does not have many models must have one of the shapes presented in Figure 12.1. Moreover in the first case we have $(0_A : \mu) \prec 1_A$. From Theorem 12.3 we know that $\mathrm{typ}\{0_A, (0_A : \mu)\} \subseteq \{\mathbf{2}\}$. In the next theorem we show that in fact $\mathrm{typ}((0_A : \mu), 1_A) = \mathbf{3}$.

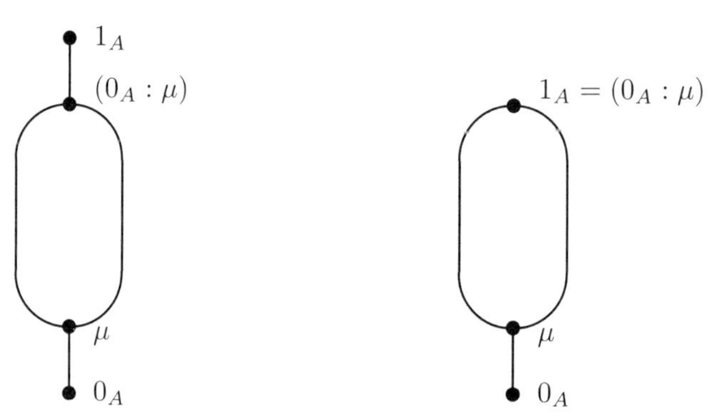

FIGURE 12.1. $\mathsf{Con}\,\mathbf{A}$ for a subdirectly irreducible $\mathbf{A}$

THEOREM 12.6. *If* $\mathbf{A}$ *is a finite subdirectly irreducible algebra with monolith* μ *of type* $\mathbf{2}$*, and if the variety generated by* $\mathbf{A}$ *does not have many models, then the solvable radical of* $\mathbf{A}$ *centralizes* μ*.*

PROOF. Let ν denote the centralizer of μ. If $\nu = 1_A$ we are done. If $\nu < 1_A$, then by previous remarks we have $\nu \prec 1_A$. Since the variety generated by $\mathbf{A}$ does not have many models, $\text{typ}(\nu, 1_A)$ is neither $\mathbf{4}$ nor $\mathbf{5}$. We show that this type cannot be $\mathbf{1}$ or $\mathbf{2}$. Suppose the contrary.

Choose and fix

$$\begin{aligned} U &\in M_{\mathbf{A}}(0_A, \mu) \\ V &\in M_{\mathbf{A}}(\nu, 1_A) \\ (a, e) &\in \mu|_U - 0_U \\ (0, 1) &\in 1_V - \nu. \end{aligned}$$

Thus $\mu = \text{Cg}^{\mathbf{A}}(a, e)$ and $1_A = \text{Cg}^{\mathbf{A}}(1, 0)$ since both μ and 1_A are join-irreducible.

We have $\text{typ}(0_A, \mu) = \mathbf{2}$ and the failure of $C(1_A, \mu; 0_A)$. By means of Theorem 11.3 and Lemma 12.2 we can find a binary polynomial $\mathbf{s}(x, y)$ on $\mathbf{A}$ with $\mathbf{s}(e, 1) = \mathbf{s}(a, 1)$ and $\mathbf{s}(e, 0) \neq \mathbf{s}(a, 0)$. (Note that we have reversed the role of 0 and 1 here.) Moreover, from Lemma 12.2 we may also assume that

$$\begin{aligned} \mathbf{s}(x, 0) &= x \text{ for all } x \in U \\ \mathbf{s}(x, 1) &= e \text{ for all } x \in U \cap e/\mu. \end{aligned}$$

Let T be a set of cardinality 2^k and let R be an arbitrary equivalence relation on T. The number of such equivalence relations that are pairwise nonisomorphic is $\Pi(2^k)$, which is an at least doubly exponential function of k. As usual we consider the Boolean algebras $\mathbf{B}$ with universe $\{0, 1\}^T$ and the subalgebra $\mathbf{S}$ of $\mathbf{B}$ each of whose atoms is the join of atoms in $\mathbf{B}$ corresponding to equivalence classes of R, with each equivalence class of R determining one atom of $\mathbf{S}$. Let $f_1, \ldots, f_k$ and $g_1, \ldots, g_k$ be generating sets for $\mathbf{B}$ and $\mathbf{S}$ respectively. For every $1 \leqslant i \leqslant k$ let f_i^{01}, f_i^{10} and $g_i^{01} \in A^T$ be as in the proof of Theorem 9.1.

Let $\mathbf{D} = \mathbf{D}_R$ be the diagonal subalgebra of $\mathbf{A}^T$ generated by $f_1^{01}, \ldots, f_k^{01}$, $f_1^{10}, \ldots, f_k^{10}$ and $g_1^{01}, \ldots, g_k^{01}$. Note that $\mathbf{D}$ is $(3k + |A|)$-generated. Since $g_i = g_i^{01}$ we suppress the superscripts in what follows.

As in Claim 2 of Lemma 12.4 we can show that every atom in $\mathsf{Con}\,\mathbf{D}$ is of the form $\text{Cg}^{\mathbf{D}}(e, a^t)$ where a^t is a on coordinate t and is e everywhere else and that every $t \in T$ corresponds to a unique atom of this form. Thus, we may identify T with the atoms of $\mathsf{Con}\,\mathbf{D}$.

As usual, η_t is the t-th projection kernel. Since $\mathbf{D}$ is a diagonal algebra, we have $\mathbf{D}/\eta_t \simeq \mathbf{A}$ by Proposition 2.1. Moreover, every congruence of $\mathbf{D}$ that lies above η_t is of the form $\{(c, d) \in D \times D : (c_t, d_t) \in \gamma\}$ for some $\gamma \in \mathsf{Con}\,\mathbf{A}$. In particular, if $\varphi \in \mathsf{Con}\,\mathbf{D}$ is a dual atom in $\mathsf{Con}\,\mathbf{D}$ and $\varphi \geqslant \eta_t$, then

$$\varphi = \{(c, d) \in D \times D : (c_t, d_t) \in \nu\}$$

since 1_A is join-irreducible in $\mathsf{Con}\,\mathbf{A}$ and $\nu \prec 1_A$. Consequently we have

$$\overline{\nu} = \bigcap_{t \in T} \{\varphi \in \mathsf{Con}\,\mathbf{D} : \eta_t \leqslant \varphi \prec 1_D\}.$$

Moreover, if $\eta_t \leqslant \varphi \prec 1_D$, then $\mathbf{D}/\varphi \simeq \mathbf{A}/\nu$. The type of this simple algebra is, by our hypothesis, $\mathbf{1}$ or $\mathbf{2}$. Thus, $\mathbf{D}/\overline{\nu}$ is isomorphic to a subalgebra of $(\mathbf{A}/\nu)^T$, that is, $\mathbf{D}/\overline{\nu}$ belongs to the (quasi)variety generated by a finite simple Abelian algebra $\mathbf{A}/\nu$. The algebra $\mathbf{D}/\overline{\nu}$ is at most $(4k+|A|)$-generated since $\mathbf{D}$ is. It is known [**4**, Corollary 7.13] that a variety generated by a finite simple Abelian algebra has free spectrum that is at most singly exponential. So $|D/\overline{\nu}|$ is at most singly exponential as a function of k.

Let $\Sigma_0 = \left\{\overline{\nu} \vee \mathrm{Cg}^{\mathbf{D}}(c,d) : c, d \in D\right\}$. The mapping

$$\overline{\nu} \vee \mathrm{Cg}^{\mathbf{D}}(c,d) \longmapsto \mathrm{Cg}^{\mathbf{D}/\overline{\nu}}(c/\overline{\nu}, d/\overline{\nu})$$

is injective by the Correspondence Theorem [**55**, Theorem 4.12]. Hence

$$|\Sigma_0| \leqslant \text{number of principal congruence relations in } \mathbf{D}/\overline{\nu} \leqslant |D/\overline{\nu}|^2,$$

so $|\Sigma_0|$ is at most singly exponential as a function of k.

For every k-element subset Σ of Σ_0 define an equivalence relation R_Σ on the set of atoms of $\mathsf{Con}\,\mathbf{D}$ by putting

$$(\alpha,\beta) \in R_\Sigma \quad \text{iff} \quad (\forall \sigma \in \Sigma)\,(C(\sigma,\alpha;0_A) \quad \text{iff} \quad C(\sigma,\beta;0_A))\,.$$

That is, two congruences are R_Σ related if and only if they are centralized by the same set of congruences in Σ. Let Λ denote the set of all such equivalence relations R_Σ. We identify the set T with the set of atoms of $\mathsf{Con}\,\mathbf{D}$. So we may view Λ as a family of equivalence relations on T. The set Σ_0 is at most singly exponential in k, so the collection of k-element subsets of Σ_0 also has this property. Therefore $|\Lambda|$ is at most singly exponential as a function of k. If we can show that an equivalence relation isomorphic to our original R is in Λ, then the proof is complete, since if R_1 and R_2 are equivalence relations on T for which the associated families Λ_1 and Λ_2 are distinct, then the k-generated algebras $\mathbf{D}_{R_1}$ and $\mathbf{D}_{R_2}$ are not isomorphic because each Λ_i is definable using algebraic properties of $\mathbf{D}_{R_i}$. Thus, if $R \in \Lambda$, then at most exponentially many, pairwise nonisomorphic equivalence relations on T can share the same Λ.

So we complete the proof by showing the relation R_Σ for

$$\Sigma = \left\{\overline{\nu} \vee \mathrm{Cg}^{\mathbf{D}}(\overline{0}, g_j) : j = 1, \ldots, k\right\}$$

is isomorphic to our original relation R on T. We do this by proving that for all $t, s \in T$,

$$(t,s) \in R \quad \text{iff} \quad \forall j\ (C(\overline{\nu} \vee \theta_j, \alpha_t; 0_D) \quad \text{iff} \quad C(\overline{\nu} \vee \theta_j, \alpha_s; 0_D))$$

where

$$\alpha_t = \mathrm{Cg}^{\mathbf{D}}(e, a^t) \quad \text{and} \quad \theta_j = \mathrm{Cg}^{\mathbf{D}}(\overline{0}, g_j).$$

Our proof is in two steps:

CLAIM 1: $C(\overline{\nu}, \alpha_t; 0_D)$ for all $t \in T$.

CLAIM 2: $(t,s) \in R$ iff $\forall j\ (C(\theta_j, \alpha_t; 0_D)$ iff $C(\theta_j, \alpha_s; 0_D))$.

For if we prove these two claims, then

$$C(\overline{\nu} \vee \theta_j, \alpha_t; 0_D) \quad \text{iff} \quad (C(\overline{\nu}, \alpha_t; 0_D) \text{ and } C(\theta_j, \alpha_t; 0_D)) \quad \text{iff} \quad C(\theta_j, \alpha_t; 0_D)$$

and

$$C(\overline{\nu} \vee \theta_j, \alpha_s; 0_D) \quad \text{iff} \quad (C(\overline{\nu}, \alpha_s; 0_D) \text{ and } C(\theta_j, \alpha_s; 0_D)) \quad \text{iff} \quad C(\theta_j, \alpha_s; 0_D).$$

Proof of Claim 1: Suppose to the contrary that there exists $t \in T$ for which $C(\overline{\nu}, \alpha_t; 0_D)$ does not hold. We map this failure on coordinate t into $\mathbf{A}$ and thereby obtain a contradiction to our hypothesis $C(\nu, \mu; 0_A)$. In particular, since $C(\overline{\nu}, \alpha_t; 0_D)$ does not hold, there exists a polynomial $\mathbf{p} \in \operatorname{Pol} \mathbf{D}$ and $(u, v) \in \overline{\nu}$ and $(x_1, y_1), \ldots, (x_n, y_n) \in \alpha_t$ such that

$$\begin{aligned} \mathbf{p}(u, x_1, \ldots, x_n) &= \mathbf{p}(u, y_1, \ldots, y_n) \\ \mathbf{p}(v, x_1, \ldots, x_n) &\neq \mathbf{p}(v, y_1, \ldots, y_n). \end{aligned}$$

For each $(x_i, y_i) \in \alpha_t$ we have $x_i(s) = y_i(s)$ for all $s \neq t$. Let $\mathbf{r}$ be the term used to obtain the polynomial $\mathbf{p}$. When we project on coordinate t we obtain

$$\begin{aligned} \mathbf{r}(u(t), x_1(t), \ldots, x_n(t), c_1, \ldots, c_m) &= \mathbf{r}(u(t), y_1(t), \ldots, y_n(t), c_1, \ldots, c_m) \\ \mathbf{r}(v(t), x_1(t), \ldots, x_n(t), c_1, \ldots, c_m) &\neq \mathbf{r}(v(t), y_1(t), \ldots, y_n(t), c_1, \ldots, c_m), \end{aligned}$$

where $c_1, \ldots, c_m$ are the t-th coordinates of the parameters in $\mathbf{D}$ used to form $\mathbf{p}$. We have $(u(t), v(t)) \in \nu$ and $(x_i(t), y_i(t)) \in \mu$, which contradicts $C(\nu, \mu; 0_A)$.

Proof of Claim 2: First suppose that $(t, s) \notin R$. Then there exists $1 \leqslant j \leqslant k$ such that $g_j(t) \neq g_j(s)$. By interchanging s and t if necessary we may assume that $g_j(t) = 0$ and $g_j(s) = 1$. Then $C(\theta_j, \alpha_s; 0_D)$ does not hold since

$$\mathbf{s}(\overline{e}, g_j) = \overline{e} = \mathbf{s}(a^s, g_j)$$

and

$$\mathbf{s}(\overline{e}, \overline{0}) = \overline{e} \neq a^s = \mathbf{s}(a^s, \overline{0}).$$

On the other hand $C(\theta_j, \alpha_t; 0_D)$ does hold since $\theta_j \wedge \alpha_t = 0_D$ when $g_j(t) = 0$.

For the opposite direction in Claim 2, suppose $(t, s) \in R$ and $C(\theta_j, \alpha_t; 0_D)$ holds while $C(\theta_j, \alpha_s; 0_D)$ fails for some $1 \leqslant j \leqslant k$. We must have $g_j(s) = 1$ for otherwise $\theta_j \wedge \alpha_s = 0_D$ and so $C(\theta_j, \alpha_s; 0_D)$ would hold. Therefore $g_j(t) = 1$ since all of the g_i are constant on equivalence classes of R.

Now

$$\mathbf{s}(\overline{e}, g_j) = \overline{e} = \mathbf{s}(a^t, g_j)$$

while

$$\mathbf{s}(\overline{e}, \overline{0}) = \overline{e} \neq a^t = \mathbf{s}(a^t, \overline{0}).$$

Since $(\overline{e}, a^t) \in \alpha_t$ and $(g_j, \overline{0}) \in \theta_j$ we see that $C(\theta_j, \alpha_t; 0_D)$ does not hold, contrary to our assumption. □

Before we proceed to investigate type **2** finite subdirectly irreducible algebras in locally finite varieties omitting type **1** we summarize our results in the following.

THEOREM 12.7. *If $\mathcal{V}$ is a locally finite variety that does not have many models, then*

(1) $\operatorname{typ}\{\mathcal{V}\} \subseteq \{\mathbf{1}, \mathbf{2}, \mathbf{3}\}$;
(2) *minimal sets of type* **2** *and* **3** *have no tails;*
(3) *every finite subdirectly irreducible algebra in $\mathcal{V}$ with type* **3** *monolith is simple;*
(4) *for every finite subdirectly irreducible algebra $\mathbf{A} \in \mathcal{V}$ with type* **2** *monolith μ and the centralizer of the monolith $\nu = (0_A : \mu)$ we have*
    (4.1) *ν is the solvable radical of $\mathbf{A}$,*
    (4.2) *ν is comparable to all congruences of $\mathbf{A}$,*
    (4.3) *the algebra $\mathbf{A}/\nu$ is either trivial or simple of type* **3**.

PROOF. That $\mathcal{V}$ omits types **4** and **5** is Theorem 8.1, while Corollary 9.2 and Theorem 11.3 rule out tails of type **3** and **2**, respectively. Theorem 9.1 shows the simplicity of type **3** finite subdirectly irreducibles in $\mathcal{V}$. Theorems 12.3 and 12.6 show that $(0_A : \mu)$ is the solvable radical of $\mathbf{A}$. For (4.2) and (4.3) we may appeal to Corollary 12.5, Theorem 12.6, and (4.1). □

Assuming that a locally finite variety omits type **1** allows us to put further restrictions on the behavior of the solvable radical.

COROLLARY 12.8. *If $\mathcal{V}$ is a locally finite variety that omits type* **1** *and does not have many models, then every solvable congruence relation on a finite algebra in $\mathcal{V}$ is nilpotent.*

PROOF. By Corollary 11.4 we know that $\mathcal{V}$ is congruence modular. Recall that $\theta^{(n)}$ and $\theta^{[n]}$ denote the n-th nilpotent and solvable power with respect to the modular commutator as described in Chapter 2.

Suppose $\mathbf{A} \in \mathcal{V}$ is a finite algebra with a solvable congruence relation β that is not nilpotent. Therefore there exist $r, n > 0$ such that

$$\beta^{[r]} = 0_A \text{ and } \beta > \beta^{(2)} > \cdots > \beta^{(n)} = \beta^{(n+1)} > 0_A.$$

Pick α to be any subcover of $\beta^{(n)}$ and let η be a congruence relation that is maximal with respect to being above α but not above $\beta^{(n)}$. Obviously η is meet irreducible. An easy induction on k shows that

$$(\beta \vee \eta)^{[k]} \vee \eta \geqslant \beta^{[k]} \vee \eta.$$

Since β is solvable, the congruence relation $(\beta \vee \eta)/\eta$ is solvable in the subdirectly irreducible algebra $\mathbf{A}/\eta$. Consequently, by Theorem 12.6 the congruence $(\beta \vee \eta)/\eta$ centralizes the monolith $(\beta^{(n)} \vee \eta)/\eta$ of $\mathbf{A}/\eta$. This gives

$$\eta \geqslant [\beta \vee \eta, \beta^{(n)} \vee \eta] \geqslant \beta^{(n+1)} = \beta^{(n)},$$

a contradiction. □

We conclude this Chapter with the following summary of our results.

THEOREM 12.9. *If $\mathcal{V}$ is a locally finite variety that omits type* **1** *and does not have many models, then*

(1) *$\mathcal{V}$ is congruence permutable;*
(2) *for every finite subdirectly irreducible algebra $\mathbf{A} \in \mathcal{V}$ with monolith μ and the centralizer of the monolith $\nu = (0_A : \mu)$ we have*
 (2.1) *ν is the solvable radical of $\mathbf{A}$,*
 (2.2) *ν is comparable to all congruences of $\mathbf{A}$,*
 (2.3) *ν is nilpotent,*
 (2.4) *the algebra $\mathbf{A}/\nu$ is either trivial or simple of type* **3**.

PROOF. That $\mathcal{V}$ is congruence permutable is Corollary 11.4 while nilpotency of $(0_A : \mu)$ follows from Corollary 12.8. The other properties of $(0_A : \mu)$ are already listed in Theorem 12.7. □

In Theorem 17.1 we refine condition (2.3) but at the expense of restricting to varieties with few models.

CHAPTER 13

Varieties with Very Few Models

This Chapter contains a characterization of locally finite, congruence modular varieties with very few models. This characterization was first obtained by P. Idziak and R. McKenzie in [**36**]. It will be needed in our characterization of varieties with few models to describe the behavior of nilpotent and Abelian congruences. For this reason we include complete proofs. We start with the following Lemma that is a generalization of one from [**36**].

LEMMA 13.1. *Suppose* $\mathbf{A}$ *is a finite subdirectly irreducible algebra from a congruence permutable variety with few models. Let* ν *be the centralizer of the monolith of* $\mathbf{A}$ *and let* $\mathbf{C}$ *be the subalgebra of* $\mathbf{A}^k$ *consisting of all* ν*-constant functions. Then*

- $\mathbf{C}$ *is* $|A|^2k$*-generated,*
- *if* ν *is not Abelian, then* $\mathbf{C}$ *has at least* $\Pi(k)$ *nonisomorphic quotients, where* $\Pi(k)$ *is the number of partitions of the integer* k.

PROOF. Since $\mathbf{A}$ generates a congruence modular variety with few models we know from Theorem 12.9 that ν is its solvable and simultaneously nilpotent radical. Moreover ν is either 1_A or the unique coatom in $\mathsf{Con}\,\mathbf{A}$ and is comparable with all congruences.

Fix a Maltsev term $\mathbf{d}(x, y, z)$ for the variety generated by $\mathbf{A}$. Let X be a set with $|X| = k$. For $f \in A^X, a \in A$ and $t \in X$ define $f[{}^t_a]$ to be the element of A^X that differs from f at the t-th coordinate where $f(t)$ is replaced by a.

We will prove the first item of the Lemma by showing that $\mathbf{C}$ is generated by the set $G = \{\overline{a}[{}^t_b] : (a, b) \in \nu \text{ and } t \in X\}$. Indeed, if $X = \{1, \ldots, k\}$ and $(a_1, \ldots, a_k) \in C$, then all elements of the form $f_i = a_1[{}^{\,i}_{a_i}]$ are in G and therefore the element

$$(a_1, \ldots, a_k) = \mathbf{d}(\mathbf{d}(\ldots(\mathbf{d}(\mathbf{d}(f_2, f_1, f_3), f_1, f_4)\ldots), f_1, f_{k-1}), f_1, f_k)$$

is in the subalgebra of $\mathbf{A}^X$ generated by G.

Since ν is not an Abelian congruence of $\mathbf{A}$ it follows that $\mu \subseteq [\nu, \nu]$. First we will prove the Lemma under the assumption that this inclusion is not proper.

CASE $[\nu, \nu] = \mu$

For an equivalence relation R on the set X we will define a congruence Θ_R of $\mathbf{C}$ and we will show that the quotients $\mathbf{C}/\Theta_{R_1}$ and $\mathbf{C}/\Theta_{R_2}$ are isomorphic only if the structures $(X; R_1)$ and $(X; R_2)$ are so.

Before defining Θ_R we choose and fix a $(0, \mu)$-minimal set U, a pair $(e, a) \in \mu|_U - 0$ and a $(0, \mu)$–trace $N = e/\mu \cap U$. The algebra $\mathbf{A}|_N$ is polynomially equivalent to a vector space over a finite field and without loss of generality we may assume that e is the zero of this vector space. Let Θ_R be the congruence relation of $\mathbf{C}$ generated by the set

$$\left\{(f,\bar{e}) \in C \times C \ : \ f \in \mathbf{C}(N) \text{ and } \sum_{t\in Z} f(t) = e \text{ for all } \ Z \in X/R\right\}.$$

Since all generators of Θ_R are $\overline{\mu}$–related we immediately get

(1) $\Theta_R \subseteq \overline{\mu}$.

Moreover, we argue that

(2) if $f^0, f^1 \in \mathbf{C}(N)$, then $(f^0, f^1) \in \Theta_R$ iff $\sum_{t\in Z} f^0(t) = \sum_{t\in Z} f^1(t)$ for all $Z \in X/R$.

One direction in (2) is easy since the assumption $\sum_{t\in Z} f^0(t) = \sum_{t\in Z} f^1(t)$ for all $Z \in X/R$ together with $f^0, f^1 \in \mathbf{C}(N)$ gives $f^0 - f^1 \overset{\Theta_R}{\equiv} \bar{e}$ and therefore $(f^0, f^1) \in \Theta_R$.

For the other direction, if $(f^0, f^1) \in \Theta_R$, then congruence permutability of the variety generated by $\mathbf{A}$ gives, by Proposition 2.2, that there is a polynomial, say n-ary, $\mathbf{h}(x_1, \ldots, x_n)$ of $\mathbf{C}$ and $f_1, \ldots, f_n \in \mathbf{C}(N)$ such that $\sum_{t\in Z} f_i(t) = e$ for all $Z \in X/R$ and

$$\begin{aligned} f^0 &= \mathbf{h}(f_1, \ldots, f_n) \\ f^1 &= \mathbf{h}(\bar{e}, \ldots, \bar{e}). \end{aligned}$$

This means that there is an $(n+m)$-ary polynomial $\mathbf{h}'(x_1, \ldots, x_n, y_1, \ldots, y_m)$ of $\mathbf{A}$ and $g_1, \ldots, g_m \in C$ with

$$\begin{aligned} f^0 &= \mathbf{h}'(f_1, \ldots, f_n, g_1, \ldots, g_m) \\ f^1 &= \mathbf{h}'(\bar{e}, \ldots, \bar{e}, g_1, \ldots, g_m). \end{aligned}$$

Since $f^0, f^1 \in \mathbf{C}(N) \subseteq \mathbf{C}(U)$ we may assume that the range of $\mathbf{h}'$ is contained in U.

Choose and fix $t_0 \in X$ and $Z = t_0/R$. We invoke the assumption $C(\mu, \nu; 0)$ to

$$\mathbf{h}'(\underline{e}, \ldots, \underline{e}, \overline{g}(t)) - \mathbf{h}'(e, \ldots, e, \overline{g}(t)) = e = \mathbf{h}'(\underline{e}, \ldots, \underline{e}, \overline{g}(t_0)) - \mathbf{h}'(e, \ldots, e, \overline{g}(t_0))$$

and replace the underlined elements to obtain

$$\mathbf{h}'(u_1, \ldots, u_n, \overline{g}(t)) - \mathbf{h}'(e, \ldots, e, \overline{g}(t)) = \mathbf{h}'(u_1, \ldots, u_n, \overline{g}(t_0)) - \mathbf{h}'(e, \ldots, e, \overline{g}(t_0))$$

for all $t \in Z$ and $u_1, \ldots, u_n \in N$. Thus

$$\mathbf{h}'(u_1, \ldots, u_n, \overline{g}(t)) = \mathbf{h}'(u_1, \ldots, u_n, \overline{g}(t_0)) - \mathbf{h}'(e, \ldots, e, \overline{g}(t_0)) + \mathbf{h}'(e, \ldots, e, \overline{g}(t)).$$

But the map

$$(u_1, \ldots, u_n) \mapsto \mathbf{h}'(u_1, \ldots, u_n, \overline{g}(t_0)) - \mathbf{h}'(e, \ldots, e, \overline{g}(t_0))$$

is a polynomial of the vector space $\mathbf{A}|_N$ that sends $(e, \ldots, e)$ to e. Thus it must be of the form

$$(u_1, \ldots, u_n) \mapsto \sum_{i=1}^{n} \lambda_i u_i$$

for some elements $\lambda_1, \ldots, \lambda_n$ of the field. Consequently

$$\mathbf{h}'(u_1, \ldots, u_n, \overline{g}(t)) = \sum_{i=1}^{n} \lambda_i u_i + \mathbf{h}'(e, \ldots, e, \overline{g}(t))$$

for all $t \in Z$ and $u_1, \ldots, u_n \in N$. In particular

$$\begin{aligned} f^0(t) &= \mathbf{h}'(f_1(t), \ldots, f_n(t), \overline{g}(t)) \\ &= \sum_{i=1}^{n} \lambda_i f_i(t) + \mathbf{h}'(e, \ldots, e, \overline{g}(t)) \\ &= \sum_{i=1}^{n} \lambda_i f_i(t) + f^1(t). \end{aligned}$$

Thus

$$\sum_{t\in Z} f^0(t) = \sum_{i=1}^{n} \lambda_i \sum_{t\in Z} f_i(t) + \sum_{t\in Z} f^1(t)$$

which together with $\sum_{t\in Z} f_i(t) = e$ gives

$$\sum_{t\in Z} f^0(t) = \sum_{t\in Z} f^1(t),$$

as required in (2).

Now for a subset Z of X we define the following congruences of $\mathbf{C}$

$$\begin{aligned} \eta_Z &= \{(f,g) \in C \times C \ : \ f_t = g_t \ \text{ for all } \ t \in Z\} \\ \eta'_Z &= \eta_{X-Z}. \end{aligned}$$

Moreover if φ is a congruence of $\mathbf{A}$ and $Z \subseteq X$ we put

$$\begin{aligned} \varphi_Z &= \{(f,g) \in C \times C \ : \ (f_t, g_t) \in \varphi \ \text{ for all } \ t \in Z\} \\ \varphi'_Z &= \varphi_Z \cap \eta'_Z. \end{aligned}$$

As usual we will write $\eta_t, \eta'_t, \varphi_t, \varphi'_t$, instead of $\eta_{\{t\}}, \eta'_{\{t\}}, \varphi_{\{t\}}, \varphi'_{\{t\}}$, respectively. Moreover for any congruence γ of $\mathbf{C}$ the congruence $(\gamma \vee \Theta_R)/\Theta_R$ of $\mathbf{C}/\Theta_R$ will be denoted by $\tilde{\gamma}$.

First we observe that

(3) μ'_t is the unique atom of $\mathsf{Con}\,\mathbf{C}$ that is below η'_t.

Suppose that $\sigma = \mathrm{Cg}^{\mathbf{C}}(f,g)$ is an atom below η'_t. Then we have $f(s) = g(s)$ for all $s \neq t$ and $f(t) \neq g(t)$. In particular $(a,e) \in \mu \subseteq \mathrm{Cg}^{\mathbf{A}}(f(t), g(t))$ so that there is a unary polynomial $\mathbf{q}(x)$ of $\mathbf{A}$ with $\mathbf{q}(f(t)) = a$ and $\mathbf{q}(g(t)) = e$. Thus $0 \neq \mathrm{Cg}^{\mathbf{C}}(\mathbf{q}(f), \mathbf{q}(g)) \subseteq \sigma$ and therefore $\sigma \subseteq \mathrm{Cg}^{\mathbf{C}}(\mathbf{q}(f), \mathbf{q}(g)) \subseteq \mu'_t$. Changing notation we may assume that $\sigma = \mathrm{Cg}^{\mathbf{C}}(f,g)$ for some $f, g \in C$ with $f(t) \stackrel{\mu}{\equiv} g(t)$ and $f(s) = g(s)$ for all $s \neq t$. Obviously $\mu'_t \cap \eta_t \subseteq \eta'_t \cap \eta_t = 0$. Moreover if $(c,d) \in \mu'_t$ then $(c(t), d(t)) \in \mu = \mathrm{Cg}^{\mathbf{A}}(f(t), g(t))$. Thus there is a unary polynomial $\mathbf{p}(x)$ of $\mathbf{A}$ with $\mathbf{p}(f(t)) = c(t)$ and $\mathbf{p}(g(t)) = g(t)$. Applying $\mathbf{p}$ coordinatewise to f and g we get

$$c \stackrel{\eta_t}{\equiv} \mathbf{p}(f) \stackrel{\sigma}{\equiv} \mathbf{p}(g) \stackrel{\eta_t}{\equiv} d,$$

i.e., $(c,d) \in \sigma \vee \eta_t$. Consequently $\mu'_t \subseteq \sigma \vee \eta_t$. Now modularity gives that $\sigma = \mu'_t$, as required.

Note that for $t \in X$ the element $a^t = \overline{e}[{}^t_a]$ lies in C and in fact $\mu'_t = \mathrm{Cg}^{\mathbf{C}}(\overline{e}, a^t)$. Since $(\overline{e}, a^t) \notin \Theta_R$ the covering pair $0 \prec \mu'_t$ projects up to $\Theta_R \prec \Theta_R \vee \mu'_t$ and therefore we have

(4) $\tilde{\mu}'_t$ is an atom in $\mathsf{Con}\,\mathbf{C}/\Theta_R$.

Moreover we have

(5) $\tilde{\mu}'_t = \tilde{\mu}'_s$ if and only if $(t,s) \in R$.

To see this, note that if $(t,s) \in R$, then (2) gives $(a^t, a^s) \in \Theta_R$. Consequently $(a^t, \overline{e}) \in \Theta_R \vee \mu'_s$, which proves the 'if' direction.

Conversely, suppose that $(t,s) \notin R$. If $\tilde{\mu}'_t = \tilde{\mu}'_s$, then $(a^t, \overline{e}) \in \Theta_R \vee \mu'_s$. This together with congruence permutability gives an element $f \in C$ with

$$a^t \overset{\Theta_R}{\equiv} f \overset{\mu'_s}{\equiv} \overline{e}.$$

Applying a unary idempotent polynomial of $\mathbf{A}$ that has range U we may assume that $f \in \mathbf{C}(U)$. In fact this gives that $f \in \mathbf{C}(N)$. Applying (2) to $a^t \overset{\Theta_R}{\equiv} f$ and $Z = s/R$ we get $\sum_{z \in Z} f(z) = e$. On the other hand from $f \overset{\mu'_s}{\equiv} \overline{e}$ we know that $f(z) = e$ for all $z \in X - \{s\}$. Consequently $f = \overline{e}$, i.e., $(a^t, \overline{e}) \in \Theta_R$, a contradiction.

We next show

(6) for $\gamma \in \mathsf{Con}\,\mathbf{C}$ and $t \in X$ either $[\gamma, \overline{\nu}] \subseteq \eta_t$ or $\mu'_t \subseteq [\gamma, \overline{\nu}]$.

Suppose that $[\gamma, \overline{\nu}] \not\subseteq \eta_t$. Using Exercise 6.6 in [**22**] we can witness this failure of centrality by a binary polynomial $\mathbf{s}(x,y)$ of $\mathbf{C}$, a pair $(c,d) \in \overline{\nu}$ and a pair $(f,g) \in \gamma$ satisfying

$$\begin{aligned} \mathbf{s}(f,c) &\overset{\eta_t}{\equiv} \mathbf{s}(f,d) \\ \mathbf{s}(g,c) &\overset{\eta_t}{\not\equiv} \mathbf{s}(g,d). \end{aligned}$$

Since $(c(t), d(t)) \in \nu$ we get that $d' = c[^{t}_{d(t)}]$ is an element of C and $(c,d') \in \overline{\nu} \cap \eta'_t$. Moreover

$$\begin{aligned} \mathbf{s}(f,c) &= \mathbf{s}(f,d') \\ \mathbf{s}(g,c) &\neq \mathbf{s}(g,d') \end{aligned}$$

so that $0 \neq [\gamma, \overline{\nu} \cap \eta'_t] \subseteq \eta'_t$. This together with (3) gives $\mu'_t \subseteq [\gamma, \overline{\nu} \cap \eta'_t] \subseteq [\gamma, \overline{\nu}]$, as required.

A congruence of $\mathbf{C}/\Theta_R$ is called *regular* if it is the only atom below $[\gamma, \overline{\nu}/\Theta_R]$ for some $\gamma \leqslant \overline{\nu}/\Theta_R$.

(7) A congruence of $\mathbf{C}/\Theta_R$ is regular if and only if it is of the form $\tilde{\mu}'_t$ for some $t \in X$.

First suppose that $\alpha, \gamma \geqslant \Theta_R$ are such that $\tilde{\alpha}$ is the unique atom below the congruence $[\gamma/\Theta_R, \overline{\nu}/\Theta_R]$. This means that α is the unique congruence of $\mathbf{C}$ with $\Theta_R \prec \alpha \subseteq [\gamma, \overline{\nu}] \vee \Theta_R$. In particular $[\gamma, \overline{\nu}] \neq 0$ so that there is a $t \in X$ with $[\gamma, \overline{\nu}] \not\subseteq \eta_t$. By (6) we get that $\mu'_t \subseteq [\gamma, \overline{\nu}]$ and therefore both α and $\mu'_t \vee \Theta_R$ cover Θ_R and are below $[\gamma, \overline{\nu}] \vee \Theta_R$. Consequently $\alpha = \mu'_t \vee \Theta_R$ and we are done with the 'only if' direction.

To prove that $\tilde{\mu}'_t$ is regular it suffices to show that $\mu'_t \vee \Theta_R$ is the only congruence α with $\Theta_R \prec \alpha \subseteq [\nu'_t, \overline{\nu}] \vee \Theta_R$. From (4) we know that $\mu'_t \vee \Theta_R$ covers Θ_R. Obviously $[\nu'_t, \overline{\nu}] \subseteq \eta'_t$ and $[\nu'_t, \overline{\nu}] \subseteq [\overline{\nu}, \overline{\nu}] \subseteq \overline{\mu}$. Thus $[\nu'_t, \overline{\nu}] \subseteq \eta'_t \cap \overline{\mu} = \mu'_t \subseteq [\nu'_t, \overline{\nu}]$. So $[\nu'_t, \overline{\nu}] \vee \Theta_R = \mu'_t \vee \Theta_R$ as desired.

Since the congruence $\overline{\nu}/\Theta_R$ is definable in $\mathbf{C}/\Theta_R$ as the largest solvable congruence, we can recover all congruences of the form $\tilde{\mu}'_t$. This together with (5) allows us to recover the number $|X/R|$ of R–cosets in X as the number of regular congruences in $\mathbf{C}/\Theta_R$.

We will show that nonisomorphic equivalence relations $(X; R)$ give rise to nonisomorphic quotients $\mathbf{C}/\Theta_R$ by proving that the sizes of R–cosets are recoverable.

First note that for $\theta = (0 : \nu)$ we have $\mu \leqslant \theta < \nu$. The following claim allows us to compare sizes of R–cosets.

(8) Two subsets Y, Z of X have the same number of elements if and only if the quotients $\mathbf{C}/(\nu_Y \cap \theta_{X-Y})$ and $\mathbf{C}/(\nu_Z \cap \theta_{X-Z})$ are isomorphic.

By permuting coordinates of $\mathbf{C}$ we may assume that Y and Z are comparable with respect to inclusion. Obviously, if $Y \subseteq Z \subseteq X$ then $\nu_Y \cap \theta_{X-Y} \subseteq \nu_Z \cap \theta_{X-Z}$, or in other words

$$\mathbf{C}/(\nu_Z \cap \theta_{X-Z}) \text{ is a homomorphic image of } \mathbf{C}/(\nu_Y \cap \theta_{X-Y}).$$

Moreover, if $Y \subset Z$, then picking $t \in Z - Y$ and $(c, d) \in \nu - \theta$ we get that $(\overline{c}, \overline{c}[^t_d]) \in (\nu_Z \cap \theta_{X-Z}) - (\nu_Y \cap \theta_{X-Y})$, i.e., $\nu_Y \cap \theta_{X-Y} < \nu_Z \cap \theta_{X-Z}$, and consequently, since $\mathbf{C}$ is finite,

$$\mathbf{C}/(\nu_Y \cap \theta_{X-Y}) \text{ is not a homomorphic image of } \mathbf{C}/(\nu_Z \cap \theta_{X-Z}).$$

Our claim (8) easily follows from the last two displays.

Obviously $\Theta_R \subseteq \nu_Z \cap \theta_{X-Z}$. We can finish the proof by showing that for $Z \in X/R$ the congruences $(\nu_Z \cap \theta_{X-Z})/\Theta_R$ are recoverable from $\mathbf{C}/\Theta_R$.

A congruence relation δ of $\mathbf{C}/\Theta_R$ will be called *co-regular* if δ is maximal with respect to the properties:

- $\delta \subseteq \overline{\nu}/\Theta_R$,
- $[\delta, \overline{\nu}/\Theta_R]$ contains exactly one atom.

To recover $(X; R)$ up to isomorphism it remains to show the following

(9) A congruence δ of $\mathbf{C}/\Theta_R$ is co-regular if and only if $\delta = (\nu_Z \cap \theta_{X-Z})/\Theta_R$ for some $Z \in X/R$.

Suppose that $\Theta_R \subseteq \gamma \subseteq \overline{\nu}$ and γ/Θ_R is co-regular. Then $[\gamma/\Theta_R, \overline{\nu}/\Theta_R]$ contains exactly one atom and, by (7), this atom must be of the form $\tilde{\mu}'_t$ for some $t \in X$. Put $Z = t/R$. We will show that $\gamma \subseteq \theta_{X-Z}$. Suppose to the contrary that there are $s \in X - Z$ and $(f, g) \in \gamma$ with $(f(s), g(s)) \notin \theta$. This means that $\left[\mathrm{Cg}^{\mathbf{A}}(f(s), g(s)), \nu\right] \neq 0$. Using Exercise 6.6 in [**22**] we can witness this failure of centrality by a binary polynomial $\mathbf{s}(x, y)$ of $\mathbf{A}$ and $(c, d) \in \nu$ such that

$$\begin{aligned} \mathbf{s}(f(s), c) &= \mathbf{s}(f(s), d) \\ \mathbf{s}(g(s), c) &\neq \mathbf{s}(g(s), d). \end{aligned}$$

In particular, if $\overline{\mathbf{s}}$ is the natural extension of $\mathbf{s}$ to $\mathbf{C}$, then

$$\begin{aligned} \overline{\mathbf{s}}(f, \overline{c}) &\overset{\eta_s}{\equiv} \overline{\mathbf{s}}(f, \overline{d}) \\ \overline{\mathbf{s}}(g, \overline{c}) &\overset{\eta_s}{\not\equiv} \overline{\mathbf{s}}(g, \overline{d}), \end{aligned}$$

i.e., $[\gamma, \overline{\nu}] \nsubseteq \eta_s$. Consequently (6) gives $\mu'_s \subseteq [\gamma, \overline{\nu}]$ and therefore $\tilde{\mu}'_s$ is an atom below $[\gamma/\Theta_R, \overline{\nu}/\Theta_R]$. Thus $\tilde{\mu}'_s = \tilde{\mu}'_t$, which by (5) gives $s \in Z$, a contradiction.

We first show that for $Z \in X/R$ the congruence $(\nu_Z \cap \theta_{X-Z})/\Theta_R$ is co-regular. This congruence is contained in $\overline{\nu}/\Theta_R$ since $\theta < \nu$. We argue that for $t \in Z$ the congruence $\tilde{\mu}'_t$ is the only atom below $[(\nu_Z \cap \theta_{X-Z})/\Theta_R, \overline{\nu}/\Theta_R] = ([\nu_Z \cap \theta_{X-Z}, \overline{\nu}] \vee \Theta_R)/\Theta_R$. We have

$$[\nu_Z \cap \theta_{X-Z}, \overline{\nu}] \subseteq [\nu_Z, \overline{\nu}] \cap [\theta_{X-Z}, \overline{\nu}] \subseteq \mu_Z \cap \eta_{X-Z} = \mu'_Z.$$

From (5) and the fact that $\mu'_Z = \bigvee_{t \in Z} \mu'_t$ we know that $\mu'_Z \vee \Theta_R = \mu'_t \vee \Theta_R$. Consequently, if $\Theta_R \prec \alpha \subseteq [\nu_Z \cap \theta_{X-Z}, \overline{\nu}] \vee \Theta_R$, then $\alpha = \mu'_t \vee \Theta_R$, as required.

To show that $(\nu_Z \cap \theta_{X-Z})/\Theta_R$ is maximal suppose that $\nu_Z \cap \theta_{X-Z} \subseteq \gamma \subseteq \overline{\nu}$ and $[\gamma/\Theta_R, \overline{\nu}/\Theta_R]$ contains exactly one atom. This atom must be $\tilde{\mu}'_t$. We have already shown that $\gamma \subseteq \theta_{X-Z}$ and so $\gamma \subseteq \nu_Z \cap \theta_{X-Z}$. This gives $\gamma = \nu_Z \cap \theta_{X-Z}$ and thus the congruence $(\nu_Z \cap \theta_{X-Z})/\Theta_R$ is co-regular.

For the converse, suppose $\delta = \gamma/\Theta_R$ is co-regular with $\Theta_R \subseteq \gamma \subseteq \overline{\nu}$. As observed above, there is a $Z \in X/R$ for which $\gamma \subseteq \theta_{X-Z}$. So $\delta \subseteq (\nu_Z \cap \theta_{X-Z})/\Theta_R$, and maximality forces equality here.

CASE $[\nu, \nu] > \mu$

If $[\nu, \nu] > \mu$, then we pick a congruence η of $\mathbf{A}$ that is maximal with the property of being above μ but not above $[\nu, \nu]$. Obviously the quotient $\mathbf{A}' = \mathbf{A}/\eta$ is subdirectly irreducible with monolith $\mu' = ([\nu, \nu] \vee \eta)/\eta$, the centralizer of the monolith is $\nu' = \nu/\eta$, and $[\nu', \nu'] = \mu'$.

By the previous Case the subalgebra $\mathbf{C}'$ of $(\mathbf{A}')^k$ consisting of all ν'–constant functions has $\Pi(k)$ nonisomorphic quotients. Since obviously $\mathbf{C}'$ is a homomorphic image of $\mathbf{C}$ we are done. □

COROLLARY 13.2. *Let $\mathcal{V}$ be a locally finite variety with $\mathbf{1} \notin \mathrm{typ}\{\mathcal{V}\}$. If $\mathcal{V}$ has very few models, then $\mathcal{V}$ is affine.*

PROOF. Combining Proposition 3.3 with Theorem 12.9 we get that $\mathcal{V}$ is congruence permutable and that the centralizer of the monolith of every finite finite subdirectly irreducible algebra $\mathbf{A}$ in $\mathcal{V}$ is 1_A. Since $\mathcal{V}(\mathbf{A})$ has very few models, Lemma 13.1 together with the fact that $\Pi(k)$ is not at most polynomial show that $\mathbf{A}$ is in fact Abelian. The variety $\mathcal{V}$ is locally finite and is therefore generated by its finite subdirectly irreducible algebras, which are Abelian. Since it is known that any congruence modular variety that is generated by Abelian algebras is itself Abelian, we see that $\mathcal{V}$ is Abelian and so we may conclude that $\mathcal{V}$ is affine. □

LEMMA 13.3. *Let $t : \omega \longrightarrow \omega$ be a nondecreasing function. Then the function*

$$t'(k) = \binom{k + t(k)}{t(k)}$$

is bounded from above by a polynomial in k if and only if t is eventually constant.

PROOF. The 'if' direction is obvious. To see the 'only if' suppose that t is not eventually constant so that $\lim_{k \to \infty} t(k) = \infty$. Then also for the function

$$g(k) = \mathrm{MIN}\left\{ \frac{t(k)}{2}, \frac{\lfloor \sqrt{k} \rfloor}{2} \right\}$$

we have $\lim_{k \to \infty} g(k) = \infty$.

We will show that $t'(k) \geqslant k^{g(k)}$ for all k. This would give a contradiction as then $t'(k)$ cannot be bounded from above by a polynomial in k.

First we note that

$$t'(k) = \frac{(k+1)\dots(k+t(k))}{t(k)!} \geqslant \left(\frac{k}{t(k)}\right)^{t(k)}.$$

Now the proof that $t'(k) \geqslant k^{g(k)}$ splits into 3 cases.

CASE 1. $t(k) \leqslant \sqrt{k}$.

We have

$$t'(k) \geqslant \left(\frac{k}{t(k)}\right)^{t(k)} \geqslant \left(\frac{k}{\sqrt{k}}\right)^{t(k)} = k^{\frac{t(k)}{2}} \geqslant k^{g(k)},$$

as required.

CASE 2. $\sqrt{k} \leqslant t(k) \leqslant k$.

In this case we have

$$t'(k) \geqslant \binom{k+t(k)}{\lfloor\sqrt{k}\rfloor} \geqslant \binom{k+\lfloor\sqrt{k}\rfloor}{\lfloor\sqrt{k}\rfloor} \geqslant \left(\frac{k}{\sqrt{k}}\right)^{\lfloor\sqrt{k}\rfloor} = \sqrt{k}^{\lfloor\sqrt{k}\rfloor} = k^{\frac{\lfloor\sqrt{k}\rfloor}{2}} \geqslant k^{g(x)}.$$

CASE 3. $k \leqslant t(k)$.

Here we have

$$t'(k) = \binom{k+t(k)}{t(k)} = \binom{k+t(k)}{k} \geqslant \binom{k+t(k)}{\lfloor\sqrt{k}\rfloor} \geqslant \binom{k+\lfloor\sqrt{k}\rfloor}{\lfloor\sqrt{k}\rfloor} \geqslant k^{g(x)}.$$

□

THEOREM 13.4. *Let $\mathcal{V}$ be a locally finite variety with $\mathbf{1} \notin \operatorname{typ}\{\mathcal{V}\}$. Then $\mathcal{V}$ has very few models if and only if $\mathcal{V}$ is a directly representable affine variety.*

PROOF. The 'if' direction is a part of Corollary 6.6.

To see the converse we first use Corollary 13.2 to get that $\mathcal{V}$ is affine. For an algebra $\mathbf{A} \in \mathcal{V}$ let $\mathbf{A}_\nabla$ be the linearization of $\mathbf{A}$ as defined in [**22**], p.114, namely $\mathbf{A}_\nabla = \mathbf{A}^2/\Delta_{1,1}$, where $\Delta_{1,1}$ is the congruence on $\mathbf{A}^2$ generated by the diagonal relation on $\mathbf{A}$, i.e., $\Delta_{1,1}$ is the smallest congruence Θ on $\mathbf{A}^2$ such that the elements of the diagonal $\Delta = \{(a,a) : a \in A\}$ are identified by Θ. For Abelian algebras one has that Δ is a coset of $\Delta_{1,1}$.

We need the following properties of this linearization. Proofs of these five statements may be found in Lemma 2 of [**16**].

(10) $\mathbf{A}_\nabla$ has a one element subalgebra,
(11) $\mathbf{A}_\nabla \in \mathcal{V}$,
(12) $(\mathbf{A} \times \mathbf{B})_\nabla$ is isomorphic to $\mathbf{A}_\nabla \times \mathbf{B}_\nabla$,
(13) $|A| = |A_\nabla|$,
(14) If $\mathbf{A}$ is directly indecomposable then so is $\mathbf{A}_\nabla$.

In particular we know that the class $\mathcal{V}_\nabla = \{\mathbf{A}_\nabla : \mathbf{A} \in \mathcal{V}\}$ is contained in the class $\mathcal{V}_0$ of all algebras from $\mathcal{V}$ that have one element subalgebras. The class $\mathcal{V}_0$ is closed under taking products and homomorphic images. Since $\mathcal{V}$ is congruence permutable we can apply a result of G. Birkhoff (see e.g., Theorem 5.3 in [**55**]) to get that finite algebras in $\mathcal{V}_0$ have the unique factorization property. This in turn will allow us to show that the class $\mathcal{V}_0$ has only finitely many finite directly indecomposable algebras.

To this end, we first show that if $\mathbf{D}_1, \ldots, \mathbf{D}_t \in \mathcal{V}_0$ are k–generated, then the product $\mathbf{D}_1 \times \ldots \times \mathbf{D}_t$ is $k \cdot t$–generated. Indeed, let $\mathbf{d}(x, y, z)$ be a Maltsev term for $\mathcal{V}$. Suppose that $\{e_i\}$ is a subuniverse of $\mathbf{D}_i$ and that $\mathbf{D}_i$ is generated by the set $\{d_i^1, \ldots, d_i^k\}$. Then the subuniverse $\{e_1\} \times \ldots \times \{e_{i-1}\} \times D_i \times \{e_{i+1}\} \times \ldots \times \{e_t\}$ of the product $\mathbf{D}_1 \times \ldots \times \mathbf{D}_t$ is generated by the set

$$G_i = \Big\{(e_1, \ldots, e_{i-1}, d_i^j, e_{i+1}, \ldots, e_t) \ : \ j = 1, \ldots, k\Big\}.$$

In particular each element $(a_1, \ldots, a_t) \in D_1 \times \ldots \times D_t$ for which there is at most one i with $a_i \neq e_i$, lies in the subalgebra of $\mathbf{D}_1 \times \ldots \times \mathbf{D}_t$ generated by $G = G_1 \cup \ldots \cup G_t$. We have $|G| = kt$. Now for an arbitrary $a = (a_1, \ldots, a_t) \in D_1 \times \ldots \times D_t$ define $a^i = (e_1, \ldots, \ldots, e_{i-1}, a_i, e_{i+1}, \ldots, e_t)$ and observe that

$$(a_1, \ldots, a_t) = \mathbf{d}(\mathbf{d}(\ldots(\mathbf{d}(\mathbf{d}(a^1, e, a^2), e, a^3)\ldots), e, a^{t-1}), e, a^t),$$

where $e = (e_1, \ldots, e_t)$. This shows that the product $\mathbf{D}_1 \times \ldots \times \mathbf{D}_t$ is generated by the set G, as desired.

Now let $t(k)$ denote the number of nonisomorphic, directly indecomposable, k–generated algebras in $\mathcal{V}_0$. Let $\mathbf{D}_1, \ldots, \mathbf{D}_{t(k)}$ be a transversal of the directly indecomposable k–generated algebras in $\mathcal{V}_0$. Since finite algebras in $\mathcal{V}_0$ have unique factorization we get that for different tuples $(n_1, \ldots, n_{t(k)})$ of nonnegative integers the corresponding products $\mathbf{D}_1^{n_1} \times \cdots \times \mathbf{D}_{t(k)}^{n_{t(k)}}$ are nonisomorphic. Moreover, if $n_1 + \cdots + n_{t(k)} \leqslant k$, then all such products are k^2–generated. The number of nonnegative integral solutions of this inequality is $\binom{k+t(k)}{t(k)}$. On the other hand the variety $\mathcal{V}$ has very few models, i.e., there is a constant c with $\mathrm{G}_{\mathcal{V}}(k) \leqslant k^c$. Thus, $\binom{k+t(k)}{t(k)} \leqslant \mathrm{G}_{\mathcal{V}_0}(k^2) \leqslant \mathrm{G}_{\mathcal{V}}(k^2) \leqslant k^{2c}$. Since $t(k)$ is a nondecreasing integer-valued function, Lemma 13.3 shows that $t(k)$ is eventually constant. This means that the class $\mathcal{V}_0$ as well as $\mathcal{V}_\nabla$ have only finitely many finite directly indecomposables.

Finally, using (14) and (13) we see that $\mathcal{V}$ and $\mathcal{V}_\nabla$ have the same bound on the size of their finite directly indecomposables, and this bound is finite. Therefore all finite directly indecomposables are homomorphic images of some fixed finite free algebra in $\mathcal{V}$. Consequently $\mathcal{V}$ has only finitely many finite directly indecomposables, i.e., $\mathcal{V}$ is directly representable. □

A ring $\mathbf{R}$ is said to be of *finite representation type* if there are only finitely many (up to isomorphism) finitely generated and directly indecomposable $\mathbf{R}$–modules. If a ring $\mathbf{R}$ is finite and has an identity element, then the variety of all unitary $\mathbf{R}$–modules is locally finite. The following is immediate from Theorem 13.4.

THEOREM 13.5. *Let $\mathbf{R}$ be a finite ring with identity. The variety $\mathcal{M}_{\mathbf{R}}$ of all unitary $\mathbf{R}$–modules has very few models if and only if the ring $\mathbf{R}$ is of finite representation type, i.e., $\mathcal{M}_{\mathbf{R}}$ is directly representable.* □

In our characterization of finitely generated congruence modular varieties with few models we will need the following specialization of the results of this Chapter.

COROLLARY 13.6. *Let $\mathbf{R}$ be a finite ring with unit. Expand the language of $\mathbf{R}$–modules by an arbitrary set of constants. Let $\mathcal{A}$ be the variety, in this new language, consisting of all unitary $\mathbf{R}$–modules with an arbitrary interpretation of these constants. Then the following conditions are equivalent:*

(1) *$\mathcal{A}$ has very few models,*

(2) $\mathcal{A}$ *is directly representable,*
(3) $\mathcal{M}_{\mathbf{R}}$ *has very few models,*
(4) *the ring* $\mathbf{R}$ *is of finite representation type.*

PROOF. The first two conditions are equivalent by Theorem 13.4, while the last two are equivalent by Corollary 13.5. To see that (1) implies (3) note that $\mathcal{M}_{\mathbf{R}}$ can be treated as a subvariety of $\mathcal{A}$ with all the new constants set to zero. To show that (4) implies (2) we argue as in the last paragraph of the proof of Theorem 13.4, using $\mathcal{A}$ in place of $\mathcal{V}$ so that $\mathcal{V}_{\nabla}$ is essentially $\mathcal{M}_{\mathbf{R}}$. $\square$

CHAPTER 14

Restricting Nilpotent Behavior

Condition (2.3) in Theorem 12.9 states that the centralizer of the monolith is nilpotent. We wish to refine this condition in congruence modular varieties. Our analysis of the subdirectly irreducible algebras in this setting will break into two cases depending on whether the algebra is nilpotent or not.

14.1. Nilpotent Congruences in non Nilpotent Algebras

We start with the following two technical Lemmas that will be needed in this Chapter as well as in Chapter 16.2.

LEMMA 14.1. *Suppose that a finite algebra* $\mathbf{E}$ *in a congruence permutable variety and a congruence* θ *of* $\mathbf{E}$ *satisfy the following conditions:*

- θ *is the unique coatom in* $\mathsf{Con}\,\mathbf{E}$,
- $[1,\alpha]=\alpha$ *for all* $\alpha\in\mathsf{Con}\,\mathbf{E}$.

Let $\mathbf{C}$ *be the subalgebra of* $\mathbf{E}^k$ *consisting of all* θ*–constant functions. Then the algebra* $\mathbf{C}^{2^k}$ *is* $k(|E|^2+1)$*–generated.*

PROOF. Let X be a set with $|X|=2^k$. We identify the algebra $\mathbf{C}^X$ with a subalgebra of $\mathbf{E}^{X\times k}$ consisting of all functions $f\in E^{X\times k}$ such that $f(t,i)\overset{\theta}{\equiv}f(t,j)$ for all $t\in X$ and $i,j\in k$.

From our assumptions on $\mathbf{E}$ we know that $[1,1]=1$, so 1 is not Abelian over θ, i.e., $\mathrm{typ}(\theta,1)=\mathbf{3}$. Let $\{0,1\}$ be a $(\theta,1)$–minimal set of $\mathbf{E}$. We use the set $\{f_1',\dots,f_k'\}$ of free generators of the Boolean algebra $\mathbf{2}^X$ to define $f_1,\dots,f_k\in E^{X\times k}$ by putting $f_i(t,j)=f_i'(t)$.

Moreover for $(a,b)\in\theta$ and $j\in k$ we define $g^j_{ab}\in E^{X\times k}$ by

$$g^j_{ab}(t,i)=\begin{cases} b, & \text{if } i=j,\\ a, & \text{otherwise.}\end{cases}$$

Now let $\mathbf{D}$ be the subalgebra of $\mathbf{E}^{X\times k}$ generated by the set

$$\{f_1,\dots,f_k\}\cup\{g^i_{ab}: i\in k \text{ and } (a,b)\in\theta\}.$$

Thus $\mathbf{D}$ is $k(|E|^2+1)$–generated and contains the diagonal of $\mathbf{E}^{X\times k}$. Our goal is to show that $\mathbf{D}=\mathbf{C}^X=\prod_{t\in X}\mathbf{C}_t$ where each $\mathbf{C}_t$ is a copy of $\mathbf{C}$. Trivially $D\subseteq\prod_{t\in X}C_t$. To prove the reverse inclusion we will show that

(1) For any $a,b\in E$ and $s\in X$ the function h^s_{ab} defined by

$$h^s_{ab}(t,i)=\begin{cases} b, & \text{if } t=s,\\ a, & \text{otherwise,}\end{cases}$$

is in D.

(2) For any $(a,b) \in \theta$ and $(s,j) \in X \times k$ the function $h_{ab}^{(s,j)}$ defined by

$$h_{ab}^{(s,j)}(t,i) = \begin{cases} b, & \text{if } (t,i) = (s,j), \\ a, & \text{otherwise}, \end{cases}$$

is in D.

Having done that we argue as follows to show that every element of $\prod_{t \in X} C_t$ is in D. For $x, y \in E^{X \times k}$ we define

$$\nabla(x,y) = \{(t,i) \in X \times k : x(t,i) \neq y(t,i)\}.$$

Moreover we define the distance $dist(x)$ between the element $x \in E^{X \times k}$ and the set D as the least value of $|\nabla(x,y)|$, where y ranges over D. Obviously $x \in D$ if and only if $dist(x) = 0$. Suppose that D is a proper subset of C^X, i.e., there is an element $x \in C^X$ with $dist(x) > 0$. Pick $y \in D$ such that $dist(x) = |\nabla(x,y)|$. Choose and fix $(s,j) \in \nabla(x,y)$. Let $a = y(s,j)$ and $b = x(s,j)$ and let $\mathbf{d}$ be a Maltsev term for the variety containing $\mathbf{D}$.

If $(a,b) \in \theta$ then, by (2), $h_{ab}^{(s,j)} \in D$ and therefore $y' = \mathbf{d}\left(y, \overline{a}, h_{ab}^{(s,j)}\right) \in D$. Moreover, $y'(s,j) = b = x(s,j)$ and $y'(t,i) = y(t,i)$ for all $(t,i) \neq (s,j)$. Therefore $|\nabla(x,y')| < |\nabla(x,y)| = dist(x)$ gives a contradiction.

Now suppose that $(a,b) \notin \theta$. From (1) we get $h_{ab}^{s} \in D$ and thus $y' = \mathbf{d}(y, \overline{a}, h_{ab}^{s}) \in D$. Obviously $y'(s,j) = b = x(s,j)$ and $y'(t,i) = y(t,i)$ for all $t \neq s$. On the other hand all elements of C^X are θ–constant on the sets of the form $\{t\} \times k$. Consequently $\{s\} \times k \subseteq \nabla(x,y)$ and therefore $\nabla(x,y') \subseteq \nabla(x,y) - \{(s,j)\}$, which together with $y' \in D$, contradicts $dist(x) = |\nabla(x,y)|$.

Therefore to finish the proof of the Lemma it suffices to show (1) and (2). To see (1) first note that from the generators of $\mathbf{D}$ of the form f_i and the diagonal elements of $\mathbf{E}^{X \times k}$, we can produce in $\mathbf{D}$ all elements of the form h_{01}^{s}. Now since $(a,b) \in 1_E = \mathrm{Cg}^{\mathbf{E}}(0,1)$ and $\mathbf{E}$ is in a congruence permutable variety, there is a unary polynomial $\mathbf{p}$ of $\mathbf{E}$ that sends 0 to a and 1 to b. Consequently $h_{ab}^{s} = \mathbf{p}(h_{01}^{s})$ is in D, as required.

Suppose that (2) fails. Pick $(a_0, b_0) \in \theta$ witnessing this failure and such that $\delta = \mathrm{Cg}^{\mathbf{E}}(a_0, b_0)$ is minimal among all $\mathrm{Cg}^{\mathbf{E}}(a,b)$ for which $(a,b) \in \theta$ and (2) fails. Choose γ to be a subcover of δ and let V be a (γ, δ)–minimal set. Then for $(c,d) \in \delta|_V - \gamma$ we have $\delta = \gamma \vee \mathrm{Cg}^{\mathbf{E}}(c,d)$. Applying congruence permutability to γ and $\mathrm{Cg}^{\mathbf{E}}(c,d)$ we get an element b_0' with $(a_0, b_0') \in \mathrm{Cg}^{\mathbf{E}}(c,d)$ and $(b_0', b_0) \in \gamma$. Since $\mathrm{Cg}^{\mathbf{E}}(b_0', b_0) \subseteq \gamma \subset \mathrm{Cg}^{\mathbf{E}}(a_0, b_0)$ our minimality condition gives $h_{b_0' b_0}^{(s,j)} \in D$. Because $\mathbf{E}$ is in a congruence permutable variety there is a unary polynomial $\mathbf{g}$ of $\mathbf{E}$ that sends the pair (c,d) to (a_0, b_0'). By [**32**, Theorem 2.8(3)] we know that $\mathbf{g}(V)$ is a (γ, δ)–minimal set of $\mathbf{E}$. So a_0 and b_0' are in a (γ, δ)–minimal set and we may apply the next Claim to prove that $h_{a_0 b_0'}^{(s,j)} \in D$. But then $h_{a_0 b_0}^{(s,j)} = \mathbf{d}\left(h_{a_0 b_0'}^{(s,j)}, \overline{b_0'}, h_{b_0' b_0}^{(s,j)}\right)$, which contradicts our choice of (a_0, b_0). Thus, to prove that (2) holds it suffices to prove the following.

CLAIM. Let $(a,b) \in \theta$ and $\delta = \mathrm{Cg}^{\mathbf{E}}(a,b)$ with γ a subcover of δ. If a, b are in a (γ, δ)–minimal set, then $h_{ab}^{(s,j)} \in D$.

Proof. Let U be the (γ, δ)–minimal set with $a, b \in U$ and let $\mathbf{e}$ be an idempotent unary polynomial with $\mathbf{e}(E) = U$. Since $\left[\mathrm{Cg}^{\mathbf{E}}(0,1), \mathrm{Cg}^{\mathbf{E}}(a,b)\right] = \mathrm{Cg}^{\mathbf{E}}(a,b) \not\leq \gamma$, there is a binary polynomial $\mathbf{t}'(x,y)$ of $\mathbf{E}$ with

$$\begin{aligned} \mathbf{t}'(0,a) &\overset{\gamma}{\equiv} \mathbf{t}'(0,b), \\ \mathbf{t}'(1,a) &\overset{\gamma}{\not\equiv} \mathbf{t}'(1,b). \end{aligned}$$

We may assume that the range of $\mathbf{t}'$ is contained in U, or replace $\mathbf{t}'$ by $\mathbf{f}\mathbf{t}'$ where $\mathbf{f}$ is a unary polynomial of $\mathbf{E}$ with $\mathbf{f}(E) = U$ and $(\mathbf{f}\mathbf{t}'(1,a), \mathbf{f}\mathbf{t}'(1,b)) \in \delta - \gamma$, (cf. 2.8.(4) of [**32**]). On the other hand the unary polynomial $\mathbf{t}'(1,y)$ does not collapse $\delta|_U$ into γ which means that it is a permutation of U. Consequently, iterating $\mathbf{t}'(x,y)$ sufficiently many times in the second variable we may arrange that $\mathbf{t}'(1,y)$ is the identity function on U.

Observe that the polynomial $\mathbf{t}(x,y) = \mathbf{e}(\mathbf{d}(\mathbf{t}'(x,y), \mathbf{t}'(x,a), \mathbf{t}'(1,a)))$ has its range contained in U and that for $a' = \mathbf{t}(0,b)$ we have

$$\begin{aligned} a &= \mathbf{t}(0,a) \overset{\gamma}{\equiv} \mathbf{t}(0,b) = a', \\ a &= \mathbf{t}(1,a) \overset{\gamma}{\not\equiv} \mathbf{t}(1,b) = b. \end{aligned}$$

Now for arbitrary $c \in E$ and $(s,j) \in X \times k$ define $h^{(s,j)}_{abc} \in E^{X \times k}$ by

$$h^{(s,j)}_{abc}(t,i) = \begin{cases} b, & \text{if } (t,i) = (s,j), \\ c, & \text{if } t \neq s \text{ and } i = j, \\ a, & \text{otherwise.} \end{cases}$$

Since $(a,a') \in \gamma|_U$ and $h^{(s,j)}_{aba'} = \mathbf{t}(h^s_{01}, g^j_{ab}) \in D$ we get

(3) There is an $a' \in U$ such that $h^{(s,j)}_{aba'} \in D$ and $\mathrm{Cg}^{\mathbf{E}}(a,a')|_U \subseteq \gamma$.

We choose $a' \in U$ satisfying (3) and such that the congruence $\mathrm{Cg}^{\mathbf{E}}(a,a')|_U$ is as small as possible.

If $\mathrm{Cg}^{\mathbf{E}}(a,a')|_U = 0|_U$ then we are done, as then $h^{(s,j)}_{ab} = h^{(s,j)}_{aba'} \in D$.

Otherwise we choose a subcover, say γ', of $\delta' = \mathrm{Cg}^{\mathbf{E}}(a,a')$ and we show that this leads to a contradiction. Since $\delta'|_U \not\subseteq \gamma'$ the (γ,δ)–minimal set U contains a (γ',δ')–minimal set V. Now, using 4.30 of [**32**], we get that $V = U$.

From $(a,a') \in \mathrm{Cg}^{\mathbf{E}}(a,b)$ and that $\mathbf{E}$ is in a congruence permutable variety, we get a unary polynomial $\mathbf{q}$ of $\mathbf{E}$ with $(\mathbf{q}(a), \mathbf{q}(b)) = (a,a')$. Composing $\mathbf{q}$ with the unary idempotent polynomial $\mathbf{e}$ whose range is U we may assume that $\mathbf{q}$ itself has range contained in U. Since $\mathbf{q}|_U$ maps a pair, namely (a,b), from outside $\gamma|_U$ into $\gamma|_U$, it cannot be a permutation of U. Therefore $\mathbf{q}(\delta'|_U) \subseteq \gamma'|_U$. Now for $a'' = \mathbf{q}(a')$ we have $h^{(s,j)}_{aba''} = \mathbf{d}(h^{(s,j)}_{aba'}, g^j_{aa'}, \mathbf{q}(h^{(s,j)}_{aba'})) \in D$. This together with $\mathrm{Cg}^{\mathbf{E}}(a,a'')|_U \subseteq \gamma'|_U < \delta'|_U = \mathrm{Cg}^{\mathbf{E}}(a,a')|_U$ contradicts our choice of a'.

This finishes the proof of the Claim and thereby establishes (2) and completes the proof of the Lemma. □

LEMMA 14.2. *Suppose that a finite algebra* $\mathbf{E}$ *is in a congruence permutable variety and* θ *is a congruence relation on* $\mathbf{D}$ *such that:*

- θ *is the unique coatom in* $\mathsf{Con}\,\mathbf{E}$,
- $[1,\alpha] = \alpha$ *for all* $\alpha \in \mathsf{Con}\,\mathbf{E}$.

For a natural number k let $\mathbf{C}_k$ be the subalgebra of $\mathbf{E}^k$ consisting of all θ–constant functions. If the variety $\mathcal{V}(\mathbf{E})$ has few models, then there is a polynomial $p(k)$ such that $\mathbf{C}_k$ has at most $p(k)$ nonisomorphic quotients.

PROOF. The congruence equation $[1, x] = x$ is preserved by homomorphic images and finite subdirect products (see e.g., Remarks 8.8 and 8.9 in [**22**]). Since $\mathbf{C}_k$ is a subdirect power of $\mathbf{E}$ we get that $\mathbf{C}_k$, as well as each algebra from $\mathsf{P}_{\mathsf{fin}}\mathsf{H}(\mathbf{C}_k)$, satisfies this congruence identity.

Moreover each algebra in $\mathsf{H}(\mathbf{C}_k)$ is directly indecomposable. To prove this it is enough to show that $\overline{\theta}$ is the unique coatom in $\mathsf{Con}\,\mathbf{C}_k$. It is obviously a coatom, as by Proposition 2.1 the quotient $\mathbf{C}/\overline{\theta}$ is isomorphic to the simple algebra $\mathbf{E}/\theta$. Let α be another coatom in $\mathsf{Con}\,\mathbf{C}_k$. Note that if $\eta_0, \dots, \eta_{k-1}$ are the kernels of the projections $\mathbf{C}_k \longrightarrow \mathbf{E}$, then $\eta_i \subseteq \overline{\theta}$. The interval $I\,[\eta_i, 1]$ is isomorphic to $\mathsf{Con}\,\mathbf{E}$ and therefore has only one coatom. Thus, for no $i < k$ does $\eta_i \leqslant \alpha$ hold. Consequently $1 = \eta_i \vee \alpha$ for all $i < k$. On the other hand

$$1 = 1^{(k)} = [\eta_0 \vee \alpha, [\eta_1 \vee \alpha, \dots [\eta_{k-2} \vee \alpha, \eta_{k-1} \vee \alpha]]] \,.$$

Distributing over joins we get that the last expression is a join of

$$[\eta_0, [\eta_1, \dots [\eta_{k-2}, \eta_{k-1}]]]$$

and commutators with at least one occurrence of α. Since the last displayed nested commutator is bounded from above by $\eta_0 \cap \dots \cap \eta_{k-1} = 0$ and all of the others summands by are bounded by α we get $1 \leqslant \alpha$, a contradiction.

If a class $\mathcal{C}$ of algebras satisfies the congruence identity $[1, x] = x$, then there are no skew congruences in finite direct products of algebras from $\mathcal{C}$. This condition on skew congruences means that for $\mathbf{D}_1, \dots, \mathbf{D}_s \in \mathcal{C}$, if $\alpha \in \mathsf{Con}\,(\mathbf{D}_1 \times \dots \times \mathbf{D}_s)$ and $\eta_1, \dots, \eta_s$ are the projection kernels, then $\alpha = \bigcap_{i=1}^{s} (\eta_i \vee \alpha)$. The proof of this fact about the congruence identity $[1, x] = x$ is contained in the proof of the implication (1) $\Rightarrow$ (5) of Theorem 8.5 in [**22**] since the argument there only requires congruence modularity.

Now let $\mathbf{D}_1, \dots, \mathbf{D}_s$ be the nontrivial homomorphic images of $\mathbf{C}_k$. From the absence of skew congruences in the product $\mathbf{D} = \mathbf{D}_1 \times \dots \times \mathbf{D}_s$ and the fact that the $\mathbf{D}_i$ are directly indecomposable we get that $\mathbf{D}_1 \times \dots \times \mathbf{D}_s$ is the unique factorization of $\mathbf{D}$ into directly indecomposable algebras.

Let $s(k)$ be the number of nonisomorphic nontrivial quotients of $\mathbf{C}_k$ and let the set $\{\varphi_1, \dots, \varphi_{s(k)}\} \subseteq \mathsf{Con}\,\mathbf{C}_k - \{1\}$ be such that the $\mathbf{D}_i = \mathbf{C}_k/\varphi_i$ are pairwise nonisomorphic. Now for each sequence $(\alpha_1, \dots, \alpha_s)$ of nonnegative integers with $\alpha_1 + \dots + \alpha_{s(k)} \leqslant 2^k$ the product $\mathbf{D}_1^{\alpha_1} \times \dots \times \mathbf{D}_{s(k)}^{\alpha_{s(k)}}$ is isomorphic to a quotient of $\mathbf{C}_k^{2^k}$. Thus, by Lemma 14.1 each such product is $k(|E|^2 + 1)$–generated. From the unique factorization property and the fact that the $\mathbf{D}_i$'s are directly indecomposable we get that different sequences give rise to nonisomorphic products. Let $t(k) = \min\{s(k), 2^k\}$. There are at least $2^{t(k)}$ solutions of the inequality $\alpha_1 + \dots + \alpha_{s(k)} \leqslant 2^k$ since there are $2^{t(k)}$ solutions using $\alpha_i \in \{0, 1\}$ for $1 \leqslant i \leqslant t(k)$. On the other hand $\mathcal{V} = \mathcal{V}(\mathbf{E})$ has few models so there is a polynomial $q(k)$ with $\mathrm{G}_{\mathcal{V}}(k) \leqslant 2^{q(k)}$. Consequently $t(k) \leqslant q(k(|E|^2 + 1))$. But if $t(k)$ is bounded by a polynomial, then for all but at most finitely many k, $t(k) = s(k)$. So $s(k)$ is also bounded by a polynomial, say $p(k)$. Then $\mathbf{C}_k$ has at most $p(k)$ nonisomorphic quotients. □

THEOREM 14.3. *Let $\mathcal{V}$ be a locally finite congruence modular variety with few models. Then every finite subdirectly irreducible algebra $\mathbf{A} \in \mathcal{V}$ is either nilpotent or the centralizer of its monolith is Abelian.*

PROOF. Let $\mathbf{A}$ be a finite subdirectly irreducible algebra in $\mathcal{V}$ that is not nilpotent. So from Theorem 12.9 we know that the variety $\mathcal{V}$ is congruence permutable and the centralizer ν of the monolith μ of $\mathbf{A}$ is the unique coatom in $\mathsf{Con}\,\mathbf{A}$.

We argue that $\mathbf{A}$ satisfies the congruence identity $[1,x]=x$. Indeed, if $[1,\alpha]<\alpha$ for some $\alpha \in \mathsf{Con}\,\mathbf{A}$, then we can pick a congruence η that is maximal with the property that it is above $[1,\alpha]$ but not above α. Obviously $\eta \subseteq \nu$ and η has a unique cover $\eta \vee \alpha$. This means that the subdirectly irreducible algebra $\mathbf{A}/\eta$ is not solvable as it has a simple non-Abelian quotient $\mathbf{A}/\nu$. On the other hand $[1,\eta\vee\alpha]=[1,\eta]\vee[1,\alpha]\subseteq\eta$, i.e., $1_{\mathbf{A}/\eta}$ centralizes the monolith $(\eta\vee\alpha)/\eta$ of $\mathbf{A}/\eta$. Therefore, by Theorem 12.9 (2.1), $\mathbf{A}/\eta$ has to be solvable. This contradiction proves that $\mathbf{A}$ satisfies the congruence identity $[1,x]=x$.

Now let $\mathbf{C}_k$ be the subalgebra of $\mathbf{A}^k$ consisting of all ν–constant functions. Lemma 14.2 supplies us with a polynomial that bounds from above the number of nonisomorphic quotients of $\mathbf{C}_k$. Therefore Lemma 13.1 gives that ν is Abelian. □

14.2. Nilpotent Algebras

The aim of this Section is to show that if $\mathbf{A}$ is a finite nilpotent algebra in a congruence modular variety that does not have many models, then $\mathbf{A}$ factors as a direct product of algebras of prime power order and that the clone of $\mathbf{A}$ is finitely generated. As we have seen in the proof of Corollary 6.11, for finitely generated, congruence modular, nilpotent varieties the G-spectrum is closely related to the free spectrum and to the prime power order factorization property.

Free spectra of finitely generated nilpotent congruence modular varieties were studied in [**5**] by J. Berman and W. Blok. A full characterization of congruence modular varieties with at most singly exponential free spectra was described by K. Kearnes in [**43**]. In his characterization the so-called commutator terms play important role. By a commutator term for an algebra $\mathbf{A}$, or a variety it generates, we mean a term $\mathbf{t}(x_1,\dots,x_r,z)$ of $\mathbf{A}$ such that

$$\mathbf{A} \models \mathbf{t}(x_1,\dots,x_{i-1},z,x_{i+1},\dots,x_r,z)=z,$$

holds for any $i=1,\dots,r$. The number r above is called the rank of the commutator term $\mathbf{t}$. A commutator term $\mathbf{t}$ is nontrivial if

$$\mathbf{A} \not\models \mathbf{t}(x_1,\dots,x_r,z)=z.$$

By combining the proofs of Theorem 3.14 and Corollary 3.15 in [**43**] we may restate the results they contain as follows.

THEOREM 14.4 (K. Kearnes). *Let $\mathbf{A}$ be a finite nilpotent algebra from a congruence modular variety. Then the following conditions are equivalent:*

(1) *The free spectrum of $\mathcal{V}(\mathbf{A})$ is at most (singly) exponential,*
(2) *$\mathbf{A}$ has a finite bound on the rank of its nontrivial commutator terms,*
(3) *$\mathbf{A}$ has a finitely generated clone and every finite algebra in $\mathcal{V}(\mathbf{A})$ factors as a direct product of algebras of prime power order.*

We will use the above Theorem to show the following

LEMMA 14.5. *Let $\mathcal{N}$ be a finitely generated, nilpotent, congruence modular variety of finite type. If there is no finite bound on the rank of nontrivial commutator terms of $\mathcal{N}$, then $\mathcal{N}$ has many models.*

PROOF. Let $\mathbf{A}$ be a minimal, with respect to cardinality, finite algebra in $\mathcal{N}$ such that there is no finite bound on the rank of commutator terms for $\mathbf{A}$. One can easily show that if $\mathbf{A}$ is a finite subdirect product of smaller algebras, then the number that bounds the ranks of nontrivial commutator terms in each stalk is also a bound for the rank of nontrivial commutator terms for $\mathbf{A}$. Thus our minimality assumption about the cardinality of $\mathbf{A}$ gives that $\mathbf{A}$ is subdirectly irreducible. Denote by μ the monolith of $\mathbf{A}$. Then by our minimality condition, $\mathbf{A}/\mu$ generates a variety with a finite bound on the rank of nontrivial commutator terms. Theorem 14.4 tells us that in particular the free spectrum of $\mathcal{V}(\mathbf{A}/\mu)$ is at most singly exponential.

Let X be a set with cardinality 2^{k+1} and let R be an equivalence relation on X with exactly 2^k blocks. There are at least $\Pi(2^k)$, i.e., at least doubly exponentially many, nonisomorphic such structures $(X; R)$. To see this, note that each of the $\Pi(2^k)$ partitions of the integer 2^k can be mapped to a unique partition of 2^{k+1} that has exactly 2^k summands simply by adding 1 to each summand in the partition and padding with a sufficient number of 1s.

By $\mathbf{B}$ we denote the (free) Boolean algebra $(2^X; \wedge, \vee, \neg)$ and we let $\mathbf{B}_R$ denote the subalgebra of $\mathbf{B}$ with atoms corresponding to the blocks of the equivalence relation R on X. Since $\mathbf{B}_R$ has 2^k atoms the algebra $\mathbf{B}_R$ is free as well. Fix the sets $\{f_0, \ldots, f_k\}$ and $\{g_0, \ldots, g_{k-1}\}$ of free generators for $\mathbf{B}$ and $\mathbf{B}_R$, respectively. For $h \in B$ we write $-1h$ for $\neg h$ and $1h$ for h. Then the mappings

$$\{-1, 1\}^{k+1} \ni (\varepsilon_0, \ldots, \varepsilon_k) \mapsto \varepsilon_0 f_0 \wedge \ldots \wedge \varepsilon_k f_k$$

and

$$\{-1, 1\}^k \ni (\sigma_0, \ldots, \sigma_{k-1}) \mapsto \sigma_0 g_0 \wedge \ldots \wedge \sigma_{k-1} g_{k-1}$$

establish the bijections between the sets $\{-1, 1\}^{k+1}$ or $\{-1, 1\}^k$ on one hand and the sets of atoms in $\mathbf{B}$ or $\mathbf{B}_R$ on the other.

We have assumed that the algebra $\mathbf{A}$ has no finite bound on the rank of its nontrivial commutator terms. For every k, by considering a $(k+3)$–ary nontrivial commutator term we know that there exist a $(k+2)$–ary polynomial $\mathbf{t}(x_0, \ldots, x_{k+1})$ of $\mathbf{A}$ and elements $a_0, \ldots, a_{k+1}, 0$ in A such that

$$\mathbf{t}(a_0, \ldots, a_{k+1}) \neq 0 \text{ and } \mathbf{t}(x_0, x_1, \ldots, x_{i-1}, 0, x_{i+1}, \ldots, x_{k+1}) = 0$$

for all $x_j \in A$ and all $0 \leqslant i \leqslant k+1$. The variety $\mathcal{N}$ is congruence regular (see e.g., Corollary 7.7 in [**22**]). Therefore there is an element $b \in A$ such that $\mu = \mathrm{Cg}^{\mathbf{A}}(0, b)$. Moreover he variety $\mathcal{N}$ is congruence permutable (see e.g., Theorem 6.3 in [**22**]) so that Proposition 2.2 applied to $\mathbf{t}(a_0, \ldots, a_{k+1}) \neq 0$ supplies us with a unary polynomial $\mathbf{p}$ of $\mathbf{A}$ such that $(\mathbf{p}(0), \mathbf{p}(\mathbf{t}(a_0, \ldots, a_{k+1}))) = (0, b)$. Replacing $\mathbf{t}$ by $\mathbf{pt}$ we get a new $(k+2)$–ary polynomial $\mathbf{t}(x_0, \ldots, x_{k+1})$ with

$$\mathbf{t}(a_0, \ldots, a_{k+1}) = b$$

and

$$\mathbf{t}(a_0, \ldots, a_{i-1}, 0, a_{i+1}, \ldots, a_{k+1}) = 0,$$

for all $i = 0, \ldots, k+1$.

For $j = 0, \ldots, k+1$ and appropriate i's we define elements f_{ij} and g_{ij} of A^X by

$$f_{ij}(t) = \begin{cases} a_j, & \text{if } f_i(t) = 1, \\ 0, & \text{if } f_i(t) = 0, \end{cases}$$

and

$$g_{ij}(t) = \begin{cases} a_j, & \text{if } g_i(t) = 1, \\ 0, & \text{if } g_i(t) = 0. \end{cases}$$

Now let $\mathbf{D}_R$ be a diagonal subalgebra of $\mathbf{A}^X$ generated by the set

$$\{f_{ij}\}_{i=0,\ldots,k}^{j=0,\ldots,k+1} \cup \{g_{ij}\}_{i=0,\ldots,k-1}^{j=0,\ldots,k+1}.$$

For a fixed i there are at most $|A|$ different elements of the form f_{ij} and at most $|A|$ different elements of the form g_{ij}. Consequently the algebra $\mathbf{D}_R$ has at most $(2k+2)|A|$ generators.

Our goal is to uniformly define two sets Φ and Ψ of sequences of congruences of $\mathbf{D}_R$ such that

- $|\Phi \times \Psi|$ is at most singly exponential as a function of k,
- there is a uniform way to associate with every $(\varphi, \psi) \in \Phi \times \Psi$ a set $At(\varphi)$ and a binary relation $R_{(\varphi,\psi)}$ on $At(\varphi)$,
- there is a pair $(\varphi, \psi) \in \Phi \times \Psi$ such that the structure $(At(\varphi); R_{(\varphi,\psi)})$ is isomorphic to $(X; R)$.

Having made these constructions we argue as follows. Let R' be another equivalence relation on X with $|X/R'| = 2^k$. We form an algebra $\mathbf{D}_{R'}$ and the sets Φ' and Ψ'. The properties listed above guarantee that if $\mathbf{D}_R$ and $\mathbf{D}_{R'}$ are isomorphic, then every structure of the form $(At(\varphi); R_{(\varphi,\psi)})$ with $(\varphi, \psi) \in \Phi \times \Psi$ is isomorphic to a structure of the form $(At(\varphi'); R_{(\varphi',\psi')})$ for some $(\varphi', \psi') \in \Phi' \times \Psi'$, and vice versa. The cardinality of $\Phi \times \Psi$ is at most singly exponential as a function of k. Hence, the number of relations on $\Phi \times \Psi$ that are isomorphic to $(X; R)$ where R is an equivalence relation on X having exactly 2^k equivalence classes is also at most singly exponential as a function of k. But each such $(X; R)$ is isomorphic to a member of some $\Phi \times \Psi$ for the algebra $\mathbf{D}_R$. Since there are at least doubly exponentially many such equivalence relations $(X; R)$, we conclude that there are at least doubly exponentially many $(2k+2)|A|$–generated, pairwise nonisomorphic algebras $\mathbf{D}_R$ in the variety generated by $\mathbf{A}$. Thus, this variety and therefore $\mathcal{N}$ have many models.

Before proceeding with our constructions we collect some information about $\mathbf{D}_R$.

Since $\mathcal{N}$ is a nilpotent congruence modular variety it has a Maltsev term, say $\mathbf{d}(x, y, z)$. Consequently $f'_{ij} = \mathbf{d}(\overline{a}_j, f_{ij}, \overline{0})$ and $g'_{ij} = \mathbf{d}(\overline{a}_j, g_{ij}, \overline{0})$ are in $\mathbf{D}_R$.

CLAIM 1: $\{0, b\}^X \subseteq D_R$.

Proof. For $t \in X$ we denote by b^t the element of $\{0, b\}^X$ whose only nonzero entry is in the t–th place. Since every element of $\{0, b\}^X$ can be obtained from the b^t's and $\overline{0}$ by using the Maltsev term $\mathbf{d}$, it suffices to show that all of the b^t's are in D_R. From our assumption on the f_i's we know that there exists $\varepsilon \in \{-1, 1\}^{k+1}$ such that $\{t\} = \bigcap_{i=0}^{k} (\varepsilon_i f_i)^{-1}(1)$. One can easily check that $b^t = \mathbf{t}(\varepsilon_0 f_{00}, \ldots, \varepsilon_k f_{kk}, \overline{a}_{k+1})$, where $1 f_{ij} = f_{ij}$ and $-1 f_{ij} = f'_{ij}$.

CLAIM 2: The congruence $\mathrm{Cg}^{\mathbf{D_R}}(\overline{0}, b^t)$ is the only atom below $\eta_{X-\{t\}}$ and satisfies $\mathrm{Cg}^{\mathbf{D_R}}(\overline{0}, b^t) = \overline{\mu} \cap \eta_{X-\{t\}}$.

Proof. The proof that $\overline{\mu} \cap \eta_{X-\{t\}}$ is the unique atom below $\eta_{X-\{t\}}$ is essentially that of (3) in the proof of Lemma 13.1. Since $(\overline{0}, b^t) \in \overline{\mu} \cap \eta_{X-\{t\}} - 0$ we get $\mathrm{Cg}^{\mathbf{D_R}}(\overline{0}, b^t) = \overline{\mu} \cap \eta_{X-\{t\}}$.

CLAIM 3: The congruence $\overline{\mu}$ is definable in $\mathbf{D}_R$ as the join of all atoms in $\mathsf{Con}\,\mathbf{D}_R$.

Proof. First note that each atom σ is contained in $\overline{\mu}$. For if otherwise, then suppose that σ is generated by a pair (f, g) with $(f(t), g(t)) \notin \mu$. Pick a unary polynomial $\mathbf{p}(x)$ of $\mathbf{A}$ such that $(\mathbf{p}(f(t)), \mathbf{p}(g(t))) \in \mu - 0$. Then the pair $(\mathbf{p}(f), \mathbf{p}(g))$ generates a nonzero congruence σ' of $\mathbf{D}_R$ that is contained in $\mu_t := \{(r,s) \in D_R \times D_R : (r_t, s_t) \in \mu\}$. Since $\sigma \not\leqslant \mu_t$ we get $0 \neq \sigma' < \sigma$, a contradiction. Let α be the join of all the atoms of $\mathbf{D}_R$. We thus have $\alpha \leqslant \overline{\mu}$. To see that $\overline{\mu} \leqslant \alpha$ suppose that $(f, g) \in \overline{\mu}$. Since for $s \in X$ we have $(f(s), g(s)) \in \mu = \mathrm{Cg}^{\mathbf{A}}(0, b)$ there is a unary polynomial $\mathbf{q}_s$ of $\mathbf{A}$ such that $\mathbf{q}_s(0) = f(s)$ and $\mathbf{q}_s(b) = g(s)$. To see that $(f, g) \in \alpha$ we let $m = 2^{k+1}$ and we enumerate the set $X = \{t_1, \ldots, t_m\}$ and then define $f_0, f_1, \ldots, f_m \in D_R$ by putting

$$\begin{aligned} f_0 &= f \\ f_\ell &= \mathbf{d}\left(f_{\ell-1}, \mathbf{q}_{t_\ell}(\overline{0}), \mathbf{q}_{t_\ell}(b^{t_\ell})\right) \qquad \text{for } \ell = 1, \ldots, m. \end{aligned}$$

Observe that

$$f_\ell = g|_{\{t_1,\ldots,t_\ell\}} \cup f|_{X-\{t_1,\ldots,t_\ell\}}.$$

Moreover $(f_{\ell-1}, f_\ell) \in \mathrm{Cg}^{\mathbf{D_R}}(\overline{0}, b^{t_l}) \subseteq \alpha$ so that $f = f_0 \stackrel{\alpha}{\equiv} f_m = g$, as required.

We have $[\mu, 1] = 0$ since $\mathbf{A}$ is nilpotent. This together with Propositions 4.4.(2) and 4.5 from [**22**] immediately give

CLAIM 4: $[\overline{\mu}, 1] = 0$.

For $i = 0, \ldots, r-1$, a sequence $\varphi = (\varphi_0, \ldots, \varphi_{2r-1})$ of $2r$ congruences of $\mathbf{D}_R$ and $\varepsilon \in \{-1,1\}^r$ define $\varphi_i^{\varepsilon_i} = \varphi_{\frac{1-\varepsilon_i}{2}\cdot r+i}$. We let Φ be the set of all sequences $\varphi = (\varphi_0, \ldots, \varphi_k, \varphi_{k+1}, \ldots, \varphi_{2k+1})$ such that

- $\mathsf{Con}\,\mathbf{D}_R \ni \varphi_i \geqslant \overline{\mu}$,
- for every $\varepsilon \in \{-1,1\}^{k+1}$ the intersection $\displaystyle\bigcap_{i=0}^{k} [\varphi_i^{\varepsilon_i}, 1]$ contains exactly one atom of $\mathsf{Con}\,\mathbf{D}_R$.

Since $\mathbf{D}_R/\overline{\mu}$ is a $(2k+2)|A|$–generated algebra from a nilpotent variety $\mathcal{V}(\mathbf{A}/\mu)$ with an at most singly exponential free spectrum, it follows from Lemma 6.8 that $|\mathsf{Con}\,(\mathbf{D}_R/\overline{\mu})|$ is at most singly exponential as a function of k. So $|\Phi| \leqslant |\mathsf{Con}\,(\mathbf{D}_R/\overline{\mu})|^{2k+2}$ is at most singly exponential in k. Finally, let Ψ be the set of all sequences $\psi = (\psi_0, \ldots, \psi_{k-1}, \psi_k, \ldots, \psi_{2k-1})$ of congruences of $\mathbf{D}_R$ with $\psi_i \geqslant \overline{\mu}$. As with Φ, we easily observe that $|\Psi|$ is at most singly exponential in k.

For $\varphi \in \Phi$ and $\varepsilon \in \{-1,1\}^{k+1}$ denote by $\varphi^{(\varepsilon)}$ the unique atom contained in $\bigcap_{i=0}^{k} [\varphi_i^{\varepsilon_i}, 1]$ and let $At(\varphi)$ denote the set $\left\{\varphi^{(\varepsilon)} : \varepsilon \in \{-1,1\}^{k+1}\right\}$ of all atoms obtained in this way. Now for every pair $(\varphi, \psi) \in \Phi \times \Psi$ we define a binary relation $R_{(\varphi,\psi)}$ on the set $At(\varphi)$ by

$$(\alpha, \beta) \in R_{(\varphi,\psi)} \text{ iff } (\forall \sigma \in \{-1,1\}^k) \left(\alpha \subseteq \bigcap_{i=0}^{k-1} [\psi_i^{\sigma_i}, 1] \text{ iff } \beta \subseteq \bigcap_{i=0}^{k-1} [\psi_i^{\sigma_i}, 1] \right).$$

For $j = 0, \ldots, k$ define

$$\begin{aligned} \theta_j &= \mathrm{Cg}^{\mathbf{D_R}}(\overline{0}, f_{jj}) \\ \theta'_j &= \mathrm{Cg}^{\mathbf{D_R}}(\overline{0}, f'_{jj}) \end{aligned}$$

and for $j = 0, \ldots, k-1$ put

$$\begin{aligned} \xi_j &= \mathrm{Cg}^{\mathbf{D_R}}(\overline{0}, g_{jj}) \\ \xi'_j &= \mathrm{Cg}^{\mathbf{D_R}}(\overline{0}, g'_{jj}). \end{aligned}$$

We will conclude the proof of the Theorem by showing that for the sequences:

$$\varphi = (\overline{\mu} \vee \theta_0, \ldots, \overline{\mu} \vee \theta_k,\ \overline{\mu} \vee \theta'_0, \ldots, \overline{\mu} \vee \theta'_k)$$

and

$$\psi = (\overline{\mu} \vee \xi_0, \ldots, \overline{\mu} \vee \xi_{k-1},\ \overline{\mu} \vee \xi'_0, \ldots, \overline{\mu} \vee \xi'_{k-1})$$

the pair (φ, ψ) is in $\Phi \times \Psi$ and the structures $(At(\varphi); R_{(\varphi,\psi)})$ and $(X; R)$ are isomorphic.

We write $1\theta_i$ for θ_i and $-1\theta_i$ for θ'_i and we adopt the same convention for the ξ's. To see that $\varphi \in \Phi$ take $\varepsilon \in \{-1,1\}^{k+1}$. Let $Z_i = (\varepsilon_i f_i)^{-1}(1)$. Since $[\overline{\mu}, 1] = 0$ we have $[\varphi_i^{\varepsilon_i}, 1] = [\overline{\mu} \vee \varepsilon_i\theta_i, 1] = [\varepsilon_i\theta_i, 1] \subseteq \varepsilon_i\theta_i \subseteq \eta_{X-Z_i}$. On the other hand the intersection $Z_0 \cap \ldots \cap Z_k$ contains exactly one point from X, say t. Therefore the intersection $\bigcap_{i=0}^k [\varphi_i^{\varepsilon_i}, 1]$ is contained in $\eta_{X-\{t\}}$. Consequently it can contain at most one atom, namely $\mathrm{Cg}^{\mathbf{D_R}}(b^t, \overline{0})$. To see that in fact $(b^t, \overline{0})$ lies in $[\varphi_i^{\varepsilon_i}, 1]$ for all $i = 0, \ldots, k$ note that

$$\mathbf{t}(\overline{0}, \ldots, \overline{0}, \varepsilon_i f_{ii}, \overline{0}, \ldots, \overline{0}, \overline{a}_{k+1}) = \overline{0} = \mathbf{t}(\overline{0}, \ldots, \overline{0}, \overline{0}, \overline{0}, \ldots, \overline{0}, \overline{a}_{k+1})$$

i.e., those elements are collapsed by $[\varphi_i^{\varepsilon_i}, 1]$. Changing each $\overline{0}$ to an appropriate $\varepsilon_l f_{ll}$ we get that the elements

$$\begin{aligned} b^t &= \mathbf{t}(\varepsilon_0 f_{00}, \ldots, \varepsilon_k f_{kk}, \overline{a}_{k+1}), \\ \overline{0} &= \mathbf{t}(\varepsilon_0 f_{00}, \ldots, \varepsilon_{i-1} f_{i-1,i-1}, \overline{0}, \varepsilon_{i+1} f_{i+1,i+1}, \ldots, \varepsilon_k f_{kk}, \overline{a}_{k+1}) \end{aligned}$$

get collapsed by $[\varphi_i^{\varepsilon_i}, 1]$, as required. This proves that $\varphi \in \Phi$. Moreover the argument above shows that for any $\varphi \in \Phi$ there is a bijection between $At(\varphi)$ and X given by

$$\varphi^{(\varepsilon)} \quad \mapsto \quad \text{the unique element } t \in X \text{ such that } (\varepsilon_i f_{ii})(t) \neq 0.$$

Now recall that the elements σ of $\{-1,1\}^k$ correspond bijectively to R–blocks by $\sigma \mapsto Z_\sigma = \bigcap_{i=0}^{k-1} (\sigma_i g_i)^{-1}(1)$. To prove that $R_{(\varphi,\psi)} = R$, all we need to show is that

CLAIM 5: $(b^t, \overline{0}) \in \bigcap_{i=0}^{k-1} [\psi_i^{\sigma_i}, 1]$ if and only if $t \in Z_\sigma$.

First observe that an argument similar to the one given for $\bigcap_{i=0}^k [\varphi_i^{\varepsilon_i}, 1]$ shows that $\bigcap_{i=0}^{k-1} [\psi_i^{\sigma_i}, 1] = \bigcap_{i=0}^{k-1} [\sigma_i\xi_i, 1] \subseteq \eta_{Z_\sigma}$. This proves the 'only if' direction of the Claim. To see the converse, fix $t \in Z_\sigma$ and $i \in \{0, \ldots, k-1\}$. We will show that

$(b^t, \bar{0}) \in [\sigma_i \xi_i, 1]$. To simplify our notation we assume that $i = 0$. Let $\varepsilon \in \{-1, 1\}^{k+1}$ be such that $\{t\} = \bigcap_{i=0}^{k} (\varepsilon_i f_i)^{-1}(1)$. As before we get

$$\left(\mathbf{t}(\sigma_0 g_{00}, \bar{0}, \ldots, \bar{0}), \mathbf{t}(\bar{0}, \bar{0}, \ldots, \bar{0})\right) = (\bar{0}, \bar{0}) \in [\sigma_0 \xi_0, 1]$$

and consequently the elements

$$\begin{aligned} b^t &= \mathbf{t}(\sigma_0 g_{00}, \varepsilon_0 f_{01}, \ldots, \varepsilon_k f_{k,k+1}), \\ \bar{0} &= \mathbf{t}(\bar{0}, \varepsilon_0 f_{01}, \ldots, \varepsilon_k f_{k,k+1}) \end{aligned}$$

are collapsed by $[\sigma_0 \xi_0, 1] = [\psi_0^{\sigma_0}, 1]$.

This proves the last Claim and completes the proof of Lemma 14.5. □

Lemma 14.5, Theorem 14.4 and Corollary 6.11 immediately give the following.

THEOREM 14.6. *Let $\mathcal{N}$ be a finitely generated, nilpotent, congruence modular variety. Then the following conditions are equivalent:*

(1) *$\mathcal{N}$ has few models,*
(2) *$\mathcal{N}$ does not have many models,*
(3) *$\mathcal{N}$ has a finitely generated clone and is generated by a finite algebra that factors as direct product of algebras of prime power order.* □

In order to understand finitely generated congruence modular varieties that do not have many models we need the following theorem.

THEOREM 14.7. *Suppose $\mathcal{V}$ is a finitely generated congruence modular variety that does not have many models. Let $\mathcal{N}$ be the class of all nilpotent algebras in $\mathcal{V}$. Then $\mathcal{N}$ is a finitely generated variety. Actually if $\mathcal{V} = \mathsf{HSP}(\mathbf{A})$ for a finite algebra $\mathbf{A}$, then $\mathcal{N} = \mathsf{HSP}(\mathsf{HS}(\mathbf{A}) \cap \mathcal{N})$.*

PROOF. From Theorem 3.10 of R.McKenzie [**51**] we know that there is a natural number n such that the variety $\mathcal{V}$ satisfies two congruence identities: $\theta^{[n+1]} = \theta^{[n]}$ and $\theta^{(n+1)} = \theta^{(n)}$. In particular

$$\mathcal{N} = \left\{ \mathbf{D} \in \mathcal{V} \; : \; \mathsf{Con}\, \mathbf{D} \models 1^{(n)} = 0 \right\}.$$

Since the congruence identity $1^{(n)} = 0$ is preserved by homomorphic images, subalgebras and direct products (cf. Propositions 4.4 and 4.5 of [**22**]), the class $\mathcal{N}$ is a variety.

Before showing that this variety is finitely generated we observe that in any algebra $\mathbf{D} \in \mathcal{V}$ the congruence

$$1^{[\omega]} = \bigcap\nolimits_{k<\omega} 1^{[k]} = 1^{[n]}$$

is the smallest congruence φ such that $\mathbf{D}/\varphi$ is solvable, while

$$1^{(\omega)} = \bigcap\nolimits_{k<\omega} 1^{(k)} = 1^{(n)}$$

is the smallest congruence φ such that $\mathbf{D}/\varphi$ is nilpotent. Obviously $1^{[\omega]} \subseteq 1^{(\omega)}$. On the other hand Corollary 12.8 gives that the solvable algebra $\mathbf{D}/1^{[\omega]}$ is nilpotent and therefore $1^{[\omega]} = 1^{(\omega)}$.

CLAIM. Let $\mathbf{D}$ be a finite algebra in $\mathcal{V}$ and $\gamma = 1^{[\omega]} = 1^{(\omega)}$. Then for $\theta_0, \ldots, \theta_{s-1} \in \mathsf{Con}\, \mathbf{D}$ we have $\gamma \vee \bigcap_{i<s} \theta_i = \bigcap_{i<s} (\gamma \vee \theta_i)$.

Proof. We will prove the Claim for $s = 2$ as the general case can be easily obtained by induction. Put $\gamma' = \gamma \vee (\theta_0 \cap \theta_1)$ and suppose, to the contrary that $\gamma' < (\theta_0 \vee \gamma) \cap (\theta_1 \vee \gamma)$ and pick δ such that $\gamma' \prec \delta \subseteq (\theta_0 \vee \gamma) \cap (\theta_1 \vee \gamma)$. Define

$$\begin{aligned}\alpha_i &= (\theta_i \cap \delta) \vee (\theta_{1-i} \cap \gamma'),\\ \varepsilon &= \alpha_0 \cap \alpha_1.\end{aligned}$$

We will need the following

(4) $\theta_i \vee \delta = \theta_i \vee \gamma'$,
(5) $\varepsilon = (\theta_0 \cap \gamma') \vee (\theta_1 \cap \gamma')$,
(6) $\alpha_0 \vee \alpha_1 \leqslant \delta$,
(7) $\alpha_i \cap \gamma' = \varepsilon$
(8) $\alpha_i \vee \gamma' = \delta = \alpha_i \vee \gamma$.

The first item is equivalent to $\delta \leqslant \theta_i \vee \gamma'$. But this follows immediately from our choice of δ as $\delta \leqslant (\theta_0 \vee \gamma) \cap (\theta_1 \vee \gamma) \leqslant \theta_i \vee \gamma \leqslant \theta_i \vee \gamma'$.

To get (5) we apply modularity twice and use the fact that $\theta_0 \cap \theta_1 \leqslant \gamma' \leqslant \delta$

$$\begin{aligned}\varepsilon &= ((\theta_0 \cap \delta) \vee (\theta_1 \cap \gamma')) \cap ((\theta_1 \cap \delta) \vee (\theta_0 \cap \gamma'))\\ &= (((\theta_0 \cap \delta) \vee (\theta_1 \cap \gamma')) \cap (\theta_1 \cap \delta)) \vee (\theta_0 \cap \gamma')\\ &= (((\theta_0 \cap \delta) \cap (\theta_1 \cap \delta)) \vee (\theta_1 \cap \gamma')) \vee (\theta_0 \cap \gamma')\\ &= (\theta_0 \cap \theta_1) \vee (\theta_1 \cap \gamma') \vee (\theta_0 \cap \gamma')\\ &= (\theta_1 \cap \gamma) \vee (\theta_0 \cap \gamma).\end{aligned}$$

From $\alpha_i \leqslant (\theta_i \cap \delta) \vee (\theta_{1-i} \cap \delta) \leqslant \delta$ we get (6).

To see (7) we apply (5) in the last equality of the following calculation

$$\begin{aligned}\alpha_i \cap \gamma' &= ((\theta_i \cap \delta) \vee (\theta_{1-i} \cap \gamma')) \cap \gamma'\\ &= (\theta_i \cap \delta \cap \gamma') \vee (\theta_{1-i} \cap \gamma')\\ &= (\theta_i \cap \gamma') \vee (\theta_{1-i} \cap \gamma') = \varepsilon.\end{aligned}$$

Finally we have

$$\begin{aligned}\alpha_i \vee \gamma &= (\theta_i \cap \delta) \vee (\theta_{1-i} \cap \gamma') \vee \gamma\\ &\geqslant (\theta_i \cap \delta) \vee \gamma\\ &= (\theta_i \vee \gamma) \cap \delta\\ &= \delta,\end{aligned}$$

where the last equality follows from $\delta \leqslant (\theta_0 \vee \gamma) \cap (\theta_1 \vee \gamma) \leqslant \theta_i \vee \gamma$. The displayed calculation together with $\alpha_i \vee \gamma \leqslant \alpha_i \vee \gamma' \leqslant \delta$ gives (8).

In particular we have

$$I[\gamma', \delta] \searrow I[\varepsilon, \alpha_i] \nearrow I[\alpha_{1-i}, \alpha_0 \vee \alpha_1]$$

and consequently $\alpha_0 \vee \alpha_1$ covers α_0. We have $\mathrm{typ}(\alpha_0, \alpha_0 \vee \alpha_1) = \mathbf{2}$ since $\gamma \leqslant \gamma'$ and $\mathrm{typ}\{\gamma, 1\} \subseteq \{\mathbf{2}\}$. Pick η to be a congruence that is maximal with respect to being above α_0 but not above $\alpha_0 \vee \alpha_1$. Then $\mathbf{D}/\eta$ is a subdirectly irreducible and $(\eta \vee \alpha_1)/\eta$ is its (Abelian) monolith. On the other hand

$$[\alpha_1, \delta] = [\alpha_1, \alpha_0 \vee \gamma'] = [\alpha_1, \alpha_0] \vee [\alpha_1, \gamma'] \subseteq \varepsilon \subseteq \alpha_0 \subseteq \eta$$

so that

$$[\eta \vee \alpha_1, \eta \vee \delta] \subseteq \eta \vee [\alpha_1, \delta] \subseteq \eta.$$

This means that $(\eta \vee \delta)/\eta$ centralizes the monolith of $\mathbf{D}/\eta$ and therefore Theorem 12.3 gives that $\eta \vee \delta$ is solvable over η. From $I[\eta, \eta \vee \delta] \searrow I[\eta \cap \delta, \delta]$ we get that δ is solvable over $\eta \cap \delta$. This together with the fact that 1 is solvable over δ shows that 1 is solvable over $\eta \cap \delta$. Consequently $\gamma \subseteq \eta \cap \delta$ and therefore $\eta \geqslant \alpha_0 \vee \gamma = \delta \geqslant \alpha_0 \vee \alpha_1$, a contradiction. This finishes the proof of the Claim.

The variety $\mathcal{V}$ and therefore $\mathcal{N}$ are locally finite. Thus to prove that $\mathcal{N} = \mathsf{HSP}(\mathsf{HS}(\mathbf{A}) \cap \mathcal{N})$ it suffices to show that every finite algebra $\mathbf{D} \in \mathcal{N}$ is also in $\mathsf{HSP}(\mathsf{HS}(\mathbf{A}) \cap \mathcal{N})$. Since $\mathbf{D} \in \mathsf{HSP}(\mathbf{A}) \cap \mathcal{N}$ we know that $\mathbf{D}$ is a homomorphic image of some quotient of the form $\mathbf{C}/1^{(\omega)}$ where $\mathbf{C}$ is a subalgebra of a finite power $\mathbf{A}^s$. Thus it suffices to show that $\mathbf{C}/1^{(\omega)} \in \mathsf{ISP}(\mathsf{HS}(\mathbf{A}) \cap \mathcal{N})$. Let η_i be the kernel of the i–th projection $\mathbf{C} \longrightarrow \mathbf{A}$. In particular $\mathbf{C}/\eta_i$ is isomorphic to a subalgebra of $\mathbf{A}$. From the Claim we know $\mathbf{C}/(1^{(\omega)} \vee \bigcap_{i<s} \eta_i)$ is embeddable into the product $\prod_{i<s} \mathbf{C}/(1^{(\omega)} \vee \eta_i)$. But $\bigcap_{i<s} \eta_i = 0_C$, so $\mathbf{C}/1^{(\omega)}$ is also embeddable into this product. Since each $\mathbf{C}/(1^{(\omega)} \vee \eta_i)$ is a homomorphic image of $\mathbf{C}/\eta_i$, which is isomorphic to a subalgebra of $\mathbf{A}$, we are done. □

Combining Theorems 14.7 and 14.6 we immediately get the following.

COROLLARY 14.8. *If a finitely generated congruence modular variety $\mathcal{V}$ does not have many models, then the class $\mathcal{N}$ of all nilpotent algebras in $\mathcal{V}$ is a variety generated by a finite algebra that factors as direct product of algebras of prime power order and the clone of $\mathcal{N}$ is finitely generated.* □

CHAPTER 15

Decomposing Finite Algebras

In this Chapter we summarize our results and give a list of necessary conditions for a locally finite variety omitting type **1** to have few models. Then we will show that every finite algebra from a variety satisfying all the conditions from our list decomposes into a direct product of a nilpotent algebra and a centerless algebra. This decomposition will be used both in Chapter 16 to isolate an additional necessary condition that the Abelian part of a centerless algebra from a locally finite variety omitting type **1** with few models has to fulfill as well as in Chapter 17 in our proof that this expanded list of necessary conditions is actually complete.

Combining Theorems 12.9, 14.3 and Corollary 14.8 we immediately get the following result.

THEOREM 15.1. *Let $\mathcal{V}$ be a locally finite variety omitting type* **1**. *If $\mathcal{V}$ has few models then*

- **(1)** *$\mathcal{V}$ is congruence permutable,*
- **(2)** *for any finite subdirectly irreducible algebra $\mathbf{A}$ in $\mathcal{V}$ with monolith μ and its centralizer $\nu = (0 : \mu)$ we have:*
  - **(2.1)** *ν is the solvable radical of $\mathbf{A}$,*
  - **(2.2)** *ν is comparable to all congruences of $\mathbf{A}$,*
  - **(2.3)** *ν is Abelian or $\mathbf{A}$ is nilpotent,*
  - **(2.4)** *the quotient $\mathbf{A}/\nu$ is either trivial or simple non-Abelian,*
- **(3)** *every finite nilpotent subdirectly irreducible algebra in $\mathcal{V}$ is of prime power order and has a finitely generated clone.* □

In the rest of this Chapter we assume that a locally finite variety $\mathcal{V}$ satisfies the conditions **(1)**–**(3)** of Theorem 15.1. We introduce a number of congruence relations for finite algebras in $\mathcal{V}$ and then use these congruences to form a canonical direct decomposition of each finite member of $\mathcal{V}$ into a nilpotent and a centerless factor.

In our analysis that leads to this decomposition we make heavy use of modular commutator theory. Recall that for a congruence α we denote the solvable and nilpotent n-th power by $\alpha^{[n]}$ and $\alpha^{(n)}$, respectively. Since $\alpha^{[2]} = \alpha^{(2)}$ we simply write α^2 in this case.

For a finite algebra $\mathbf{A}$ from the variety $\mathcal{V}$ we define the following congruences:

$$\begin{aligned}
\sigma_{\mathbf{A}} &= \bigcap \{\theta \in \mathsf{Con}\,\mathbf{A} : [1_{\mathbf{A}}, 1_{\mathbf{A}}] \not\leq \theta \prec 1_{\mathbf{A}}\}, \\
1^{\omega}_{\mathbf{A}} &= \bigcap_{n<\omega} 1^{[n]}_{\mathbf{A}}, \\
\delta_{\mathbf{A}} &= \left(\sigma^2_{\mathbf{A}} : 1^{\omega}_{\mathbf{A}}\right), \\
\rho_{\mathbf{A}} &= \bigvee \left\{\alpha \in \mathsf{Con}\,\mathbf{A} : \alpha^{[n]} = 0 \text{ for some } n < \omega\right\}.
\end{aligned}$$

We will omit the subscript **A** if it is clear from the context. Note that 1^ω is the smallest congruence φ of **A** for which the quotient $\mathbf{A}/\varphi$ is solvable. Solvable congruences in (finite) algebras from $\mathcal{V}$ are nilpotent so $1^\omega = \bigcap_{n<\omega} 1^{(n)}$ and consequently 1^ω is also the smallest congruence φ of **A** such that $\mathbf{A}/\varphi$ is nilpotent. On the other hand ρ is the join of all solvable congruences of **A**, i.e., ρ is the solvable radical of **A**.

Our goal is to show that the algebra **A** decomposes into the direct product $\mathbf{A}/\delta \times \mathbf{A}/1^\omega$.

The congruence relation 1 is solvable over 1^ω and thus

(1) For $\theta \prec 1$ we have: $1^\omega \leqslant \theta$ iff $1^2 \leqslant \theta$ iff $\operatorname{typ}(\theta, 1) = \mathbf{2}$.

Before proving an analogous characterization of coatoms above δ we need some preparatory remarks. We start with

(2) $[1^\omega, \varphi] = 1^\omega \wedge \varphi$ for every $\varphi \in \mathsf{Con}\,\mathbf{A}$.

Obviously $[1^\omega, \varphi] \leqslant 1^\omega \wedge \varphi$. If this inequality is strict, then we can find a congruence β with $[1^\omega, \varphi] \prec \beta \leqslant 1^\omega \wedge \varphi$. Next we consider a congruence η that is maximal with respect to being above $[1^\omega, \varphi]$ but not above β. Then $\mathbf{A}/\eta$ is subdirectly irreducible. This algebra cannot be nilpotent, as then $\eta \geqslant 1^\omega \geqslant 1^\omega \wedge \varphi \geqslant \beta$. Therefore $\mathbf{A}/\eta$ has a type **3** subcover of 1 by virtue of **(2.3)** and **(2.4)**. This gives that $\eta \vee 1^\omega = 1$ for otherwise there is a subcover of 1 that is above both η and 1^ω and this would force the existence of a type **3** coatom above 1^ω, contrary to (1). On the other hand $[1^\omega, \beta] \leqslant [1^\omega, \varphi]$, i.e., $1^\omega \leqslant ([1^\omega, \varphi] : \beta)$. By our choice of η we know that the interval $I\left[[1^\omega, \varphi], \beta\right]$ projects up to $I\left[\eta, \eta^+\right]$ and therefore $([1^\omega, \varphi] : \beta) = (\eta : \eta^+)$. Consequently $1 = 1^\omega \vee \eta \leqslant (\eta : \eta^+)$, which means that the centralizer of the monolith of $\mathbf{A}/\eta$ is not solvable, contradicting **(2.1)**.

From (2) we easily infer the following

(3) $(\delta : 1^\omega) = \delta$.

Indeed, $(\delta : 1^\omega) \geqslant \delta$ is obvious. Using (2) we get

$$\begin{aligned}
&\left[1^\omega, \left((\sigma^2 : 1^\omega) : 1^\omega\right)\right] = 1^\omega \wedge \left((\sigma^2 : 1^\omega) : 1^\omega\right) = \\
&1^\omega \wedge 1^\omega \wedge \left((\sigma^2 : 1^\omega) : 1^\omega\right) = \left[1^\omega, \left[1^\omega, \left((\sigma^2 : 1^\omega) : 1^\omega\right)\right]\right] \leqslant \\
&\left[1^\omega, (\sigma^2 : 1^\omega)\right] \leqslant \sigma^2,
\end{aligned}$$

and therefore $(\delta : 1^\omega) = \left((\sigma^2 : 1^\omega) : 1^\omega\right) \leqslant \left(\sigma^2 : 1^\omega\right) = \delta$, as desired.

(4) $\sigma^2 \leqslant \delta \leqslant \sigma$.

The first inequality in (4) is immediate from the definition of δ. To prove the second one assume to the contrary that there is $\theta \prec 1$ with $1^2 \not\leqslant \theta$ and $\delta \not\leqslant \theta$. In particular we have $\delta \vee \theta = 1$. From $\delta = \left(\sigma^2 : 1^\omega\right)$ we have $\delta \leqslant \left(\sigma^2 : 1^\omega\right)$ and thus $[1^\omega, \delta] \leqslant \sigma^2$. This gives $[1^\omega, \delta] \leqslant \sigma^2 \leqslant \sigma \leqslant \theta$. Thus

$$1^\omega = [1^\omega, 1] = [1^\omega, \delta \vee \theta] = [1^\omega, \delta] \vee [1^\omega, \theta] \leqslant \theta \vee [1^\omega, \theta] = \theta.$$

This however contradicts the fact that 1 is not Abelian over θ.

Now we are ready to prove the following

(5) For $\theta \prec 1$ we have: $\delta \leqslant \theta$ iff $1^2 \not\leqslant \theta$ iff $\operatorname{typ}(\theta, 1) = \mathbf{3}$.

Obviously the second two conditions are equivalent and imply the first one, since $1^2 \not\leqslant \theta$ gives $\sigma \leqslant \theta$ and then (4) gives $\delta \leqslant \theta$.

Suppose that (5) fails and that $\mathbf{A}$ is a smallest, with respect to cardinality, counterexample. Thus in $\mathbf{A}$ we have $\delta \leqslant \theta$ and $\operatorname{typ}(\theta, 1) = 2$. By (1) we have $1^\omega \leqslant \theta$ as well. So $\delta \vee 1^\omega \leqslant \theta \prec 1$.

From the minimality of $\mathbf{A}$ we get that

(5.1) $\delta = 0 = \sigma^2$.

In order to prove (5.1) it suffices to show that $\delta_A = 0$. To this end put $\mathbf{B} = \mathbf{A}/\delta_{\mathbf{A}}$. Note that $1^\omega_{\mathbf{B}} = (1^\omega_{\mathbf{A}} \vee \delta_{\mathbf{A}})/\delta_{\mathbf{A}}$. By (4) we know that $\sigma_{\mathbf{B}} = \sigma_{\mathbf{A}}/\delta_{\mathbf{A}}$ so that $\sigma^2_{\mathbf{B}} = (\sigma^2_{\mathbf{A}} \vee \delta_{\mathbf{A}})/\delta_{\mathbf{A}} = \delta_{\mathbf{A}}/\delta_{\mathbf{A}} = 0_{\mathbf{B}}$. In particular, $\delta_{\mathbf{B}} = (0_{\mathbf{B}} : 1^\omega_{\mathbf{B}})$. Let $\varphi \geqslant \delta_{\mathbf{A}}$ be such that $\delta_{\mathbf{B}} = (0_{\mathbf{B}} : 1^\omega_{\mathbf{B}}) = \varphi/\delta_{\mathbf{A}}$. This gives that $0_{\mathbf{B}} \geqslant [\varphi/\delta_{\mathbf{A}}, (1^\omega_{\mathbf{A}} \vee \delta_{\mathbf{A}})/\delta_{\mathbf{A}}] = ([\varphi, 1^\omega_{\mathbf{A}} \vee \delta_{\mathbf{A}}] \vee \delta_{\mathbf{A}})/\delta_{\mathbf{A}}$, i.e., $\delta_{\mathbf{A}} \geqslant [\varphi, 1^\omega_{\mathbf{A}} \vee \delta_{\mathbf{A}}] \geqslant [\varphi, 1^\omega_{\mathbf{A}}]$. Thus $\varphi \leqslant (\delta_{\mathbf{A}} : 1^\omega_{\mathbf{A}}) = \delta_{\mathbf{A}}$. Consequently $\varphi = \delta_{\mathbf{A}}$ which means that $\delta_{\mathbf{B}} = 0_{\mathbf{B}}$. However $\delta_{\mathbf{A}} \leqslant \theta$ gives that in the algebra $\mathbf{A}$ the congruence $\theta/\delta_{\mathbf{A}}$ witnesses the fact that $\mathbf{B}$ is also a counterexample to (5). Thus $|B| = |A|$ and consequently $\delta_{\mathbf{A}} = 0$.

In the next step we show that $\mathbf{A}$ is subdirectly irreducible. Suppose to the contrary that α_1, α_2 are two different atoms in $\mathsf{Con}\, \mathbf{A}$. Obviously

(5.2) $\alpha_i \leqslant 1^\omega$,

as otherwise $[\alpha_i, 1^\omega] \leqslant \alpha_i \wedge 1^\omega = 0$, so that $\alpha_i \leqslant (0 : 1^\omega) = (\delta : 1^\omega) = \delta = 0$, a contradiction.

Since $\sigma_{\mathbf{A}/\alpha_i} = (\sigma \vee \alpha_i)/\alpha_i$ we get

$$(\sigma_{\mathbf{A}/\alpha_i})^2 = ((\sigma \vee \alpha_i)^2 \vee \alpha_i)/\alpha_i = (\sigma^2 \vee [\sigma, \alpha_i] \vee \alpha_i^2 \vee \alpha_i)/\alpha_i = \alpha_i/\alpha_i = 0_{\mathbf{A}/\alpha_i}.$$

Moreover $(1_{\mathbf{A}/\alpha_i})^\omega = 1^\omega/\alpha_i$. Thus $\delta_{\mathbf{A}/\alpha_i} = \left((\sigma_{\mathbf{A}/\alpha_i})^2 : (1_{\mathbf{A}/\alpha_i})^\omega\right) = (\alpha_i : 1^\omega)/\alpha_i$. The quotients $\mathbf{A}/\alpha_i$ are smaller than $\mathbf{A}$ and therefore satisfy (5). In particular every coatom above $\delta_{\mathbf{A}/\alpha_i}$ is of type **3**. On the other hand every coatom above $(1_{\mathbf{A}/\alpha_i})^\omega$ is of type **2** by (1). Thus, there is no coatom above both of these congruences. So they join to $1_{\mathbf{A}/\alpha_i}$. Let ξ_i denote $(\alpha_i : 1^\omega)$. Note that $\xi_i \geqslant \alpha_i$ and from (5.2) we know $1^\omega \geqslant \alpha_i$. We have $1_{\mathbf{A}/\alpha_i} = \xi_i/\alpha_i \vee 1^\omega/\alpha_i = (\xi_i \vee 1^\omega)/\alpha_i$. Thus we have

(5.3) $\xi_i \vee 1^\omega = 1$.

Consequently

(5.4) $\alpha_i < \xi_i$,

as otherwise $\alpha_i = \xi_i$ and so $1 = \xi_i \vee 1^\omega = \alpha_i \vee 1^\omega = 1^\omega \leqslant 1^2$, contrary to our assumption that $1^2 \leqslant \theta$ for some $\theta \prec 1$.

Moreover (5.2), (5.4), and (2) give $\alpha_i = 1^\omega \wedge \alpha_i \leqslant 1^\omega \wedge \xi_i = [1^\omega, \xi_i]$ while from the definition of ξ_i we have $[1^\omega, \xi_i] \leqslant \alpha_i$. Therefore

(5.5) $1^\omega \wedge \xi_i = [1^\omega, \xi_i] = \alpha_i$.

Now we have the following two equations

(5.6) $1^\omega \wedge (\xi_1 \vee \xi_2) = [1^\omega, \xi_1 \vee \xi_2] = [1^\omega, \xi_1] \vee [1^\omega, \xi_2] = \alpha_1 \vee \alpha_2$

and

(5.7) $1^\omega \vee (\alpha_1 \vee \xi_2) = (1^\omega \vee \xi_2) \vee \alpha_1 = 1$

from which we infer

$$\alpha_1 \vee \xi_2 = \xi_1 \vee \xi_2,$$

for if otherwise there would be an $[\alpha_1 \vee \xi_2,\ \xi_1 \vee x_2,\ 1^\omega]$-pentagon by virtue of (5.4), (5.7) and (5.6).

Using modularity again we get

$$\xi_1 = \xi_1 \wedge (\xi_1 \vee \xi_2) = \xi_1 \wedge (\alpha_1 \vee \xi_2) = \alpha_1 \vee (\xi_1 \wedge \xi_2).$$

But $\xi_1 \wedge \xi_2 = (\alpha_1 \wedge \alpha_2 : 1^\omega) = (0 : 1^\omega) = 0$, and consequently the last display gives $\xi_1 = \alpha_1$, contradicting (5.4).

This finishes the proof that $\mathbf{A}$ is subdirectly irreducible. By our assumption there is a coatom θ in $\mathsf{Con}\,\mathbf{A}$ with $\mathrm{typ}(\theta, 1) = \mathbf{2}$. Therefore $\mathbf{A}$ must be nilpotent. Hence $1^\omega = 0$ and so $\delta = \left(\sigma^2_{\mathbf{A}} : 0\right) = 1$. This contradicts (5.1) and thereby proves (5).

We must have

(6) $1^\omega \vee \delta = 1$

for if otherwise there is a coatom θ above $1^\omega \vee \delta$. Since θ is above 1^ω it is of type $\mathbf{2}$ because of (1), while θ must be of type $\mathbf{3}$ because of (5).

Now we show that

(7) $1^\omega \wedge \delta = 0.$

Suppose otherwise. Pick a congruence α with $0 \prec \alpha \leqslant 1^\omega \wedge \delta$. Let η be a congruence of $\mathbf{A}$ that is maximal with respect to not being above α. Then $\mathbf{A}/\eta$ is subdirectly irreducible. This algebra cannot be nilpotent as then $\eta \geqslant 1^\omega \geqslant \alpha$. Therefore $I\,[\eta, 1] = I\,[\eta, \theta] \cup \{1\}$ for some $\theta \prec 1$ with $\mathrm{typ}(\theta, 1) = \mathbf{3}$ and all other prime quotients above η have type $\mathbf{2}$. This gives $\sigma \leqslant \theta$ and so $\sigma^2 \leqslant \theta^2$. It is also the case that θ is solvable over η. Consequently in $\mathbf{A}/\eta$ the congruence θ/η is the centralizer of the monolith of $\mathbf{A}/\eta$. The algebra $\mathbf{A}/\eta$ is not nilpotent since it has a type $\mathbf{3}$ prime quotient at the top. It follows from condition (**2.3**) that θ/η is Abelian over η/η. This gives $\theta^2 \leqslant \eta$. From $\alpha \leqslant \delta = \left(\sigma^2 : 1^\omega\right)$ it follows that $[\alpha, 1^\omega] = [1^\omega, \alpha] \leqslant \sigma^2$. These observations together with (2) and $\alpha \leqslant \delta$ yield $\alpha = \alpha \wedge 1^\omega = [\alpha, 1^\omega] \leqslant \sigma^2 \leqslant \theta^2 \leqslant \eta$, contrary to our choice of η.

Let $\mathcal{N}$ denote the class of all nilpotent algebras in $\mathcal{V}$. Moreover let $\mathcal{C}$ be the class of all finite algebras $\mathbf{A}$ in $\mathcal{V}$ for which $1^2 = 1$, or equivalently such that $\mathrm{typ}(\theta, 1) = \mathbf{3}$, for every coatom θ in $\mathsf{Con}\,\mathbf{A}$. From $1^2 = 1$ we immediately get $1^\omega = 1$, so that (2) yields that each algebra in $\mathcal{C}$ is centerless.

From (1) and (5) we get that $\mathbf{A}/1^\omega \in \mathcal{N}$ and $\mathbf{A}/\delta \in \mathcal{C}$. Congruence permutability, (6) and (7) immediately give the following.

THEOREM 15.2. *Let a locally finite variety $\mathcal{V}$ satisfy conditions* **(1)**–**(3)** *of Theorem* 15.1. *Then every finite algebra $\mathbf{A}$ in $\mathcal{V}$ can be decomposed into a direct product $\mathbf{A} = \mathbf{N} \times \mathbf{C}$ for a nilpotent algebra $\mathbf{N} \in \mathcal{N}$ and a centerless algebra $\mathbf{C} \in \mathcal{C}$.* □

CHAPTER 16

Restricting Affine Behavior

In Theorem 14.3 every finite subdirectly irreducible algebra in a locally finite conguence modular variety with few models was shown to be either nilpotent or to have an Abelian centralizer of its monolith. Corollary 14.8 handles the nilpotent case. This Chapter is devoted to the study of the Abelian case. For a finite algebra $\mathbf{D}$ in a congruence modular variety $\mathcal{V}$ and an Abelian congruence β of $\mathbf{D}$ we describe an action of $\mathbf{D}/\beta$ on a module $\mathbf{M}(\beta)$ derived from the congruence β. An analysis of this action allows us, for any fixed algebra $\mathbf{S}$, to form a ring $\mathbf{R}_{\mathbf{S}}^{\mathcal{V}}$ and to uniformly recover every finite algebra $\mathbf{D} \in \mathcal{V}$ from $\mathbf{S}$ and an $\mathbf{R}_{\mathbf{S}}^{\mathcal{V}}$–module $M(\beta)$ whenever $\mathbf{D}$ has an Abelian congruence β with $\mathbf{D}/\beta \simeq \mathbf{S}$. This representation is described in Section 16.1. In Section 16.2 we work in congruence modular varieties having few models and we apply this analysis to every simple non-Abelian algebra $\mathbf{S}$ that is the quotient of a subdirectly irreducible algebra in $\mathcal{V}$ by the centralizer of its monolith. In particular we show that the corresponding rings $\mathbf{RR}_{\mathbf{S}}^{\mathcal{V}}$ obtained as quotients of $\mathbf{R}_{\mathbf{S}}^{\mathcal{V}}$ are of finite representation type.

Before reading this Chapter, or simultaneously while doing so, the reader may want to read the proof of Theorem 18.2, where a similar construction is performed for the varieties of commutative rings with unit. In this situation the action mentioned above reduces to the action of the quotient field of a local commutative ring $\mathbf{D}$ modulo its Jacobson radical on the Jacobson radical $J(\mathbf{D})$. Both the action and its description are much more transparent in that setting.

16.1. Expanded Modules

In their study of congruence modular varieties R. Freese and R. McKenzie [**22**] defined a family of rings $R(\mathcal{V}, \lambda)$, where λ is a nonzero cardinal number and $\mathcal{V}$ is an arbitrary congruence modular variety, such that a great deal of information about the algebras in $\mathcal{V}$ can be inferred from the behavior of the $R(\mathcal{V}, \lambda)$–modules. In this Chapter we briefly recall the definition of the rings $R(\mathcal{V}, \lambda)$ and the properties of $R(\mathcal{V}, \lambda)$–modules that we will need in our considerations. For our purposes it suffices to restate these results only for finite λ. The material in this Section is an adaptation, for our needs, of Section 2 from [**35**].

Fix a congruence modular variety $\mathcal{V}$ with a Gumm term $\mathbf{d}(x, y, z)$. Let $\mathbf{A}$ be any algebra from $\mathcal{V}$ and let β be an Abelian congruence of $\mathbf{A}$. Then for any element $c \in A$ the equivalence class c/β can be endowed with operations:

$$\begin{aligned} a + b &= \mathbf{d}(a, c, b), \\ -a &= \mathbf{d}(c, a, c), \end{aligned}$$

such that the algebra $M_{\mathbf{A}}(\beta, c) = (d/\beta; +, -, c)$ is a commutative group.

Now let n be any positive integer and let $\mathbf{F}$ be the free algebra in $\mathcal{V}$ freely generated by the set $X = \{u, v_1, \dots, v_n\}$. Denote by θ_i the principal congruence of $\mathbf{F}$ generated by the pair (u, v_i). Since the congruence θ_i is Abelian over $[\theta_i, \theta_i]$ it follows that $\overline{H}_{ij} = \{r/[\theta_i, \theta_i] : r \in v_j/\theta_i\}$ is an Abelian group, where $v_j/[\theta_i, \theta_i]$ is the zero element and the operations $+$ and $-$ are defined by

$$\begin{aligned} (r/[\theta_i, \theta_i]) + (s/[\theta_i, \theta_i]) &= \mathbf{d}(r, v_j, s)/[\theta_i, \theta_i], \\ -(r/[\theta_i, \theta_i]) &= \mathbf{d}(v_j, r, v_j)/[\theta_i, \theta_i], \end{aligned}$$

for any $r = r(u, \mathbf{v}), s = s(u, \mathbf{v}) \in v_j/\theta_i$.

Moreover the set $R(\mathcal{V}, n)$ of all $(n \times n)$–matrices (m_{ij}) with $m_{ij} \in \overline{H}_{ji}$, treated as a product of the family $\{\overline{H}_{ij} : i, j < n\}$ of Abelian groups, is an Abelian group where the zero is the matrix whose (i, j)–th element is $v_i/[\theta_j, \theta_j]$.

The second step Freese and McKenzie take here is to introduce the multiplication of matrices from $R(\mathcal{V}, n)$ in such a way that $R(\mathcal{V}, n)$ becomes a ring with unit. This multiplication is the ordinary matrix multiplication where the product of $t/[\theta_j, \theta_j]$ and $r/[\theta_i, \theta_i]$, for any $t = t(u, \mathbf{v}) \in v_k/\theta_j$ and $r = r(u, \mathbf{v}) \in v_j/\theta_i$, is given by

$$(t/[\theta_j, \theta_j]) \cdot (r/[\theta_i, \theta_i]) = t(r, \mathbf{v})/[\theta_i, \theta_i].$$

The reader is referred to Chapter 9 of [**22**] for details.

In our work we will also need the matrices e_l, where $1 \leqslant l \leqslant n$, whose (i, j)-th element is $u/[\theta_l, \theta_l]$ for $i = j = l$ and $v_i/[\theta_j, \theta_j]$ otherwise. It should be clear that the e_l's are pairwise orthogonal idempotents that sum to the unit of the ring $R(\mathcal{V}, n)$.

Now let $\mathbf{A}$ be a finite algebra in $\mathcal{V}$ and β be an Abelian congruence of $\mathbf{A}$ with $A/\beta = \{c_1/\beta, \dots, c_n/\beta\}$. The last step is to determine the action of the ring $R(\mathcal{V}, n)$ on the direct sum $M_{\mathbf{A}}(\beta)$ of the Abelian groups $M_{\mathbf{A}}(\beta, c_i)$, for $1 \leqslant i \leqslant n$. Take $\mathbf{a} = (a_1, \dots, a_n) \in M_{\mathbf{A}}(\beta)$ and $\mathbf{r} = (r_{ij}/[\theta_j, \theta_j]) \in R(\mathcal{V}, n)$ with $r_{ij} = r_{ij}(u, v_1, \dots, v_n) \in v_i/\theta_j$. The i–th element of $\mathbf{r} \cdot \mathbf{a}$ is

$$\sum_{j=1}^{n} r_{ij}(a_j, c_1, \dots, c_n).$$

Once more we refer to [**22**] for details of this action under which $\mathbf{M}_{\mathbf{A}}(\beta)$ is a unitary $R(\mathcal{V}, n)$–module.

As was said at the very beginning, this action carries a great deal of information about the algebra $\mathbf{A}$ itself.

First of all note that for any $i, j = 1, \dots, n$ and any group homomorphism

$$h : M_{\mathbf{A}}(\beta, c_i) \longrightarrow M_{\mathbf{A}}(\beta, c_j)$$

given by the restriction of a unary polynomial of $\mathbf{A}$, there is an element $r = r(u, \mathbf{v}) \in v_j/\theta_i$ such that $h(a) = r(a, c_1, \dots, c_n)$ for all $a \in M_{\mathbf{A}}(\beta, c_i)$.

Now observe that for any fundamental operation $\mathbf{f}$, say k–ary, and all elements $a_1, \dots, a_k \in A$ with $a_j \in M_{\mathbf{A}}(\beta, c_{i_j})$ we have

$$\begin{aligned} \mathbf{f}(a_1, \dots, a_k) &= \mathbf{f}(c_{i_1}, \dots, c_{i_k}) \\ &+ \sum_{j=1}^{k} \left(\mathbf{f}(c_{i_1}, \dots, c_{i_{j-1}}, a_j, c_{i_{j+1}}, \dots, c_{i_k}) - \mathbf{f}(c_{i_1}, \dots, c_{i_k}) \right). \end{aligned}$$

Then for $c_{i_0} \in \{c_1, \ldots, c_n\}$ chosen such that $\mathbf{f}(c_{i_1}, \ldots, c_{i_k}) \in M_{\mathbf{A}}(\beta, c_{i_0})$, the mapping $M_{\mathbf{A}}(\beta, c_{i_j}) \longrightarrow M_{\mathbf{A}}(\beta, c_{i_0})$ that associates with each x from $M_{\mathbf{A}}(\beta, c_{i_j})$ the element

$$\mathbf{f}(c_{i_1}, \ldots, c_{i_{j-1}}, x, c_{i_{j+1}}, \ldots, c_{i_k}) - \mathbf{f}(c_{i_1}, \ldots, c_{i_k})$$

of $M_{\mathbf{A}}(\beta, c_{i_0})$ is a group homomorphism given by a unary polynomial. Identifying each element $a \in c_i/\beta$ with the tuple $(c_1, \ldots, c_{i-1}, a, c_{i+1}, \ldots, c_n)$ from the $R(\mathcal{V}, n)$–module $\mathbf{M_A}(\beta)$ and using the previous remark we know that there are mappings

$$\mathbf{f}^{(\mathbf{M_A}(\beta))} : \{c_1/\beta, \ldots, c_n/\beta\}^k \longrightarrow \mathbf{M_A}(\beta)$$

and for each $j = 1, \ldots, k$

$$\mathbf{f}^{(j)} : \{c_1/\beta, \ldots, c_n/\beta\}^k \longrightarrow R(\mathcal{V}, n)$$

such that

$$\mathbf{f}(a_1, \ldots, a_k) = \left(\sum_{j=1}^{k} \mathbf{f}^{(j)}(c_{i_1}/\beta, \ldots, c_{i_k}/\beta) \cdot a_j \right) + \mathbf{f}^{(\mathbf{M_A}(\beta))}(c_{i_1}/\beta, \ldots, c_{i_k}/\beta),$$

whenever $a_1 \stackrel{\beta}{\equiv} c_{i_1}$, ... , $a_k \stackrel{\beta}{\equiv} c_{i_k}$.

Note that if $\mathbf{f}(c_{i_1}, \ldots, c_{i_k}) \in M_{\mathbf{A}}(\beta, c_j)$ then $\mathbf{f}^{(\mathbf{M_A}(\beta))}(c_{i_1}/\beta, \ldots, c_{i_k}/\beta)$ is a vector $(u_1, \ldots, u_n)$ in the product $M_{\mathbf{A}}(\beta) = M_{\mathbf{A}}(\beta, c_1) \times \ldots \times M_{\mathbf{A}}(\beta, c_n)$ for which $u_j = \mathbf{f}(c_{i_1}, \ldots, c_{i_k})$ and $u_i = c_i$ for $i \neq j$.

We are going to recover the algebra $\mathbf{A}$ from its quotient $\mathbf{S} = \mathbf{A}/\beta$ and the $R(\mathcal{V}, |S|)$–module $\mathbf{M_A}(\beta)$. Obviously the constants

$$\mathbf{f}^{(j)}(c_{i_1}/\beta, \ldots, c_{i_k}/\beta) \in R(\mathcal{V}, |D|)$$

as well as

$$\mathbf{f}^{(\mathbf{M_A}(\beta))}(c_{i_1}/\beta, \ldots, c_{i_k}/\beta) \in M_{\mathbf{A}}(\beta)$$

are a big help here since they allow us to recover the fundamental operations of the algebra $\mathbf{A}$. However, first we need to recover the universe of $\mathbf{A}$. To do this, note that the universe of $\mathbf{A}$ can be identified with the following subset of $S \times M_{\mathbf{A}}(\beta)$

$$\{(c_l/\beta, \mathbf{m}) \in S \times M_{\mathbf{A}}(\beta) : \mathbf{m}_j = 0 \text{ for all } j \neq l\}.$$

A careful inspection of the action of the ring $R(\mathcal{V}, |S|)$ on $\mathbf{M_A}(\beta)$ allows us to claim that for the matrix $e_l \in R(\mathcal{V}, |S|)$ previously defined, we have that for all $\mathbf{m} \in M_{\mathbf{A}}(\beta)$

$$e_l \cdot \mathbf{m} = \mathbf{m} \text{ iff } \mathbf{m}_j = 0 \text{ for all } j \neq l.$$

This leads to the following definitions.

DEFINITION 16.1. *Fix a finite algebra* $\mathbf{S} \in \mathcal{V}$ *with universe* $\{c_1, \ldots, c_n\}$. *Define*

$$\mathbf{e} : S \longrightarrow R(\mathcal{V}, n)$$

by putting $\mathbf{e}[c_i] = e_i$.

Now, for every k*–ary fundamental operation* $\mathbf{f}$ *of* $\mathcal{V}$ *and* $j = 1, \ldots, k$ *we define a mapping*

$$\mathbf{f}^{(j)} : S^k \longrightarrow R(\mathcal{V}, n)$$

as follows. For $(c_{i_1}, \ldots, c_{i_k}) \in S^k$ *we need to define an element* $\mathbf{f}^{(j)}[c_{i_1}, \ldots, c_{i_k}]$ *of the ring* $R(\mathcal{V}, n)$, *i.e., an* $n \times n$ *matrix*

$$\left(\mathbf{f}^{(j)}_{p,q}[c_{i_1}, \ldots, c_{i_k}]/[\theta_q, \theta_q] \right)_{p,q=1,\ldots,n},$$

where

$$\mathbf{f}^{(j)}_{p,q}[c_{i_1},\dots,c_{i_k}] = \mathbf{f}^{(j)}_{p,q}[c_{i_1},\dots,c_{i_k}](u,v_1,\dots,v_n)$$

is a term from v_p/θ_q, i.e., $\mathcal{V}$ satisfies the identity

$$\mathbf{f}^{(j)}_{p,q}[c_{i_1},\dots,c_{i_k}](v_q,v_1,\dots,v_n) \approx v_p.$$

Put

$$\mathbf{f}^{(j)}_{p,q}[c_{i_1},\dots,c_{i_k}](u,v_1,\dots,v_n) =$$

$$\begin{cases} \mathbf{d}(\mathbf{f}(v_{i_1},\dots,u,\dots,v_{i_k}),\mathbf{f}(v_{i_1},\dots,v_{i_k}),v_p), & \text{if } q=i_j \\ & \text{and } \mathbf{f}(c_{i_1},\dots,c_{i_k})=c_p, \\ v_p, & \text{otherwise,} \end{cases}$$

with u occurring at the j-th place.

From this definition we know that the matrix $\mathbf{f}^{(j)}[c_{i_1},\dots,c_{i_k}]$ has at most one nonzero entry which can occur at the position with coordinates (ℓ,i_j) where $c_\ell = \mathbf{f}(c_{i_1},\dots,c_{i_k})$. This gives

$$\mathbf{e}[\mathbf{f}(c_{i_1},\dots,c_{i_k})]\cdot\mathbf{f}^{(j)}[c_{i_1},\dots,c_{i_k}] = \mathbf{f}^{(j)}[c_{i_1},\dots,c_{i_k}] = \mathbf{f}^{(j)}[c_{i_1},\dots,c_{i_k}]\cdot\mathbf{e}[c_{i_j}].$$

The reader should be warned here that the mappings

$$\mathbf{e}: S \longrightarrow R(\mathcal{V},|S|)$$

and

$$\mathbf{f}^{(j)}: S^k \longrightarrow R(\mathcal{V},|S|)$$

depend on the order in which the elements of $\mathbf{S}$ are listed, but once a function $\mathbf{e}$ is chosen as a bijection between S and the set $\{e_1,\dots,e_n\}$ (which is an orthogonal decomposition of the unit of the ring $R(\mathcal{V},|S|)$ into idempotents) then this enumeration of elements of S is fixed and is used in the definition of the $\mathbf{f}^{(j)}$'s.

Henceforth we assume that

($\star$) with every finite algebra $\mathbf{S}$ from $\mathcal{V}$ we are given mappings $\mathbf{e}$ and the $\mathbf{f}^{(j)}$ for $1 \leqslant j \leqslant k$ that satisfy the following

$$\begin{aligned} \mathbf{e}[d]\cdot\mathbf{e}[d] &= \mathbf{e}[d], \\ \mathbf{e}[d_1]\cdot\mathbf{e}[d_2] &= 0 \qquad \text{whenever } d_1 \neq d_2, \\ \sum_{d\in S}\mathbf{e}[d] &= 1, \\ \mathbf{f}^{(j)}[d_1,\dots,d_k] &= \mathbf{f}^{(j)}[d_1,\dots,d_k]\cdot\mathbf{e}[d_j], \\ \mathbf{f}^{(j)}[d_1,\dots,d_k] &= \mathbf{e}[\mathbf{f}(d_1,\dots,d_k)]\cdot\mathbf{f}^{(j)}[d_1,\dots,d_k], \end{aligned}$$

where $d, d_1, \dots, d_k$ range over the elements of $\mathbf{S}$.

Now we ready to give our second definition.

DEFINITION 16.2. *Fix a finite algebra $\mathbf{S}$ in a congruence modular variety $\mathcal{V}$.*

- *By $\mathbf{R}^{\mathcal{V}}_{\mathbf{S}}$ we denote the ring $R(\mathcal{V},|S|)$. For $d, d_1,\dots,d_k \in S$ the ring elements of the form $\mathbf{e}[d]$ and $\mathbf{f}^{(j)}[d_1,\dots,d_k]$ are fixed as in ($\star$).*

- *By an* $\mathbf{S}$*-expanded* $\mathbf{R}_{\mathbf{S}}^{\mathcal{V}}$*-module we mean any* $\mathbf{R}_{\mathbf{S}}^{\mathcal{V}}$*-module* $\mathbf{M}$ *with new constants (added to the language of* $\mathbf{R}_{\mathbf{S}}^{\mathcal{V}}$*-modules) of the form* $\mathbf{f}^{(\mathbf{M})}(d_1,\ldots,d_k)$ *(with* $\mathbf{f}$ *ranging over all fundamental operations of* $\mathcal{V}$ *and* $(d_1,\ldots,d_k) \in S^k$*), such that all identities of the following form hold in* $\mathbf{M}$

$$\mathbf{f}^{(\mathbf{M})}(d_1,\ldots,d_k) \quad = \quad \mathbf{e}[\mathbf{f}(d_1,\ldots,d_k)]\cdot\mathbf{f}^{(\mathbf{M})}(d_1,\ldots,d_k),$$

- *Given an* $\mathbf{S}$*-expanded* $\mathbf{R}_{\mathbf{S}}^{\mathcal{V}}$*-module* $\mathbf{M}$ *put*

$$S[\mathbf{M}] = \{(d,\mathbf{m}) \in S\times M \ : \ \mathbf{e}[d]\cdot\mathbf{m} = \mathbf{m}\}$$

and let $\mathbf{S}[\mathbf{M}]$ *denotes the algebra (of the same type as* $\mathbf{S}$*) with the universe* $S[\mathbf{M}]$ *and the operations defined by*

$$\mathbf{f}\left((d_1,\mathbf{m}^1),\ldots,(d_k,\mathbf{m}^k)\right) = \left(\mathbf{f}(d_1,\ldots,d_k)\ ,\ \sum_{j=1}^{k}\mathbf{f}^{(j)}[d_1,\ldots,d_k]\cdot\mathbf{m}^j + \mathbf{f}^{(\mathbf{M})}(d_1,\ldots,d_k)\right).$$

From now on it is convenient to change notation so that the ring $\mathbf{R}_{\mathbf{S}}^{\mathcal{V}}$ is obtained from a $\mathcal{V}$-free algebra $\mathbf{F}$ generated by the set $\{u\}\cup\{v_a : a\in S\}$ with the help of congruences $\theta_a = \mathrm{Cg}^{\mathbf{F}}(u,v_a)$.

Let $\mathbf{t} = \mathbf{t}(x_1,\ldots,x_k)$ be an arbitrary term of $\mathcal{V}$ and let $\mathbf{M}$ be an $\mathbf{S}$-expanded $\mathbf{R}_{\mathbf{S}}^{\mathcal{V}}$-module. For fundamental operations $\mathbf{f}(x_1,\ldots,x_l)$, by using the ring scalars $\mathbf{f}^{(j)}[d_1,\ldots,d_l]\in\mathbf{R}_{\mathbf{S}}^{\mathcal{V}}$ and constants $\mathbf{f}^{(\mathbf{M})}(d_1,\ldots,d_l)$ in the language of $\mathbf{S}$-expanded $\mathbf{R}_{\mathbf{S}}^{\mathcal{V}}$-modules, we can compute scalars $\mathbf{t}^{(j)}[d_1,\ldots,d_k]\in\mathbf{R}_{\mathbf{S}}^{\mathcal{V}}$ and constant terms $\mathbf{t}^{(\mathbf{M})}(d_1,\ldots,d_k)$ (in the language of $\mathbf{S}$-expanded $\mathbf{R}_{\mathbf{S}}^{\mathcal{V}}$-modules) such that in $\mathbf{S}[\mathbf{M}]$ we have

$$\mathbf{t}\left((d_1,\mathbf{m}^1),\ldots,(d_k,\mathbf{m}^k)\right) = \left(\mathbf{t}(d_1,\ldots,d_k)\ ,\ \sum_{j=1}^{k}\mathbf{t}^{(j)}[d_1,\ldots,d_k]\cdot\mathbf{m}^j + \mathbf{t}^{(\mathbf{M})}(d_1,\ldots,d_k)\right).$$

It is easily seen that if $\mathbf{t}(x_1,\ldots,x_k) = \mathbf{f}(\mathbf{s}_1(x_1,\ldots,x_k),\ldots,\mathbf{s}_l(x_1,\ldots,x_k))$ where $\mathbf{f}$ is a fundamental operation and the $\mathbf{s}_i$ are terms of $\mathcal{V}$, then

$$\mathbf{t}^{(j)}[d_1,\ldots,d_k] = \sum_{i=1}^{l}\mathbf{f}^{(i)}[\mathbf{s}_1(d_1,\ldots,d_k),\ldots,\mathbf{s}_l(d_1,\ldots,d_k)]\cdot\mathbf{s}_i^{(j)}[d_1,\ldots,d_k]$$

and

$$\mathbf{t}^{(\mathbf{M})}(d_1,\ldots,d_k) = \sum_{i=1}^{l}\mathbf{f}^{(i)}[\mathbf{s}_1(d_1,\ldots,d_k),\ldots,\mathbf{s}_l(d_1,\ldots,d_k)]\cdot\mathbf{s}_i^{(\mathbf{M})}(d_1,\ldots,d_k).$$

By inducting on the complexity of a term $\mathbf{t}$ one can show the following:

LEMMA 16.3. *Let* $\mathbf{S}$ *be a finite algebra in a congruence modular variety* $\mathcal{V}$*. For any (k-ary) term* $\mathbf{t} = \mathbf{t}(x_1,\ldots,x_k)$ *of* $\mathcal{V}$ *and* $j = 1,\ldots,k$ *the* $(|S|\times|S|)$*-matrix* $\mathbf{t}^{(j)}[d_1,\ldots,d_k]\in\mathbf{R}_{\mathbf{S}}^{\mathcal{V}}$ *can be represented by*

$$\left(\mathbf{t}_{p,q}^{(j)}[d_1,\ldots,d_k]/[\theta_q,\theta_q]\right)_{p,q\in S},$$

where $\mathbf{t}^{(j)}_{p,q}[d_1,\ldots,d_k] = \mathbf{t}^{(j)}_{p,q}[d_1,\ldots,d_k](u,\overline{v})$ *is a term from* v_p/θ_q, *(i.e.,* $\mathcal{V} \models \mathbf{t}^{(j)}_{p,q}[d_1,\ldots,d_k](v_q,\overline{v}) \approx v_p$*) given by*

$$\mathbf{t}^{(j)}_{p,q}[d_1,\ldots,d_k](u,\overline{v}) =$$

$$\begin{cases} \mathbf{d}(\mathbf{t}(v_{d_1},\ldots,u,\ldots,v_{d_k}),\mathbf{t}(v_{d_1},\ldots,v_{d_k}),v_p), & \text{if } q = d_j \\ & \text{and } \mathbf{t}(d_1,\ldots,d_k) = d_p, \\ v_p, & \text{otherwise.} \end{cases}$$ □

From this Lemma we see that the scalars $\mathbf{t}^{(j)}[d_1,\ldots,d_k]$ from the ring $\mathbf{R}^{\mathcal{V}}_{\mathbf{S}}$ do not depend on the $\mathbf{S}$–expanded module $\mathbf{M}$. Also the constants $\mathbf{t}^{(\mathbf{M})}(d_1,\ldots,d_k)$ depend only on the $\mathbf{f}^{(\mathbf{M})}(d_1,\ldots,d_l)$'s for basic operations $\mathbf{f}$ and the way the term $\mathbf{t}$ is composed from the basic operations. In fact, one can easily check that if the identity $\mathbf{t}(x_1,\ldots,x_k) \approx \mathbf{s}(x_1,\ldots,x_k)$ holds in $\mathcal{V}$, then the constants $\mathbf{t}^{(\mathbf{M})}(d_1,\ldots,d_k)$ and $\mathbf{s}^{(\mathbf{M})}(d_1,\ldots,d_k)$ are equal in any $\mathbf{S}$–expanded $\mathbf{R}^{\mathcal{V}}_{\mathbf{S}}$–module $\mathbf{M}$. The fact that $\mathbf{t}^{(\mathbf{M})}(d_1,\ldots,d_k) = \mathbf{s}^{(\mathbf{M})}(d_1,\ldots,d_k)$ can be expressed by a variable-free identity in the language of $\mathbf{S}$–expanded $\mathbf{R}^{\mathcal{V}}_{\mathbf{S}}$–modules. Thus for any identity $\mathbf{t} \approx \mathbf{s}$ that holds in $\mathcal{V}$ we form a set, say $\Sigma(\mathbf{t} \approx \mathbf{s})$, consisting of $|S|^k$-many variable-free identities and expressing the fact that for each k–tuple $(d_1,\ldots,d_k) \in S^k$ the constants $\mathbf{t}^{(\mathbf{M})}(d_1,\ldots,d_k)$ and $\mathbf{s}^{(\mathbf{M})}(d_1,\ldots,d_k)$ are equal.

We let $\Sigma^{\mathcal{V}}_{\mathbf{S}} = \bigcup\{\Sigma(\mathbf{t} \approx \mathbf{s}) : \mathcal{V} \models \mathbf{t} \approx \mathbf{s}\}$ and by $\mathcal{M}^{\mathcal{V}}_{\mathbf{S}}$ we denote the variety of all $\mathbf{S}$-expanded $\mathbf{R}^{\mathcal{V}}_{\mathbf{S}}$–modules satisfying $\Sigma^{\mathcal{V}}_{\mathbf{S}}$.

From the last display before Lemma 16.3 we immediately get the following

LEMMA 16.4. *If* $\mathbf{M}$ *is an ordinary* $\mathbf{R}^{\mathcal{V}}_{\mathbf{S}}$*–module then its trivial* $\mathbf{S}$*–expansion, i.e., an* $\mathbf{S}$*–expansion that interprets all constants of the form* $\mathbf{f}^{(\mathbf{M})}(d_1,\ldots,d_k)$ *by the zero element of* $\mathbf{M}$*, belongs to* $\mathcal{M}^{\mathcal{V}}_{\mathbf{S}}$*.* □

We summarize our considerations in the following:

THEOREM 16.5. *Let* $\mathbf{D}$ *be a finite algebra in a congruence modular variety* $\mathcal{V}$ *If* $\mathbf{S} = \mathbf{D}/\beta$ *for some Abelian congruence* β *of* $\mathbf{D}$*, then there is an* $\mathbf{S}$*–expanded* $\mathbf{R}^{\mathcal{V}}_{\mathbf{S}}$*–module* $\mathbf{M} \in \mathcal{M}^{\mathcal{V}}_{\mathbf{S}}$ *such that* $\mathbf{D}$ *is isomorphic to* $\mathbf{S}[\mathbf{M}]$*. Moreover if* $\mathbf{D}$ *is* ℓ*–generated, then* $\mathbf{M}$ *can be chosen to be* ℓ*–generated (as an* $\mathbf{S}$*–expanded* $\mathbf{R}^{\mathcal{V}}_{\mathbf{S}}$*–module).*

PROOF. Assume that $n = |S| = |D/\beta|$ and let $S = \{d_1/\beta,\ldots,d_n/\beta\}$. Then the product $M_{\mathbf{D}}(\beta) = M_{\mathbf{D}}(\beta,d_1) \times \ldots \times M_{\mathbf{D}}(\beta,d_n)$ has the natural structure of an $\mathbf{R}^{\mathcal{V}}_{\mathbf{S}}$–module as described at the beginning of this Chapter.

By $\overline{\mathbf{M}}_{\mathbf{D}}(\beta)$ denote the module $\mathbf{M}_{\mathbf{D}}(\beta)$ that is $\mathbf{S}$–expanded by

$$\mathbf{f}^{(\overline{\mathbf{M}}_{\mathbf{D}}(\beta))}(d_{i_1}/\beta,\ldots,d_{i_k}/\beta) = (u_1,\ldots,u_n)$$

with

$$u_p = \begin{cases} \mathbf{f}(d_{i_1},\ldots,d_{i_k}), & \text{if } \mathbf{f}(d_{i_1},\ldots,d_{i_k}) \in d_p/\beta, \\ d_p, & \text{otherwise.} \end{cases}$$

In particular we have

$$\mathbf{e}[\mathbf{f}(d_{i_1}/\beta,\ldots,d_{i_k}/\beta)] \cdot \mathbf{f}^{(\overline{\mathbf{M}}_{\mathbf{D}}(\beta))}(d_{i_1}/\beta,\ldots,d_{i_k}/\beta) = \mathbf{f}^{(\overline{\mathbf{M}}_{\mathbf{D}}(\beta))}(d_{i_1}/\beta,\ldots,d_{i_k}/\beta),$$

i.e., $\overline{\mathbf{M}}_{\mathbf{D}}(\beta)$ is an $\mathbf{S}$–expanded $\mathbf{R}_{\mathbf{S}}^{\mathcal{V}}$–module. Moreover, for a term $\mathbf{t}(x_1, \ldots, x_k)$ of $\mathcal{V}$ one can easily compute that the constant $\mathbf{t}^{(\overline{\mathbf{M}}_{\mathbf{D}}(\beta))}(d_{i_1}/\beta, \ldots, d_{i_k}/\beta)$ is a vector $(u_1, \ldots, u_n)$ with

$$u_p = \begin{cases} \mathbf{t}(d_{i_1}, \ldots, d_{i_k}), & \text{if} \quad \mathbf{t}(d_{i_1}, \ldots, d_{i_k}) \in d_p/\beta, \\ d_p, & \text{otherwise.} \end{cases}$$

Therefore $\mathbf{t}^{(\overline{\mathbf{M}}_{\mathbf{D}}(\beta))}(d_{i_1}/\beta, \ldots, d_{i_k}/\beta) = \mathbf{s}^{(\overline{\mathbf{M}}_{\mathbf{D}}(\beta))}(d_{i_1}/\beta, \ldots, d_{i_k}/\beta)$ whenever the identity $\mathbf{t} \approx \mathbf{s}$ holds in $\mathbf{S}$. This gives that $\overline{\mathbf{M}}_{\mathbf{D}}(\beta) \in \mathcal{M}_{\mathbf{S}}^{\mathcal{V}}$.

To see that $\mathbf{D} \simeq \mathbf{S}\left[\overline{\mathbf{M}}_{\mathbf{D}}(\beta)\right]$ we represent each element a of D in the module $\mathbf{M}_{\mathbf{D}}(\beta)$ by a tuple $\overrightarrow{a} = (\overrightarrow{a}_1, \ldots, \overrightarrow{a}_n)$ with

$$\overrightarrow{a}_i = \begin{cases} a, & \text{if} \quad a \in d_i/\beta, \\ d_i, & \text{otherwise.} \end{cases}$$

One can easily check that the mapping

$$h : D \ni a \mapsto (a/\beta, \overrightarrow{a}) \in S\left[\overline{\mathbf{M}}_{\mathbf{D}}(\beta)\right]$$

is a well-defined bijection. Moreover, it should be obvious that for every k–ary fundamental operation $\mathbf{f}$ of $\mathbf{D}$ and $a^1, \ldots, a^k \in D$ we have

$$\overrightarrow{\mathbf{f}(a^1, \ldots, a^k)} = \sum_{j=1}^{n} \mathbf{f}^{(j)}[a^1/\beta, \ldots, a^k/\beta] \cdot \overrightarrow{a^j} + \mathbf{f}^{(\overline{\mathbf{M}}_{\mathbf{D}}(\beta))}(a^1/\beta, \ldots, a^k/\beta).$$

This in turn gives

$$\begin{aligned} &\mathbf{f}^{\mathbf{S}[\overline{\mathbf{M}}_{\mathbf{D}}(\beta)]}(h(a^1), \ldots, h(a^k)) = \\ &= \mathbf{f}^{\mathbf{S}[\overline{\mathbf{M}}_{\mathbf{D}}(\beta)]}\left((a^1/\beta, \overrightarrow{a}^1), \ldots, (a^k/\beta, \overrightarrow{a}^k)\right) \\ &= \left(\mathbf{f}^{\mathbf{S}}(a^1/\beta, \ldots, a^k/\beta), \sum_{j=1}^{n} \mathbf{f}^{(j)}[a^1/\beta, \ldots, a^k/\beta] \cdot \overrightarrow{a}^j + \mathbf{f}^{(\overline{\mathbf{M}}_{\mathbf{D}}(\beta))}(a^1/\beta, \ldots, a^k/\beta)\right) \\ &= \left(\mathbf{f}^{\mathbf{D}}(a^1, \ldots, a^k)/\beta, \overrightarrow{\mathbf{f}(a^1, \ldots, a^k)}\right) \\ &= h\left(\mathbf{f}^{\mathbf{D}}(a^1, \ldots, a^k)\right) \end{aligned}$$

showing that h is a homomorphism between $\mathbf{D}$ and $\mathbf{S}\left[\overline{\mathbf{M}}_{\mathbf{D}}(\beta)\right]$. This proves the first part of our Theorem.

We conclude our proof by showing that if $\mathbf{D}$ is generated by the set $\{a^1, \ldots, a^\ell\}$, then $\overline{\mathbf{M}}_{\mathbf{D}}(\beta)$ is generated by $\{\overrightarrow{a}^1, \ldots, \overrightarrow{a}^\ell\}$. Obviously $\overline{\mathbf{M}}_{\mathbf{D}}(\beta)$ is generated by the set $\{\overrightarrow{a} : a \in D\}$, even as a group. Therefore it suffices to show that every element of the form $\overrightarrow{a}$ belongs to an $\mathbf{S}$–expanded $\mathbf{R}_{\mathbf{S}}^{\mathcal{V}}$–submodule of $\overline{\mathbf{M}}_{\mathbf{D}}(\beta)$ generated by $\{\overrightarrow{a}^1, \ldots, \overrightarrow{a}^\ell\}$.

Since $\mathbf{D}$ is generated by $\{a^1, \ldots, a^\ell\}$, there is an ℓ–ary term $\mathbf{t}$ with $a = \mathbf{t}(a^1, \ldots, a^\ell)$. Thus we have $\overrightarrow{a} = \overrightarrow{\mathbf{t}(a^1, \ldots, a^\ell)}$. On the other hand we have

$$\overrightarrow{\mathbf{t}(a^1, \ldots, a^\ell)} = \sum_{j=1}^{n} \mathbf{t}^{(j)}[a^1/\beta, \ldots, a^\ell/\beta] \cdot \overrightarrow{a}^j + \mathbf{t}^{(\overline{\mathbf{M}}_{\mathbf{D}}(\beta))}(a^1/\beta, \ldots, a^\ell/\beta).$$

This shows that $\overrightarrow{a}$ can be obtained from $\overrightarrow{a}^1, \ldots, \overrightarrow{a}^\ell$ by using addition, multiplication by scalars from the ring $\mathbf{R}_{\mathbf{S}}^{\mathcal{V}}$ and constants of $\overline{\mathbf{M}}_{\mathbf{D}}(\beta)$. $\square$

THEOREM 16.6. *Let* $\mathbf{S}$ *be a finite algebra in a congruence modular variety* $\mathcal{V}$. *If* $\mathbf{M} \in \mathcal{M}_{\mathbf{S}}^{\mathcal{V}}$ *then:*

- $\mathbf{S}[\mathbf{M}] \in \mathcal{V}$,
- $\sigma = \{((d_1, m_1), (d_2, m_2)) \in S[\mathbf{M}] \times S[\mathbf{M}] \ : \ d_1 = d_2\}$ *is an Abelian congruence of* $\mathbf{S}[\mathbf{M}]$,
- $\mathbf{S}[\mathbf{M}]/\sigma$ *is isomorphic to* $\mathbf{S}$,
- $\overline{\mathbf{M}}_{\mathbf{S}[\mathbf{M}]}(\sigma)$ *is isomorphic to* $\mathbf{M}$,
- *the congruence lattice* $\mathsf{Con}\,\mathbf{M}$ *is isomorphic to the interval* $I[0, \sigma]$ *in the lattice* $\mathsf{Con}\,\mathbf{S}[\mathbf{M}]$.

PROOF. Suppose that the identity $\mathbf{t}(x_1, \ldots, x_k) \approx \mathbf{s}(x_1, \ldots, x_k)$ holds in $\mathcal{V}$. Let $(d_1, m_1), \ldots, (d_k, m_k) \in S[\mathbf{M}]$. Lemma 16.3 shows that the matrices $\mathbf{t}^{(j)}[d_1, \ldots, d_k]$ and $\mathbf{s}^{(j)}[d_1, \ldots, d_k]$ are equal. On the other hand $\mathbf{M} \in \mathcal{M}_{\mathbf{S}}^{\mathcal{V}}$ so that $\mathbf{M} \models \Sigma(\mathbf{t} \approx \mathbf{s})$. Thus $\mathbf{t}^{(\mathbf{M})}(d_1, \ldots, d_k) = \mathbf{s}^{(\mathbf{M})}(d_1, \ldots, d_k)$. Applying the formula that describes the action of the terms of $\mathcal{V}$ in the algebra $\mathbf{S}[\mathbf{M}]$ we are done with the first item.

One easily verifies that the mapping

$$\mathbf{S}[\mathbf{M}] \ni (d, m) \mapsto d \in \mathbf{S}$$

is a surjective homomorphism with kernel σ. This proves the third item and the first part of the second. Moreover, from the description of how the definition of fundamental operations in $\mathbf{S}[\mathbf{M}]$ extends to arbitrary terms it immediately follows that σ satisfies the term condition, and therefore is Abelian.

Now we will show that $\overline{\mathbf{M}}_{\mathbf{S}[\mathbf{M}]}(\sigma) \simeq \mathbf{M}$. Suppose that $S = \{d_1, \ldots, d_n\}$ and consider the mapping

$$h : M \ni m \mapsto ((d_1, \mathbf{e}[d_1] \cdot m), \ldots, (d_n, \mathbf{e}[d_n] \cdot m)) \in M_{\mathbf{S}[\mathbf{M}]}(\sigma),$$

where

$$M_{\mathbf{S}[\mathbf{M}]}(\sigma) = M_{\mathbf{S}[\mathbf{M}]}(\sigma, (d_1, 0)) \times \ldots \times M_{\mathbf{S}[\mathbf{M}]}(\sigma, (d_n, 0)).$$

Since the $\mathbf{e}[d]$'s are idempotent we get that $(d_i, \mathbf{e}[d_i] \cdot m) \in S[\mathbf{M}]$ and therefore $(d_i, \mathbf{e}[d_i] \cdot m) \in (d_i, 0)/\sigma$, which means that h is well defined. Using the fact that the $\mathbf{e}[d]$'s are pairwise orthogonal and sum to the unit of the ring $\mathbf{R}_{\mathbf{S}}^{\mathcal{V}}$ we get that h is bijective. It should be obvious that h preserves module addition. The proof that it preserves multiplication by scalars from the ring $\mathbf{R}_{\mathbf{S}}^{\mathcal{V}}$ needs a little bit more effort. For any $r \in \mathbf{R}_{\mathbf{S}}^{\mathcal{V}}$ we have $r = \sum_{i,j=1}^{n} \mathbf{e}[d_i] \cdot r \cdot \mathbf{e}[d_j]$ and that matrices of the form $\mathbf{e}[d_i] \cdot r \cdot \mathbf{e}[d_j]$ have at most one nonzero entry. Thus it suffices to show that $h(\bar{\mathbf{t}} \cdot m) = \bar{\mathbf{t}} \cdot h(m)$, where $\bar{\mathbf{t}}$ denotes a matrix from $\mathbf{R}_{\mathbf{S}}^{\mathcal{V}}$ all of whose entries but one, say the (p, q)–th, are zero and the (p, q)–th entry is $\mathbf{t}/[\theta_q, \theta_q]$ for some term $\mathbf{t} = \mathbf{t}(u, \bar{v}) \in v_p/\theta_q$. In this situation we have

$$\begin{aligned}
\bar{\mathbf{t}} \cdot h(m) &= \bar{\mathbf{t}} \cdot ((d_1, \mathbf{e}[d_1] \cdot m), \ldots, (d_n, \mathbf{e}[d_n] \cdot m)) \\
&= \left(\sum_{j=1}^{n} (\bar{\mathbf{t}})_{1j} \cdot (d_j, \mathbf{e}[d_j] \cdot m) \ , \ldots, \ \sum_{j=1}^{n} (\bar{\mathbf{t}})_{nj} \cdot (d_j, \mathbf{e}[d_j] \cdot m) \right) \\
&= \Big((d_1, 0), \ldots, (d_{p-1}, 0), (\bar{\mathbf{t}})_{pq} \cdot (d_q, \mathbf{e}[d_q] \cdot m), (d_{p+1}, 0), \ldots, (d_n, 0) \Big).
\end{aligned}$$

Moreover, calculating in $\mathbf{S}[\mathbf{M}]$ we get that the element

$$(\bar{\mathbf{t}})_{pq} \cdot (d_q, \mathbf{e}[d_q] \cdot m) = \mathbf{t}((d_q, \mathbf{e}[d_q] \cdot m), (d_1, 0), \ldots (d_n, 0))$$

is equal to

$$\left(\mathbf{t}(d_q, d_1, \ldots, d_n),\ \mathbf{t}^{(0)}[d_q, d_1, \ldots, d_n] \cdot \mathbf{e}[d_q] \cdot m + \mathbf{t}^{(\mathbf{M})}(d_q, d_1, \ldots, d_n)\right).$$

Since $\mathbf{t}(v_q, v_1, \ldots, v_n) = v_p$ then Lemma 16.3 yields

$$\mathbf{t}^{(0)}[d_q, d_1, \ldots, d_n] \cdot \mathbf{e}[d_q] = \overline{\mathbf{t}}.$$

On the other hand $\mathcal{V} \models \mathbf{t}(v_q, v_1, \ldots, v_n) \approx v_p$ gives that in $\mathbf{S}[\mathbf{M}] \in \mathcal{V}$ we have

$$\begin{aligned}
(d_p, 0) &= \mathbf{t}((d_q, 0), (d_1, 0), \ldots, (d_n, 0)) \\
&= \left(\mathbf{t}(d_q, d_1, \ldots, d_n),\ \sum_{j=0}^{n} \mathbf{t}^{(j)}[d_q, d_1, \ldots, d_n] \cdot 0 + \mathbf{t}^{(\mathbf{M})}(d_q, d_1, \ldots, d_n)\right) \\
&= \left(d_p, \mathbf{t}^{(\mathbf{M})}(d_q, d_1, \ldots, d_n)\right),
\end{aligned}$$

so that $\mathbf{t}^{(\mathbf{M})}(d_q, d_1, \ldots, d_n) = 0$. Consequently

$$(\overline{\mathbf{t}})_{pq} \cdot (d_q, \mathbf{e}[d_q] \cdot m) = (d_p, \overline{\mathbf{t}} \cdot m).$$

Since $\mathbf{e}[d_p] \cdot \overline{\mathbf{t}} = \overline{\mathbf{t}}$ and $\mathbf{e}[d_j] \cdot \overline{\mathbf{t}} = 0$ for all $j \neq p$ then we get

$$\begin{aligned}
\overline{\mathbf{t}} \cdot h(m) &= \left((d_1, 0), \ldots, (d_{p-1}, 0), (d_p, \overline{\mathbf{t}} \cdot m), (d_{p+1}, 0), \ldots, (d_n, 0)\right) \\
&= \left((d_1, \mathbf{e}[d_1] \cdot \overline{\mathbf{t}} \cdot m), \ldots, (d_n, \mathbf{e}[d_n] \cdot \overline{\mathbf{t}} \cdot m)\right) \\
&= h(\overline{\mathbf{t}} \cdot m),
\end{aligned}$$

as required.

This shows that h preserves multiplication by scalars from $\mathbf{R}_{\mathbf{S}}^{\mathcal{V}}$ and so that h is an isomorphism of $\mathbf{R}_{\mathbf{S}}^{\mathcal{V}}$–modules.

The reader should find no difficulty in checking that h preserves the constants of the form $\mathbf{f}^{(\mathbf{M})}(d_{i_1}, \ldots, d_{i_k})$ which is the last step in proving that h is an isomorphism of $\mathbf{S}$–expanded $\mathbf{R}_{\mathbf{S}}^{\mathcal{V}}$–modules.

Finally, to see that $\mathsf{Con}\,\mathbf{M}$ is isomorphic to the interval $I[0, \sigma]$ note that in the view of the previous item $\mathsf{Con}\,\mathbf{M}$ is isomorphic to $\mathsf{Con}\,\overline{\mathbf{M}}_{\mathbf{S}[\mathbf{M}]}(\sigma) = \mathsf{Con}\,\mathbf{M}_{\mathbf{S}[\mathbf{M}]}(\sigma)$. Invoking Theorem 9.9 of [**22**] we get that this lattice is isomorphic to $I[0, \sigma]$. $\square$

LEMMA 16.7. *Let $\mathbf{S}$ be a finite algebra in a congruence modular variety $\mathcal{V}$.*

- *If $\mathbf{N}$ is a homomorphic image of an $\mathbf{S}$–expanded $\mathbf{R}_{\mathbf{S}}^{\mathcal{V}}$–module $\mathbf{M}$, then the algebra $\mathbf{S}[\mathbf{N}]$ is a homomorphic image of $\mathbf{S}[\mathbf{M}]$.*
- *If the module $\mathbf{M}$ is a subdirect product of the family $\{\mathbf{M}_t\}_{t\in T}$ of $\mathbf{S}$–expanded $\mathbf{R}_{\mathbf{S}}^{\mathcal{V}}$–modules, then the algebra $\mathbf{S}[\mathbf{M}]$ is a subdirect product of the family $\{\mathbf{S}[\mathbf{M}_t]\}_{t\in T}$.*
- *If $\mathbf{M}$ is an $\mathbf{S}$–expanded $\mathbf{R}_{\mathbf{S}}^{\mathcal{V}}$–module, then the algebra $\mathbf{S}[\mathbf{M}^k]$ is isomorphic to the subalgebra of $\mathbf{S}[\mathbf{M}]^k$ consisting of all σ–constant sequences.*

PROOF. Suppose $h : \mathbf{M} \longrightarrow \mathbf{N}$ is a surjective homomorphism of $\mathbf{S}$–expanded $\mathbf{R}_{\mathbf{S}}^{\mathcal{V}}$–modules. Then one easily checks that the mapping

$$\overline{h} : \mathbf{S}[\mathbf{M}] \ni (d, m) \mapsto (d, h(m)) \in \mathbf{S}[\mathbf{N}]$$

is a surjective homomorphism.

Now, for a subdirect embedding $\mathbf{M} \subseteq \prod_{t\in T} \mathbf{M}_t$ of $\mathbf{S}$–expanded $\mathbf{R}^{\mathcal{V}}_{\mathbf{S}}$–modules one easily observes that the mapping

$$\mathbf{S}[\mathbf{M}] \ni (d, (m_t)_{t\in T}) \mapsto ((d, m_t))_{t\in T} \in \prod_{t\in T} \mathbf{S}[\mathbf{M}_t]$$

is an injective homomorphism. To see that this embedding is subdirect all we need to check is that all projections of the form $S[\mathbf{M}] \longrightarrow S[\mathbf{M}_t]$ are surjective. This means that given $(d, m) \in S[\mathbf{M}_t]$ there is an $\mathbf{m} = (m_x)_{x\in T} \in M$ such that $(d, \mathbf{m}) \in S[\mathbf{M}]$ and $m_t = m$. Since $\mathbf{M}$ is a subdirect product of the $\mathbf{M}_x$'s there is an $\mathbf{n} = (n_x)_{x\in T} \in M$ with $n_t = m$. Put $\mathbf{m} = \mathbf{e}[d]\cdot\mathbf{n}$ and observe, using $(d, m) \in S[\mathbf{M}_t]$, that $m_t = (\mathbf{e}[d] \cdot \mathbf{n})_t = \mathbf{e}[d] \cdot n_t = \mathbf{e}[d] \cdot m = m$. Moreover, $\mathbf{e}[d]$ is an idempotent of the ring $\mathbf{R}^{\mathcal{V}}_{\mathbf{S}}$ so $\mathbf{e}[d] \cdot \mathbf{m} = \mathbf{m}$, which gives $(d, \mathbf{m}) \in S[\mathbf{M}]$, as required.

For the last item note that we have just shown that the mapping

$$\mathbf{S}[\mathbf{M}^k] \ni (d, (m_1, \ldots, m_k)) \mapsto ((d, m_1), \ldots, (d, m_k)) \in \mathbf{S}[\mathbf{M}]^k$$

is an embedding. Moreover it should be clear that the image of this embedding is the subalgebra of $\mathbf{S}[\mathbf{M}]^k$ consisting of all σ–constant sequences. □

16.2. Forcing Finite Representation Type.

In this Chapter we establish our last necessary condition for a locally finite variety $\mathcal{V}$ omitting type **1** to have few models. Namely, we will show that for each finite simple non-Abelian algebra $\mathbf{S} \in \mathcal{V}$ a certain canonically determined subvariety of the variety $\mathcal{M}^{\mathcal{V}}_{\mathbf{S}}$ of $\mathbf{S}$-expanded $\mathbf{R}^{\mathcal{V}}_{\mathbf{S}}$–modules has to be directly representable, i.e., that this subvariety has only finitely many directly indecomposables.

To define this subvariety we need to obtain some additional understanding of algebras of the form $\mathbf{S}[\mathbf{M}]$, where $\mathbf{M}$ ranges over the variety $\mathcal{M}^{\mathcal{V}}_{\mathbf{S}}$. We start with the following lemma, assuming in this Section that the variety $\mathcal{V}$ is locally finite and satisfies conditions **(1)**–**(3)** of Theorem 15.1.

LEMMA 16.8. *If* $\mathbf{M}$ *is a finite* $\mathbf{S}$*–expanded* $\mathbf{R}^{\mathcal{V}}_{\mathbf{S}}$*–module and* σ *is the congruence relation of the algebra* $\mathbf{S}[\mathbf{M}]$ *defined in Theorem* 16.6*, then we have*

(1) *the interval* $I[0, \sigma]$ *is isomorphic to* $\mathsf{Con}\,\mathbf{M}$,
(2) σ *is the unique coatom of type* **3** *in* $\mathsf{Con}\,\mathbf{S}[\mathbf{M}]$,
(3) $\sigma^2 = 0$,
(4) $1^\omega = 1^2$,
(5) $[1^2, \varphi] = 1^2 \wedge \varphi$ *for every* $\varphi \in \mathsf{Con}\,\mathbf{S}[\mathbf{M}]$,
(6) $(0 : 1^2) \leqslant \sigma$,
(7) $(0 : 1^2) \vee 1^2 = 1$,
(8) $(0 : 1^2) \wedge 1^2 = 0$.

PROOF. By Theorem 16.6 we know that σ is an Abelian congruence of $\mathbf{S}[\mathbf{M}]$ and that the quotient $\mathbf{S}[\mathbf{M}]/\sigma$ is isomorphic to the simple algebra $\mathbf{S}$ and consequently σ is a coatom in $\mathsf{Con}\,\mathbf{S}[\mathbf{M}]$. Moreover, the congruence lattice $\mathsf{Con}\,\mathbf{M}$ is isomorphic to $I[0, \sigma]$.

Now suppose that θ is another coatom in $\mathsf{Con}\,\mathbf{S}[\mathbf{M}]$. Then $\theta \wedge \sigma < \sigma$ and since σ is Abelian we get that 1 is Abelian over θ. Therefore σ is the unique coatom of type **3**. This shows the first three items of our Lemma.

Note also that $1^\omega \vee \sigma = 1$ as $1^\omega \not\leqslant \sigma$. Therefore the interval $I[\sigma \wedge 1^\omega, \sigma]$ projects up to $I[1^\omega, 1]$. Since σ is Abelian (over $\sigma \wedge 1^\omega$) we get that 1 is Abelian over 1^ω, and consequently $1^\omega = 1^2$.

Items (5), (6), (7) and (8) of the Lemma follow immediately from the fourth one and (2), (4), (6), (7) of Chapter 15, respectively, since the congruence $\delta = (\sigma^2 : 1^\omega)$ defined there is equal in the present setting to $(0 : 1^2)$. □

From Lemma 16.8 we know that in the algebra $\mathbf{S}[\mathbf{M}]$ the congruence σ lies above $(0 : 1^2)$ and that the algebra $\mathbf{S}[\mathbf{M}]$ decomposes into a direct product of an Abelian algebra $\mathbf{S}[\mathbf{M}]/1^2$ and a centerless algebra $\mathbf{S}[\mathbf{M}]/(0 : 1^2)$. Moreover the congruence lattice of the last factor has a unique coatom $\sigma/(0 : 1^2)$ and, by Theorem 16.5 the quotient $\mathbf{S}[\mathbf{M}]/(0 : 1^2)$ is of the form $\mathbf{S}[\mathbf{M}']$ for some $\mathbf{M}' \in \mathcal{M}_{\mathbf{S}}^{\mathcal{V}}$.

We are interested in those finite $\mathbf{S}$-expanded $\mathbf{R}_{\mathbf{S}}^{\mathcal{V}}$–modules $\mathbf{M} \in \mathcal{M}_{\mathbf{S}}^{\mathcal{V}}$ for which the corresponding algebras $\mathbf{S}[\mathbf{M}]$ have exactly one coatom in their congruence lattices. From what has been already said this is equivalent to the fact that $1^2 = 1$ or in other words that $\mathbf{S}[\mathbf{M}]$ is centerless. This leads to the following definition and our next two lemmas.

DEFINITION 16.9. *By a* reduced $\mathbf{S}$*-expanded* $\mathbf{R}_{\mathbf{S}}^{\mathcal{V}}$*–module we mean a finite* $\mathbf{S}$*-expanded* $\mathbf{R}_{\mathbf{S}}^{\mathcal{V}}$*–module* $\mathbf{M} \in \mathcal{M}_{\mathbf{S}}^{\mathcal{V}}$ *such that the corresponding algebra* $\mathbf{S}[\mathbf{M}] \in \mathcal{V}$ *is centerless. Let* $\mathcal{RM}_{\mathbf{S}}^{\mathcal{V}}$ *be the subvariety of* $\mathcal{M}_{\mathbf{S}}^{\mathcal{V}}$ *generated by the class of all finite reduced* $\mathbf{S}$*-expanded* $\mathbf{R}_{\mathbf{S}}^{\mathcal{V}}$*–modules.*

LEMMA 16.10. *The class of finite reduced* $\mathbf{S}$*-expanded* $\mathbf{R}_{\mathbf{S}}^{\mathcal{V}}$*–modules is closed under the formation of homomorphic images and finite subdirect products.*

PROOF. If $\mathbf{N}$ is a homomorphic image of $\mathbf{M}$, then Lemma 16.7 tells us that $\mathbf{S}[\mathbf{N}]$ is a homomorphic image of $\mathbf{S}[\mathbf{M}]$. Therefore it has a unique coatom in its congruence lattice whenever $\mathbf{S}[\mathbf{M}]$ has this property.

To prove that the class is closed under finite subdirect products it suffices to show this for subdirect products of two factors. Let $\mathbf{M}$ be a subdirect product of two reduced $\mathbf{S}$-expanded $\mathbf{R}_{\mathbf{S}}^{\mathcal{V}}$–modules $\mathbf{M}_1$ and $\mathbf{M}_2$. Denote by η_1 and η_2 the projection kernels onto the first and the second coordinates, and by σ_t the unique coatom in the congruence lattice of $\mathbf{S}[\mathbf{M}_t]$. Moreover let σ be the congruence of $\mathbf{S}[\mathbf{M}]$ consisting of all pairs $((d,(m_1,m_2)),(d,(n_1,n_2))) \in \mathbf{S}[\mathbf{M}] \times \mathbf{S}[\mathbf{M}]$. Obviously we have $\eta_t \leqslant \sigma$. Since σ is the unique coatom of type $\mathbf{3}$ it suffices to show that in $\mathbf{S}[\mathbf{M}]$ we have $1^2 = 1$. On the other hand, since $\mathbf{S}[\mathbf{M}_i] \simeq \mathbf{S}[\mathbf{M}]/\eta_i$ and $\mathsf{Con}\, \mathbf{S}[\mathbf{M}_i]$ is assumed to have only one coatom, we have $\eta_i \vee 1^2 = 1$. Put $\theta = (\eta_1 \wedge 1^2) \vee (\eta_2 \wedge 1^2)$ and $\delta = \eta_1 \vee \eta_2$. By Lemma 16.8 (5) we have

$$\delta \wedge 1^2 = [\delta, 1^2] = [\eta_1, 1^2] \vee [\eta_2, 1^2] = (\eta_1 \wedge 1^2) \vee (\eta_2 \wedge 1^2) = \theta.$$

Consequently, using modularity, we get

$$\delta = \delta \wedge 1 = \delta \wedge (\eta_i \vee 1^2) = \eta_i \vee (\delta \wedge 1^2) = \eta_i \vee \theta$$

and

$$\begin{aligned}
\delta &= (\eta_1 \vee \theta) \wedge (\eta_2 \vee \theta) \\
&= (\eta_1 \vee (\eta_2 \wedge 1^2)) \wedge ((\eta_1 \wedge 1^2) \vee \eta_2) \\
&= (\eta_1 \wedge ((\eta_1 \wedge 1^2) \vee \eta_2)) \vee (\eta_2 \wedge 1^2) \\
&= (\eta_1 \wedge \eta_2) \vee (\eta_1 \wedge 1^2) \vee (\eta_2 \wedge 1^2) \\
&= 0 \vee \theta = \theta.
\end{aligned}$$

This, together with the first display gives $\eta_i \leqslant \delta = \theta = \delta \wedge 1^2 \leqslant 1^2$ and consequently $1 = \eta_i \vee 1^2 = 1^2$, as required. □

LEMMA 16.11. *Every finite module from the variety $\mathcal{RM}^{\mathcal{V}}_{\mathbf{S}}$ is reduced.*

PROOF. We let $\mathcal{M}'$ denote the class of all finite reduced $\mathbf{S}$-expanded $\mathbf{R}^{\mathcal{V}}_{\mathbf{S}}$-modules, i.e., a generating class for the variety $\mathcal{RM}^{\mathcal{V}}_{\mathbf{S}}$. So $\mathcal{M}'$ is a subset of a locally finite affine variety $\mathcal{M}^{\mathcal{V}}_{\mathbf{S}}$. By Theorems 9.16(5) and 10.15 of [22] every locally finite affine variety has a finite bound on the size of its subdirectly irreducible algebras. Therefore there is a finite upper bound, say m, for the sizes of subdirectly irreducibles in $\mathcal{M}'$. On the other hand, since $\mathcal{M}'$ is closed under homomorphic images, we know that each identity $p \approx q$ that can be falsified in $\mathcal{RM}^{\mathcal{V}}_{\mathbf{S}}$, or equivalently in $\mathcal{M}'$, can be actually falsified in an algebra from $\mathcal{M}'$ of size at most m. Therefore, for $k \geqslant m$ the k–freely generated algebra in $\mathcal{RM}^{\mathcal{V}}_{\mathbf{S}}$ is a subdirect product of algebras from $\mathcal{M}'$. As $\mathcal{M}'$ is closed under finite subdirect products such a free algebra belongs to $\mathcal{M}'$. Consequently all finite algebras in $\mathcal{RM}^{\mathcal{V}}_{\mathbf{S}}$ are in $\mathcal{M}'$, as claimed. □

The variety $\mathcal{RM}^{\mathcal{V}}_{\mathbf{S}}$ is a subvariety of an affine variety $\mathcal{M}^{\mathcal{V}}_{\mathbf{S}}$ and therefore it is polynomially equivalent to the variety of modules over some quotient of $\mathbf{R}^{\mathcal{V}}_{\mathbf{S}}$. We will denote this quotient by $\mathbf{RR}^{\mathcal{V}}_{\mathbf{S}}$.

THEOREM 16.12. *Let $\mathcal{V}$ be a locally finite congruence modular variety with few models. Then for each finite simple non-Abelian algebra $\mathbf{S} \in \mathcal{V}$ the ring $\mathbf{RR}^{\mathcal{V}}_{\mathbf{S}}$ is of finite representation type.*

PROOF. Lemma 16.4 allows us to treat an ordinary $\mathbf{R}^{\mathcal{V}}_{\mathbf{S}}$–module, after identifying it with its trivial $\mathbf{S}$–expansion, as a member of the variety $\mathcal{M}^{\mathcal{V}}_{\mathbf{S}}$.

Let $\mathbf{F}(k)$ be the k–generated free algebra in the variety $\mathcal{RM}^{\mathcal{V}}_{\mathbf{S}}$ and let $\mathbf{F}_0(k)$ be the free $\mathbf{RR}^{\mathcal{V}}_{\mathbf{S}}$–module. From Lemma 9.13 of [22] we know that $\mathbf{F}_0(k)$, or rather its trivial $\mathbf{S}$–expansion, is a homomorphic image of $\mathbf{F}(k+1)$. Consequently, each finite $\mathbf{RR}^{\mathcal{V}}_{\mathbf{S}}$–module, after forming its trivial $\mathbf{S}$–expansion, belongs to $\mathcal{RM}^{\mathcal{V}}_{\mathbf{S}}$. In particular we get that the algebra $\mathbf{E} = \mathbf{S}[\mathbf{F}_0(1)]$ belongs to $\mathcal{V}$ and σ is the unique coatom in its congruence lattice. Since $\mathbf{E}/\sigma \simeq \mathbf{S}$ is not Abelian we know that $1^2 = 1$ holds in $\mathsf{Con}\,\mathbf{E}$. Therefore, by Lemma 16.8 (5), the algebra $\mathbf{E}$ satisfies the congruence identity $[1, x] = x$.

Now let $\mathbf{C}_k$ be the subalgebra of $\mathbf{E}^k$ consisting of all σ–constant sequences. By Lemma 16.7 the algebra $\mathbf{C}_k$ is isomorphic to $\mathbf{S}[\mathbf{F}_0(1)^k]$, i.e., to $\mathbf{S}[\mathbf{F}_0(k)]$. Then Lemma 14.2 supplies us with a polynomial $q(k)$ such that the number of nonisomorphic quotients of $\mathbf{S}[\mathbf{F}_0(k)]$ is not bigger than $q(k)$.

On the other hand each k–generated $\mathbf{RR}^{\mathcal{V}}_{\mathbf{S}}$–module $\mathbf{M}$ is a homomorphic image of $\mathbf{F}_0(k)$. Using Lemma 16.7 we get that $\mathbf{S}[\mathbf{M}]$ is (isomorphic to) a quotient of $\mathbf{S}[\mathbf{F}_0(k)]$. Moreover it should be clear that such an $\mathbf{M}$ is k-generated as an $\mathbf{RR}^{\mathcal{V}}_{\mathbf{S}}$–module if and only if it is k–generated as a (trivially) $\mathbf{S}$–expanded $\mathbf{R}^{\mathcal{V}}_{\mathbf{S}}$–module. Also two $\mathbf{RR}^{\mathcal{V}}_{\mathbf{S}}$–modules are isomorphic if and only if they are isomorphic as $\mathbf{S}$–expanded $\mathbf{R}^{\mathcal{V}}_{\mathbf{S}}$–modules. Therefore Theorem 16.6 guarantees that if $\mathbf{M}$ and $\mathbf{N}$ are not isomorphic, then $\mathbf{S}[\mathbf{M}] \not\simeq \mathbf{S}[\mathbf{N}]$, as otherwise $\mathbf{M} \simeq \overline{\mathbf{M}}_{\mathbf{S}[\mathbf{M}]}(\sigma_{\mathbf{S}[\mathbf{M}]}) \simeq \overline{\mathbf{M}}_{\mathbf{S}[\mathbf{N}]}(\sigma_{\mathbf{S}[\mathbf{N}]}) \simeq \mathbf{N}$. This gives that the variety of $\mathbf{RR}^{\mathcal{V}}_{\mathbf{S}}$–modules has very few models.

From Theorem 13.5 we conclude that the ring $\mathbf{RR}^{\mathcal{V}}_{\mathbf{S}}$ is of finite representation type. □

CHAPTER 17

A Characterization Theorem

We first summarize our results and give a list of necessary conditions for a locally finite variety omitting type **1** to have few models. We then prove that this list is complete if we restrict to finitely generated varieties. Our characterization substantially generalizes the one for congruence meet semi-distributive varieties given in Corollary 10.3. As observed just before this Corollary, our list of necessary conditions is not sufficient for arbitrary locally finite varieties.

17.1. Locally Finite Varieties with Few Models

Combining Theorems 15.1 and 16.12 we immediately get the following result.

THEOREM 17.1. *Let $\mathcal{V}$ be a locally finite variety omitting type* **1**. *If $\mathcal{V}$ has few models then*

(1) *$\mathcal{V}$ is congruence permutable,*

(2) *for any finite subdirectly irreducible algebra $\mathbf{A}$ in $\mathcal{V}$ with monolith μ and its centralizer $\nu = (0 : \mu)$ we have:*

- **(2.1)** *ν is the solvable radical of $\mathbf{A}$,*
- **(2.2)** *ν is comparable to all congruences of $\mathbf{A}$,*
- **(2.3)** *ν is Abelian or $\mathbf{A}$ is nilpotent,*
- **(2.4)** *the quotient $\mathbf{A}/\nu$ is either trivial or simple non-Abelian,*

(3) *every finite nilpotent subdirectly irreducible algebra in $\mathcal{V}$ is of prime power order and has a finitely generated clone,*

(4) *for any finite simple non-Abelian algebra $\mathbf{S}$ in $\mathcal{V}$ the ring $\mathbf{RR}_{\mathbf{S}}^{\mathcal{V}}$ is of finite representation type.* □

17.2. Finitely Generated Varieties with Few Models

Among finitely generated varieties of finite type omitting type **1** the conditions from Theorem 17.1 fully characterize those having few models.

THEOREM 17.2. *Let $\mathcal{V}$ be a finitely generated variety omitting type* **1**. *Then $\mathcal{V}$ has few models if and only if the following conditions hold:*

(1) *$\mathcal{V}$ is congruence permutable,*
(2) *for any finite subdirectly irreducible algebra $\mathbf{A}$ in $\mathcal{V}$ with monolith μ and its centralizer $\nu = (0 : \mu)$ we have:*
  (2.1) *ν is the solvable radical of $\mathbf{A}$,*
  (2.2) *ν is comparable to all congruences of $\mathbf{A}$,*
  (2.3) *ν is Abelian or $\mathbf{A}$ is nilpotent,*
  (2.4) *the quotient $\mathbf{A}/\nu$ is either trivial or simple non-Abelian,*
(3) *the variety $\mathcal{N}$ of all nilpotent algebras in $\mathcal{V}$ has a finitely generated clone and $\mathcal{N}$ itself is generated by finitely many finite algebras each being of prime power order,*
(4) *for any finite simple non-Abelian algebra $\mathbf{S}$ in $\mathcal{V}$ the ring $\mathbf{RR}_{\mathbf{S}}^{\mathcal{V}}$ is of finite representation type.*

For one direction we invoke Theorem 17.1 and Corollary 14.8.

To prove sufficiency of the conditions we start by recalling the decomposition obtained in Theorem 15.2. An analysis of the factors in this decomposition allows us to apply upper bound results from Chapter 6 to show that $\mathcal{V}$ has few models.

By $\mathcal{N}$ we denote the variety of all nilpotent algebras in $\mathcal{V}$, while by $\mathcal{C}$ the class of all finite algebras $\mathbf{A}$ in $\mathcal{V}$ for which $[1,1] = 1$, or equivalently such that $\mathrm{typ}(\theta, 1) = \mathbf{3}$, for every coatom θ in $\mathsf{Con}\,\mathbf{A}$. Then our conditions and Theorem 15.2 give the following.

LEMMA 17.3. *Every finite algebra $\mathbf{A}$ in $\mathcal{V}$ can be decomposed into a direct product $\mathbf{A} = \mathbf{N} \times \mathbf{C}$ for some $\mathbf{N} \in \mathcal{N}$ and $\mathbf{C} \in \mathcal{C}$.* □

From Lemma 17.3 we get that $\mathrm{G}_{\mathcal{V}}(k) \leqslant \mathrm{G}_{\mathcal{N}}(k) \cdot \mathrm{G}_{\mathcal{C}}(k)$. Since $\mathcal{N}$ is a nilpotent variety generated by a finite product of algebras of prime power order and $\mathcal{N}$ is term equivalent to a variety of finite type, Corollary 6.11 implies that $\mathrm{G}_{\mathcal{N}}(k)$ is at most exponential.

We will show that the class $\mathcal{C}$ satisfies the condition of Theorem 6.4 and therefore $\mathrm{G}_{\mathcal{C}}(k)$ is at most exponential. This will finish the proof of the main theorem of this Chapter.

In order to be able to apply Theorem 6.4 we have to understand directly indecomposable algebras in $\mathcal{C}$.

LEMMA 17.4. *Suppose $\mathbf{D}$ is a finite directly indecomposable algebra in $\mathcal{C}$. Then* $\mathsf{Con}\,\mathbf{D}$ *has exactly one coatom and this coatom is Abelian.*

PROOF. Let σ be the intersection of all coatoms in $\mathsf{Con}\,\mathbf{D}$. Then the interval $I[\sigma, 1]$ is a complemented modular lattice. Moreover for every $\beta \succ \alpha \geqslant \sigma$ there is a $\theta \prec 1$ such that the interval $I[\alpha, \beta]$ is projective to $I[\theta, 1]$. From the definition of the class $\mathcal{C}$ we know that $\mathrm{typ}(\theta, 1) = \mathbf{3}$. Consequently $\mathrm{typ}(\alpha, \beta) = \mathbf{3}$ and therefore $\mathrm{typ}\{\sigma, 1\} = \{\mathbf{3}\}$. This gives that the interval $I[\sigma, 1]$ is a Boolean lattice (as otherwise this interval contains a diamond, which by Lemma 6.6 of [**32**] would give $\mathbf{2} \in \mathrm{typ}\{\sigma, 1\}$).

If there is more than one coatom in $\mathsf{Con}\,\mathbf{D}$, then there are $\alpha, \beta \in \mathsf{Con}\,\mathbf{D} - \{\sigma, 1\}$ with $\alpha \wedge \beta = \sigma$ and $\alpha \vee \beta = 1$. We will show that

(1) $\alpha^2 \vee \beta^2 = 1$,
(2) $\alpha^2 \wedge \beta^2 = 0$,
(3) $\alpha^2 \neq 0 \neq \beta^2$,

which would contradict the assumption that $\mathbf{D}$ is directly indecomposable.

To see (1) assume to the contrary that there is a θ with $\alpha^2 \vee \beta^2 \leqslant \theta \prec 1$. It suffices to show that $\theta \geqslant \alpha$ and $\theta \geqslant \beta$, as then $\theta \geqslant \alpha \vee \beta = 1$. If $\alpha \nleqslant \theta$ then $\alpha \vee \theta = 1$. Moreover from the choice of θ we have $\alpha^2 \leqslant \theta$ so that

$$1^2 = (\alpha \vee \theta)^2 = \alpha^2 \vee \theta^2 \vee [\alpha, \theta] \leqslant \theta,$$

a contradiction. This shows $\alpha \leqslant \theta$ and similar arguments give that $\beta \leqslant \theta$.

For (2) assume to the contrary that $0 < \alpha^2 \wedge \beta^2$ and pick a congruence η that is maximal with respect to the property of not being above $\alpha^2 \wedge \beta^2$. Then the quotient $\mathbf{D}/\eta$ is subdirectly irreducible. Since all subcovers of 1 are of type **3** condition **(2)** gives that in the interval $I[\eta, 1]$ there is a unique coatom, say θ. Since θ is a meet irreducible congruence of the distributive lattice $I[\sigma, 1]$ and $\theta \geqslant \alpha \wedge \beta$ then $\theta \geqslant \alpha$ or $\theta \geqslant \beta$. In any case, $\theta^2 \geqslant \alpha^2 \wedge \beta^2$. On the other hand $\theta^2 \leqslant \eta$, since in the quotient $\mathbf{D}/\eta$ the coatom θ/η is Abelian. This gives $\alpha^2 \wedge \beta^2 \leqslant \theta^2 \leqslant \eta$, in contradiction to our choice of η.

To see (3) it suffices, by symmetry, to show that $\alpha^2 \neq 0$. Suppose $\alpha^2 = 0$. By (1) we get $1 = \beta^2 \leqslant \beta$, contradicting our choice of β.

This shows that $\mathbf{D}$ has exactly one coatom. Arguing analogously as in the proof of (2) one can show that this coatom squares to zero, i.e., it is Abelian. □

For a finite simple non-Abelian algebra $\mathbf{S} \in \mathcal{V}$ let $\mathcal{C}_{\mathbf{S}}$ denote the class of all directly indecomposables in $\mathcal{C}$ that have $\mathbf{S}$ as a homomorphic image.

LEMMA 17.5. *For a finite simple non-Abelian algebra $\mathbf{S} \in \mathcal{V}$ there is a polynomial $p_{\mathbf{S}}(k)$ and a constant $c_{\mathbf{S}}$ such that the number of k-generated directly indecomposables in $\mathcal{C}_{\mathbf{S}}$ is bounded above by $p_{\mathbf{S}}(k)$ and the size of any k-generated algebra from $\mathcal{C}_{\mathbf{S}}$ does not exceed $c_{\mathbf{S}} \cdot |\mathbf{R}_{\mathbf{S}}^{\mathcal{V}}|^k$.*

PROOF. Let $\mathbf{D}$ be a k-generated algebra in $\mathcal{C}_{\mathbf{S}}$. From Lemma 17.4 and Theorem 16.5 we know that $\mathbf{D}$ is isomorphic to $\mathbf{S}[\mathbf{M}]$ for some k-generated $\mathbf{S}$-expanded $\mathbf{R}_{\mathbf{S}}^{\mathcal{V}}$-module $\mathbf{M}$ in the subvariety $\mathcal{RM}_{\mathbf{S}}^{\mathcal{V}}$ of $\mathcal{M}_{\mathbf{S}}^{\mathcal{V}}$. The variety $\mathcal{M}_{\mathbf{S}}^{\mathcal{V}}$ is affine and therefore, by Theorem 9.16.(6) of [**22**], there is a constant c such that its k-freely generated algebra has exactly $c \cdot |\mathbf{R}_{\mathbf{S}}^{\mathcal{V}}|^k$ elements. Consequently $|M| \leqslant c \cdot |\mathbf{R}_{\mathbf{S}}^{\mathcal{V}}|^k$. Since $D \subseteq S \times M$ we get the required upper bound for $|D|$ with $c_{\mathbf{S}} = c \cdot |S|$.

Since the ring $\mathbf{RR}_{\mathbf{S}}^{\mathcal{V}}$ is of finite representation type it follows from Corollary 13.6 that the variety $\mathcal{RM}_{\mathbf{S}}^{\mathcal{V}}$ is directly representable and has very few models. In particular there is a polynomial $p_{\mathbf{S}}(k)$ that bounds the number of k-generated $\mathbf{S}$-expanded $\mathbf{R}_{\mathbf{S}}^{\mathcal{V}}$-modules in $\mathcal{RM}_{\mathbf{S}}^{\mathcal{V}}$. The same polynomial gives an upper bound for the number of k-generated algebras in $\mathcal{C}_{\mathbf{S}}$. □

LEMMA 17.6. *The number of k-generated algebras in $\mathcal{C}$ is at most exponential.*

PROOF. Since $\mathcal{V}$ is a finitely generated variety Theorem 14.5 of [**32**] guarantees that $\mathcal{V}$ has only finitely many simple algebras of type **3**. (Actually if $\mathcal{V} = \mathsf{HSP}(\mathbf{B})$, where $\mathbf{B}$ is a finite algebra, then Lemma 14.4 of [**32**] yields that all simple algebras of type **3** are in $\mathsf{HS}(\mathbf{B})$, so that they can be effectively computed from $\mathbf{B}$.) Let $\{\mathbf{S}_1, \ldots, \mathbf{S}_n\}$ be a set of all, up to isomorphism, simple non-Abelian algebras in $\mathcal{V}$. Take $p(k)$ to be a polynomial that dominates all of the polynomials $p_{\mathbf{S}_i}(k)$ described in Lemma 17.5. Analogously let c be the largest among the numbers $c_{\mathbf{S}_1}, \ldots, c_{\mathbf{S}_n}$ from that Lemma. Finally let r be the largest of the numbers $|\mathbf{R}^{\mathcal{V}}_{\mathbf{S}_1}|, \ldots, |\mathbf{R}^{\mathcal{V}}_{\mathbf{S}_n}|$.

If $\mathbf{D}$ is a finite directly indecomposable algebra in $\mathcal{C}$, then Lemma 17.4 yields that $\mathbf{D} \in \mathcal{C}_{\mathbf{S}_i}$ for some $i = 1, \ldots, n$. If moreover $\mathbf{D}$ is k–generated, then from Lemma 17.5 we get that $|D| \leqslant c \cdot r^k$. The same Lemma shows that $\mathrm{G}_{\mathcal{C}_{\mathbf{S}_i}}(k) \leqslant p_{\mathbf{S}_i}(k) \leqslant p(k)$. Applying Theorem 6.4 we get that $\mathrm{G}_{\mathcal{C}}(k)$ is at most exponential. □

Now we are ready to conclude the proof of Theorem 17.2.

From Lemma 17.3 we know that $\mathrm{G}_{\mathcal{V}}(k) \leqslant \mathrm{G}_{\mathcal{N}}(k) \cdot \mathrm{G}_{\mathcal{C}}(k)$. Since $\mathcal{N}$ is a nilpotent variety essentially of finite type and is generated by a finite product of algebras of prime power order, Corollary 6.11 gives that $\mathrm{G}_{\mathcal{N}}(k)$ is at most exponential. On the other hand Lemma 17.6 shows that $\mathrm{G}_{\mathcal{C}}(k)$ is at most exponential and therefore $\mathcal{V}$ has few models.

Part 3

Conclusions

CHAPTER 18

Application to Groups and Rings

In this Chapter we illustrate the characterization Theorems 13.4 and 17.2 by considering two common examples: groups and commutative rings. Note that these examples are congruence modular and therefore, by Theorem 6.14, their finitely generated varieties have at most doubly exponential generative complexity. Therefore we are left with considering, for these varieties, three main clusters of generative complexity: polynomial, singly exponential, and doubly exponential.

If a congruence modular variety has few models, then according to condition **(2)** of Theorem 17.2 the finite subdirectly irreducibles in $\mathcal{V}$ come in two mutually exclusive forms. One is nilpotent while the other is simple non–Abelian modulo its unique maximal proper congruence relation. The further description of the behavior of those two kinds of algebras is given in conditions **(3)** and **(4)**, respectively. It is interesting to point out that in finitely generated varieties of groups the subdirectly irreducibles are nilpotent, while in finitely generated varieties of rings with unit all subdirectly irreducibles are finite and of the second form. Thus in the group case condition **(3)** is automatically fulfilled and condition **(4)** does not apply at all. On the other hand, in the commutative ring case, condition **(3)** is vacuous and we will show that condition **(4)** gives no further restrictions.

The characterization of the finitely generated varieties of groups with a restricted generative complexity is especially beautiful.

THEOREM 18.1. *Let $\mathcal{V}$ be a finitely generated variety of groups. Then*

(1) *$\mathcal{V}$ has at most doubly exponentially many models,*
(2) *the following conditions are equivalent:*
  (2.1) *$\mathcal{V}$ has few models*
  (2.2) *$\mathcal{V}$ does not have many models*
  (2.3) *$\mathcal{V}$ is nilpotent,*
(3) *$\mathcal{V}$ has very few models iff $\mathcal{V}$ is Abelian.*

PROOF. Since every variety of groups is congruence modular, (1) follows directly from Theorem 6.14.

That (2.1) implies (2.2) is trivial while (2.3) implies (2.1) is Corollary 6.12. Suppose $\mathcal{V}$ does not have many models and $\mathcal{V}$ contains a finite group that is not nilpotent. Let $\mathbf{G} \in \mathcal{V}$ be a finite group of least cardinality that is not nilpotent. So every proper subgroup of $\mathbf{G}$ is nilpotent. It is known that if every proper subgroup of a finite group is nilpotent, then the group is solvable, e.g., [**63**, page 148]. Thus, Corollary 12.8 applies to show that $\mathbf{G}$ is nilpotent, contrary to our assumption. So every finite group in $\mathcal{V}$ is nilpotent. Since $\mathcal{V}$ is finitely generated, it is generated by a nilpotent group, and therefore every group in $\mathcal{V}$ is nilpotent.

To see (3) note that Example 6.7 shows that every finite Abelian group generates a variety with very few models. On the other hand Corollary 13.2 implies that every locally finite variety of groups with very few models is Abelian □

Note that in the above characterization of finitely generated varieties of groups with few models only nilpotent structures are allowed. An opposite phenomena occurs for another family of classical algebraic structures – commutative rings with unit. Here the only condition for a finitely generated variety $\mathcal{V}$ to have few models is that the Jacobson radical $J(\mathbf{R})$ of a generating ring $\mathbf{R}$ for $\mathcal{V} = \mathcal{V}(\mathbf{R})$ squares to zero, i.e., $J(\mathbf{R})^2 = \{0\}$. Recall that in a finite commutative ring $\mathbf{R}$ with unit the Jacobson radical is the intersection of all maximal ideals of $\mathbf{R}$, and simultaneously the largest solvable/nilpotent ideal. One can easily see that the property $J(\mathbf{R})^2 = \{0\}$ is preserved by homomorphic images as well as by finite subdirect products. Thus the understanding of finite rings with this property reduces to understanding subdirectly irreducibles among them.

The list of finite subdirectly irreducible rings with unit in which the Jacobson radical squares to zero can be obtained from Wilson's [**68**] representation of finite local rings by the so-called Szele matrices. This list is based on Galois rings $\mathbf{GR}(p^n, r)$ that generalize both finite fields and the cyclic rings $\mathbf{Z}_{p^n}$. If p is a prime number and n, r are positive integers, then by $\mathbf{GR}(p^n, r)$ we denote the quotient ring $\mathbf{Z}_{p^n}[x]/(f)$, where f is a monic polynomial of degree r that is irreducible modulo p. In particular $\mathbf{GR}(p, r)$ is the Galois field $\mathbf{GF}(p^r)$. Note that the Jacobson radical of the Galois ring $\mathbf{GR}(p^n, r)$ is $p \cdot \mathbf{GR}(p^n, r)$ and that the quotient ring $\mathbf{GR}(p^n, r)/p \cdot \mathbf{GR}(p^n, r)$ is isomorphic to the Galois field $\mathbf{GF}(p^r)$.

In the commutative setting the list of finite subdirectly irreducible rings with Jacobson radical squaring to zero reduces to the following

- Galois fields $\mathbf{GF}(p^r) = \mathbf{GR}(p, r)$,
- Galois rings $\mathbf{GR}(p^2, r)$,
- rings of matrices of the form $\begin{pmatrix} a & b \\ 0 & a \end{pmatrix}$ over a Galois field $\mathbf{GF}(p^r)$.

THEOREM 18.2. *Let $\mathcal{V}$ be a finitely generated variety of commutative rings with unit. Then*

(1) *$\mathcal{V}$ has at most doubly exponentially many models,*
(2) *$\mathcal{V}$ has few models iff the Jacobson radical of a generating ring of $\mathcal{V}$ squares to zero,*
(3) *$\mathcal{V}$ has very few models iff $\mathcal{V}$ is trivial.*

PROOF. As in the previous Theorem item (1) follows directly from Theorem 6.14. Also item (3) is an immediate consequence of Corollary 8.3, since in every nontrivial ring with unit the pair $(0, 1)$ is a 2-snag.

From now assume that $\mathbf{R}$ is a finite commutative ring with unit and $\mathcal{V}$ is the variety generated by $\mathbf{R}$.

To see that $J(\mathbf{R})^2 = \{0\}$ is a necessary condition in (2) note only that $\mathbf{R}$ is a subdirect product of finitely many subdirectly irreducible rings $\mathbf{S}_1, \ldots, \mathbf{S}_m \in \mathcal{V}$ and that Theorem 17.1 implies that $J(\mathbf{S}_i)^2 = \{0\}$ holds for each $i = 1, \ldots, m$.

Now suppose that $J(\mathbf{R})^2 = \{0\}$. This implies that $J(\mathbf{A})^2 = \{0\}$ holds for any (finite) ring $\mathbf{A}$ in $\mathcal{V}$. One way to show that this gives singly exponential generative complexity is to use Theorem 17.2. However, we choose to provide a direct proof which is more transparent and which can be viewed as a prototype for the proof of Theorem 17.2 as well as serve as a warmup for Chapter 16.

Since every finite commutative ring is a direct product of directly indecomposable rings, we start by listing the following well-known or easily provable facts:

(1) A finite ring $\mathbf{A}$ in $\mathcal{V}$ is directly indecomposable if and only if it is local. In particular, $J(\mathbf{A})$ is the unique maximal ideal of $\mathbf{A}$ and $\mathbf{A}/J(\mathbf{A})$ is a simple ring, i.e., a field.

(2) If $\mathbf{D} \in \mathcal{V}$ is local, then there is a unique prime p such that
- $\mathbf{p_D} \in J(\mathbf{D})$, where for an integer n the symbol $\mathbf{n_D}$ denotes the corresponding element of $\mathbf{D}$,
- the characteristic of $\mathbf{D}$ is either p or p^2, depending on whether $\mathbf{p_D} = 0$,
- $\mathbf{D}/J(\mathbf{D})$ is a finite field of cardinality p^k for some $k \geqslant 1$,
- $J(\mathbf{D})$ is a vector space over $\mathbf{D}/J(\mathbf{D})$ under addition and the scalar multiplication $(x + J(\mathbf{D})) \cdot a = xa$,
- the proper ideals of $\mathbf{D}$ are precisely the vector subspaces of $J(\mathbf{D})$.

The characteristic of a ring $\mathbf{D}$ will be denoted $\mathsf{char}(\mathbf{D})$, while the prime p from the above claim will sometimes be denoted $\mathsf{char}^*(\mathbf{D})$.

(3) Let $\mathbf{D} \in \mathcal{V}$ be local and $p = \mathsf{char}^*(\mathbf{D})$. Then for $a, b \in D$ we have:
- $a \overset{J(\mathbf{D})}{\equiv} b$ iff $a^p = b^p$,
- $a \in J(\mathbf{D})$ iff $a^p = 0$,
- $a^p \overset{J(\mathbf{D})}{\equiv} a$.

Now, let $\mathbf{F}$ be a field and $\mathbf{M}$ a vector space over $\mathbf{F}$. Suppose that the mapping

$$c : F \times F \longrightarrow M$$

satisfies the following conditions for all $x, y, z \in F$:

(c1) $c(x, y) = c(y, x)$,
(c2) $c(0, x) = 0$,
(c3) $c(x, -x) = 0$,
(c4) $z \cdot c(x, y) = c(zx, zy)$,
(c5) $c(x + y, z) = c(x, z) + c(y, z)$.

The reader will find no difficulties in checking that the set $F \times M$ endowed with the following operations

$$\begin{aligned}
(x_1, m_1) + (x_2, m_2) &= (x_1 + x_2, m_1 + m_2 + c(x_1, x_2)) \\
(x_1, m_1) \cdot (x_2, m_2) &= (x_1 x_2, x_1 m_2 + x_2 m_1) \\
-(x, m) &= (-x, -m) \\
0 &= (0, 0) \\
1 &= (1, 0)
\end{aligned}$$

is a commutative ring with unit. We will denote this ring by $\mathbf{F}[\mathbf{M}]$. On the other hand the $\mathbf{F}$–vector space $\mathbf{M}$, in the language expanded with $|F|^2$ many constants $\langle c(x, y) : x, y \in F \rangle$ satisfying (c1)–(c5), is to be called an $\mathbf{F}$*–expanded vector space.* The class of all $\mathbf{F}$–expanded vector spaces forms a variety, which we denote by $\mathcal{M}_{\mathbf{F}}$.

(4) Every finite local ring $\mathbf{D}$ in $\mathcal{V}$ is isomorphic to a ring $\mathbf{F}[\mathbf{M}]$, with the field $\mathbf{F} = \mathbf{D}/J(\mathbf{D})$ and for some $\mathbf{F}$–expansion of the vector space $\mathbf{M} = J(\mathbf{D})$.

Indeed, suppose that $p = \mathsf{char}^*(\mathbf{D})$ and define $c : F \times F \longrightarrow M$ by putting $c(x/J(\mathbf{D}), y/J(\mathbf{D})) = x^p + y^p - (x+y)^p$. One can easily check that this map is well defined, takes its values in $J(\mathbf{D})$ and satisfies (c1)–(c5).

We will show that the map

$$h : D \ni a \mapsto (a/J(\mathbf{D}), a - a^p) \in F \times M$$

establishes the required isomorphism.
Obviously h preserves the zero and unit of the rings. Moreover

$$\begin{aligned} h(a) + h(b) &= (a/J(\mathbf{D}),\ a - a^p) + (b/J(\mathbf{D}),\ b - b^p) \\ &= ((a+b)/J(\mathbf{D}),\ (a - a^p + b - b^p + c(a/J(\mathbf{D}), b/J(\mathbf{D})) \\ &= ((a+b)/J(\mathbf{D}),\ (a - a^p + b - b^p + a^p + b^p - (a+b)^p)) \\ &= ((a+b)/J(\mathbf{D}),\ (a+b) - (a+b)^p) \\ &= h(a+b), \end{aligned}$$

and

$$\begin{aligned} h(a) \cdot h(b) &= (a/J(\mathbf{D}),\ a - a^p) \cdot (b/J(\mathbf{D}),\ b - b^p) \\ &= (ab/J(\mathbf{D}),\ a(b - b^p) + b(a - a^p)) \\ &= (ab/J(\mathbf{D}),\ a(b - b^p) + b(a - a^p) - ab + a^p b^p + ab - a^p b^p) \\ &= (ab/J(\mathbf{D}),\ (a - a^p)(b - b^p) + ab - a^p b^p) \\ &= (ab/J(\mathbf{D}),\ ab - a^p b^p) \\ &= h(ab), \end{aligned}$$

as required. Note that the second to the last equality in the above display follows from the fact that both $a - a^p$ and $b - b^p$ are in $J(\mathbf{D})$ and $J(\mathbf{D})^2 = \{0\}$.

This shows that h is a homomorphism. To see that it is injective suppose that $a/J(\mathbf{D}) = b/J(\mathbf{D})$ and $a - a^p = b - b^p$. This immediately yields $a = b$.

To see that h is surjective take $(x/J(\mathbf{D}), m) \in F \times M$. Put $a = x^p + m$. Obviously $a/J(\mathbf{D}) = x^p/J(\mathbf{D}) = x/J(\mathbf{D})$. Since $\binom{p}{i}$ is divisible by p for $i = 1, \ldots, p-1$ and $p \cdot J(\mathbf{D}) = \{0\}$ we have $(x^p + m)^p = \sum_{i=0}^{p} \binom{p}{i} x^{pi} m^{p-i} = m^p + x^{p^2}$. Moreover $x^{p^2} = x^p$ so that $a - a^p = x^p + m - (x^p + m)^p = x^p + m - (m^p + x^{p^2}) = m$, and consequently $h(a) = (x/J(\mathbf{D}), m)$. This proves the surjectivity of h and also (4).

We will also need the following

(5) If a finite local ring $\mathbf{D} \in \mathcal{V}$ is k–generated, then so is $J(\mathbf{D})$ as a $\mathbf{D}/J(\mathbf{D})$–vector space.

Suppose that $\mathsf{char}^*(\mathbf{D}) = p$ and that $\mathbf{D}$ is generated by $\{a_1, \ldots, a_k\}$. Then obviously $\mathbf{D}$ is generated by the set $\{a_i^p : i = 1, \ldots, k\} \cup \{b_i : i = 1, \ldots, k\}$, where $b_i = a_i - a_i^p$. Note that $b_i \in J(\mathbf{D})$. Now for $m \in J(\mathbf{D})$ there is a ring term $t(x_1, \ldots, x_k, y_1, \ldots, y_k)$ such that $m = t(a_1^p, \ldots, a_k^p, b_1, \ldots, b_k)$. We can view this term as a ring polynomial that is a sum of monomials in variables $y_1, \ldots, y_k$ with coefficients that are of the form c^p with arbitrary $c \in D$. Moreover, since $J(\mathbf{D})^2 = \{0\}$ then every monomial of this polynomial depends on at most one y_i and this y_i occurs with degree 1. Consequently m is a special affine combination of $b_1, \ldots, b_k$, namely $m = c_0^p + c_1^p b_1 + \cdots + c_k^p b_k$. Now, since m as well as all of the b_i's are in $J(\mathbf{D})$, we get that $c_0^p \in J(\mathbf{D})$. Consequently $c_0^p = 0$ and $m = c_1^p b_1 + \cdots + c_k^p b_k$. This shows that each element of $J(\mathbf{D})$ is a linear combination of $b_1, \ldots, b_k$ and by the same token proves that the $\mathbf{D}/J(\mathbf{D})$–vector space $J(\mathbf{D})$ is generated by the set $\{b_1, \ldots, b_k\}$.

Since the variety $\mathcal{V}$ is finitely generated, Theorem 14.5 of [**32**] guarantees that, up to isomorphism, there are only finitely many fields in $\mathcal{V}$ (i.e., simple rings with unit). Let $\{\mathbf{F}_1, \ldots, \mathbf{F}_s\}$ be the set of all, up to isomorphism, finite fields in $\mathcal{V}$.

For a finite field $\mathbf{F} \in \mathcal{V}$ denote by $\mathcal{D}_{\mathbf{F}}$ the class of finite local rings $\mathbf{D} \in \mathcal{V}$ such that $\mathbf{D}/J(\mathbf{D})$ is isomorphic to $\mathbf{F}$. From (4) and (5) we know that every k–generated ring in $\mathcal{D}_{\mathbf{F}}$ is of the form $\mathbf{F}[\mathbf{M}]$ for some k-generated (as a vector space) $\mathbf{M} \in \mathcal{M}_{\mathbf{F}}$. Therefore $\mathrm{G}_{\mathcal{D}_{\mathbf{F}}}(k)$ is bounded from above by $k+1$, the G-spectrum of the variety of vector spaces over $\mathbf{F}$. Moreover the size of such a k–generated ring $\mathbf{F}[\mathbf{M}]$ in $\mathcal{D}_{\mathbf{F}}$ is bounded by $|F \times M| = |F|^{k+1}$.

Now, if f is the maximum of $|F_1|, \ldots, |F_s|$, then the number of nonisomorphic k–generated directly indecomposable rings in $\mathcal{V}$ is bounded from above by $s(k+1)$ and the size of any such ring is not greater than f^{k+1}. Applying Theorem 6.4 we get that $\mathrm{G}_{\mathcal{V}}(k) \leqslant f^{s(k+1)^3}$ i.e., the variety $\mathcal{V}$ has few models. □

In Example 6.5 we have shown that for the prime p the variety generated by the ring $\mathbf{Z}_{p^2}$ has few models. Theorem 18.2 allows us to characterize cyclic rings $\mathbf{Z}_m$ generating varieties with few models. Indeed, the Jacobson radical of the ring $\mathbf{Z}_m$ squares to zero if and only if m is cube free, i.e., there is no prime p with $p^3|m$. Thus we have:

COROLLARY 18.3. *A cyclic ring $\mathbf{Z}_m$ generates a variety with few models if and only if m is cube free.* □

CHAPTER 19

Open Problems

Since every finitely generated variety has at most triply exponential generative complexity the most natural questions for such varieties are the following.

PROBLEM 1. *Characterize finitely generated varieties with very few models.*

PROBLEM 2. *Characterize finitely generated varieties with few models.*

PROBLEM 3. *Characterize finitely generated varieties with at most doubly exponential generative complexity.*

We find Problem 3 to be especially difficult since the techniques we have developed are not sufficient to the task. That is, we have basically two methods of construction for building finitely generated varieties with many models. In our primary method, suppose a finite algebra $\mathbf{A}$ generates a variety $\mathcal{V}$. We consider a function $c(k)$ and for each k we form a diagonal subalgebra $\mathbf{D}$ of $\mathbf{A}^{c(k)}$ that has a generating set of cardinality $\ell(k)$, where $\ell(k)$ is some linear function of k. We code partitions of the integer $c(k)$ as $\ell(k)$-generated algebras in $\mathcal{V}$. This construction will produce at most $\Pi(c(k))$ distinct $\ell(k)$-generated algebras in $\mathcal{V}$. For a finite algebra $\mathbf{A}$, if the representation of $\mathbf{D}$ in $\mathbf{A}^{c(k)}$ is to be irredundant, then $c(k) \leqslant |A|^k$. Thus we can obtain only at most $\Pi(|A|^k)$ many nonisomorphic $\ell(k)$-generated algebras by this method. In particular, this method will not give finitely generated varieties of triply exponential generative complexity.

Our second method of constructing finitely generated varieties $\mathcal{V}$ with many models and the only way we are able to produce finitely generated varieties of triply exponential generative complexity is basically the following. We choose $\mathbf{A}$ so that in $\mathcal{V} = \mathcal{V}(\mathbf{A})$ the free algebra $\mathbf{F} = \mathbf{F}_{\mathcal{V}}(k)$ for every finite k is such that

(1) $\mathbf{F}$ is uniquely generated by $X = \{x_1, \dots, x_k\}$. (See Definition 3.4.)
(2) The universe F contains a subset U such that for every $T \subseteq U$ the congruence $\theta_T = \mathrm{Cg}^{\mathbf{F}}(T \times T)$ satisfies
   (a) $x_i/\theta_T = \{x_i\}$ for all $1 \leqslant i \leqslant k$, i.e., $\theta_T \in \mathrm{C_s}(\mathbf{F}, X)$.
   (b) For $u_1, u_2 \in U$ if $(u_1, u_2) \in \theta_T$ with $u_1 \neq u_2$, then $u_1, u_2 \in T$.

If these conditions are in force, then $|\mathrm{C_s}(\mathbf{F}, X)| \geqslant 2^{|U|}$ and so $\mathrm{G}_{\mathcal{V}}(k) \geqslant 2^{|U|}/k!$ by Corollary 3.8.

- For our examples of finitely generated varieties of triply exponential generative complexity we are able to find sets U that have cardinality that is doubly exponential as a function of k.
- For semilattices, in Theorem 4.1, we choose a set U with cardinality $\binom{k}{\lfloor k/2 \rfloor}$ to show that this variety has doubly exponential generative complexity.
- In our argument for the variety of distributive lattices, Theorem 4.2, we work with $U \subseteq F^2$ rather than $U \subseteq F$ and then for $T \subseteq U$ we let $\theta_T = \mathrm{Cg}^{\mathbf{F}}(T)$. In this case $|U| = 2^k - 2k$.

In order to obtain varieties with triply exponential G-spectra by this method we need the set U to have doubly exponential cardinality as a function of k. In our examples, the construction of such $\mathbf{A}$ and U is ad hoc. Thus, we do not have any general understanding of algebraic properties that force triply exponential generative complexity.

Note that Part 2 of our work solves Problems 1 and 2 for varieties that omit type $\mathbf{1}$.

It seems that a good starting point for attacking these problems in the general setting might be finitely generated varieties of semigroups. Recently M. Bilski [**9**] characterized such varieties with very few models. In his characterization the following varieties play a crucial role:

- $\mathcal{Z}$, the variety of semigroups with zero multiplication ($xy = uv$),
- $\mathcal{L}$, the variety of left-zero semigroups ($xy = x$),
- $\mathcal{R}$, the variety of right-zero semigroups ($xy = y$),
- $\mathcal{A}_m$, the variety of Abelian groups of exponent m treated as semigroups ($x^m = y^m$, $x^{m+1} = x$).

THEOREM 19.1 (Bilski). *A finitely generated variety $\mathcal{V}$ of semigroups has very few models if and only if $\mathcal{V} \subseteq \mathcal{Z}$ or there is an m with $\mathcal{V} \subseteq \mathcal{L} \vee \mathcal{R} \vee \mathcal{A}_m$.*

Note that the varieties $\mathcal{L}, \mathcal{R}$ and $\mathcal{A}_m$ are pairwise independent, therefore a finite semigroup $\mathbf{S}$ generates a variety with very few models if and only if either $\mathbf{S}$ is a zero semigroup or $\mathbf{S}$ is a direct product $\mathbf{L} \times \mathbf{R} \times \mathbf{A}$ for some left zero semigroup $\mathbf{L}$, a right zero semigroup $\mathbf{R}$ and an Abelian group $\mathbf{A}$.

In Corollary 8.3 we noticed that a locally finite variety $\mathcal{V}$ with very few models has to be locally solvable, i.e., $\text{typ}\{\mathcal{V}\} \subseteq \{\mathbf{1}, \mathbf{2}\}$. We actually conjecture that such a variety has to be Abelian. If so then Problem 1 reduces to the following one:

PROBLEM 4. *Characterize locally finite Abelian varieties with very few models.*

In view of Theorem 19.1 and the work done in [**29**] it is natural to make the following conjecture.

CONJECTURE 5. *A locally finite variety $\mathcal{V}$ has very few models if and only if $\mathcal{V}$ is a product $\mathcal{S} \otimes \mathcal{A}$ of a strongly Abelian variety $\mathcal{S}$ and an affine variety $\mathcal{A}$ both having very few models.*

Theorem 13.4 provides a full characterization of the affine part $\mathcal{A}$ in this decomposition so that we are left with the problem of understanding locally finite, strongly Abelian varieties with very few models. Combining our characterization of locally finite, multi-unary varieties having very few models (see Example 3.9) and the strongly Abelian part $\mathcal{L} \otimes \mathcal{R}$ in Bilski's result we are led to the following question.

PROBLEM 6. *Does every locally finite strongly Abelian variety $\mathcal{S}$ with very few models decompose into a product $\mathcal{S}_1 \otimes \ldots \otimes \mathcal{S}_n$ of finitely many varieties each categorically equivalent to a variety of multiunary algebras in which each operation is either constant or a permutation?*

Recall that at the end of Chapter 7 we show that the varieties $\mathcal{S}_i$ in Problem 6 are just matrix powers of varieties of multiunary algebras in which all operations are constant or permutations.

A characterization of locally finite varieties with very few models obtained in Theorem 13.4 is partially based in one direction on the fact that locally finite Abelian varieties are finitely generated.

In trying to characterize locally finite varieties with few models one meets difficulties already at the non-Abelian level. In Corollary 10.3 we characterized finitely generated, congruence meet-semidistributive varieties with few models as those that are semisimple arithmetical. However, this characterization is not sufficient for locally finite varieties as Examples 5.12, 5.14 and 5.15 show. On the other hand, Example 6.18 shows that there are locally finite, but not finitely generated, semisimple arithmetical varieties with few models. This leads to the following:

PROBLEM 7. *Characterize locally finite semisimple arithmetical varieties with few models.*

The G-spectrum of a variety is a numerical invariant that is independent of the similarity type, language, and identities used to present the variety. As with the free spectrum, fine spectrum, and residual bound, the G-spectrum is a natural topic of investigation when studying a particular variety or family of varieties. So, given a variety $\mathcal{V}$ one might strive to provide a closed form or asymptotic estimate for the G-spectrum $\mathrm{G}_{\mathcal{V}}(k)$. In various examples throughout this work we have presented explicit formulas for the G-spectra of specific varieties. Our paper [**6**] provides closed form expressions for the G-spectra of many of the varieties generated by 2-element algebras. Empirical evidence suggests there may exist gaps in the growth rates of G-spectra of (finitely generated) varieties. We note the existence of gaps for free spectra functions. Thus, for a finitely generated variety $\mathcal{V}$ the function $|\mathrm{F}_{\mathcal{V}}(k)|$ is either at most polynomial or at least exponential, (see [**32**, Theorem 12.2]). Our example 3.10 shows that there is no such gap for the G-spectra of finitely generated varieties but perhaps the following is true.

PROBLEM 8. *Is it the case that for every (finitely generated) variety $\mathcal{V}$ the* G-*spectrum* $\mathrm{G}_{\mathcal{V}}(k)$ *is bounded above by a polynomial in k or bounded below by $b2^{k^{\varepsilon}}$ for some positive constants b and ε?*

If we restrict this problem to locally finite varieties that omit type **1**, then any $\mathcal{V}$ that has its G-spectrum between the bounds specified in Problem 8 would have few models and so $\mathcal{V}$ would be congruence permutable by Corollary 11.4. Moreover, $\mathcal{V}$ would be Abelian by Lemma 13.1, and therefore an affine variety. It can be argued that if $\mathbf{R}$ is the finite ring associated with $\mathcal{V}$, then the G-spectrum of the variety of $\mathbf{R}$-modules would also fall within the bounds of Problem 8. So a positive solution to Problem 8 for all varieties of modules over finite rings would imply a positive solution for all finitely generated varieties that omit type **1**.

Similar problems concerning gaps can be posed for other levels in our hierarchy.

PROBLEM 9. *Is it the case that for every finitely generated variety $\mathcal{V}$, the* G-*spectrum* $\mathrm{G}_{\mathcal{V}}(k)$ *is either bounded above by an exponential function in k or is bounded below by $b2^{2^{k^{\varepsilon}}}$ for some positive constants b and ε?*

Problem 9 has a positive solution for finitely generated, congruence meet semidistributive varieties, (see Corollary 10.3). The problem also has a positive solution for varieties of groups, (see Theorem 18.1). In [**6**] it is shown that Problems 8 and

9 have positive solutions if the varieties are restricted to those generated by 2-element algebras. Indeed, for these varieties there exist gaps in the G-spectra since all are either of polynomial, exponential or at least doubly exponential generative complexity.

We know that for any locally finite variety $\mathcal{V}$

$$\mathrm{G}_{\mathcal{V}}(k) \leqslant |\mathsf{Con}\,(\mathbf{F}_{\mathcal{V}}(k))| \leqslant \mathrm{Bell}(|\mathrm{F}_{\mathcal{V}}(k)|),$$

and so, if the free spectrum of $\mathcal{V}$ is at most m-fold exponential, then the G-spectrum is at most $(m+1)$-fold exponential. If $\mathcal{V}$ is such that the cardinality of $\mathbf{F}_{\mathcal{V}}(k)$ is at least as large as the cardinality of the congruence lattice of $\mathbf{F}_{\mathcal{V}}(k)$, then $|\mathrm{G}_{\mathcal{V}}(k)| \leqslant |\mathrm{F}_{\mathcal{V}}(k)|$ and the level of exponential complexity of the G-spectrum is no higher than that of the free spectrum.

PROBLEM 10. *For which (finitely generated) varieties $\mathcal{V}$ does the inequality $|\mathsf{Con}\,(\mathbf{F}_{\mathcal{V}}(k))| \leqslant |\mathrm{F}_{\mathcal{V}}(k)|$ hold for all but finitely many integers k?*

We restate a related problem raised after Proposition 3.2.

PROBLEM 11. *If the free spectrum of a variety is at least m-fold exponential must the* G*-spectrum be at least $(m-1)$-fold exponential?*

Note that Example 5.16 is a variety $\mathcal{V}$ for which $|\mathrm{F}_{\mathcal{V}}(k)|$ can be made to be arbitrarily larger than $\mathrm{G}_{\mathcal{V}}(k)$ for all k. However, an examination of the proof in this construction shows that $\mathrm{G}_{\mathcal{V}}(k) \geqslant |\mathrm{F}_{\mathcal{V}}(k-1)|$ and so if the free spectrum of $\mathcal{V}$ is at least m-fold exponential, then the G-spectrum is as well.

We present some problems involving decidability issues for the generative complexity of finitely generated varieties.

PROBLEM 12. *Is there an algorithm that decides for a finite algebra* **A** *of finite similarity type whether the variety generated by* **A** *has* G*-spectrum that is at most polynomial? at most exponential? at most doubly exponential?*

If we relativize Problem 12 to those **A** for which $\mathcal{V} = \mathcal{V}(\mathbf{A})$ is congruence meet semi-distributive, that is, $\mathbf{1}, \mathbf{2} \notin \mathrm{typ}\{\mathcal{V}\}$, then there is an algorithm to determine if $\mathrm{G}_{\mathcal{V}}(k)$ is at most exponential. Namely, by Corollary 10.3 we need only check that **A** generates a semisimple arithmetical variety — and algorithms to perform the checking for these properties are known. However, if we move up to the next level of exponentiation we do not know the answer.

PROBLEM 13. *Is there an algorithm that decides for a finite algebra of finite similarity type that generates a congruence meet semi-distributive variety $\mathcal{V}$ whether or not $\mathcal{V}$ is of at most doubly exponential generative complexity?*

With regard to this problem we note that in Example 9.4 there is a finite algebra generating a congruence meet semi-distributive variety of triply exponential generative complexity.

If we consider the problem of deciding for algebras **A** that generate varieties that omit type **1** whether or not the variety $\mathcal{V}$ generated by **A** has G-spectrum that is at most polynomial, then by Theorem 13.4 we need only test if $\mathcal{V}$ is a directly representable affine variety. To see if **A** generates an affine variety it suffices to check if **A** has a Maltsev term and is Abelian, and there exist algorithms for doing this. If $\mathcal{V}$ is affine, then the finite ring **R** associated with $\mathcal{V}$ can be constructed using the algorithm given in Chapter 9 of [**22**]. Thus, the problem reduces to deciding if **R** is of finite representation type.

PROBLEM 14. *Does there exist an algorithm that determines whether a finite ring* **R** *is of finite representation type, i.e., whether there exist only finitely many finitely generated, directly indecomposable* **R**-*modules, up to isomorphism.*

Added in proof. Very recently P. Idziak, R. McKenzie and M. Valeriote [**37**] have been able to go along with the ideas sketched in Conjecture 5 and Problem 6. As a result they solved both Problems 1 and 4 by proving the following:

THEOREM 19.2 (P. Idziak, R. McKenzie and M. Valeriote). *A locally finite variety has at most polynomially many (in k) non-isomorphic k-generated algebras if and only if it decomposes into a varietal product of an affine variety over a ring of finite representation type, and a finite sequence of strongly Abelian varieties equivalent to matrix powers of varieties of G-sets, with constants, for various finite groups G.*

CHAPTER 20

Tables

This Chapter lists the most important results and examples in this book. They are organized into several tables. These tables also contain references to the main text, so that they can serve as navigation guides throughout the book.

The first table contains a list of results. Tables 2, 3 and 4 collect examples of varieties with very few models, few models and many models, respectively. In Table 4 we have also included examples of locally finite varieties with very large G–spectra.

According to Propositions 3.1 and 3.2 if $\mathcal{V}$ is a finitely generated variety with m–fold exponential free spectrum, then the G–spectrum of $\mathcal{V}$ could be $(m-1)$–fold, m–fold or $(m+1)$–fold exponential. In Table 5 we show that all these possibilities can actually occur.

Table 1: **List of results**

4.1	The variety of semilattices has many models
4.3	Every nontrivial variety of lattices has many models
5.1	The G-spectrum of a finitely generated variety is at most triply exponential
6.2	The G-spectrum of a variety (of finite type) with a finite residual bound is at most doubly exponential
6.3	The G-spectrum of a finitely generated quasivariety is at most doubly exponential
6.14	The G-spectrum of any finitely generated congruence modular variety is at most doubly exponential
3.9	A characterization of multiunary locally finite varieties with very few models
8.3	Every locally finite variety with very few models is locally solvable
6.12	Every locally finite variety of nilpotent groups has few models
5.5	The G-spectrum of a locally finite strongly Abelian variety is at most singly exponential
6.13	The G-spectrum of a locally finite, congruence modular, Abelian variety is at most singly exponential
6.6	Every directly representable variety has few models
6.6	A characterization of directly representable varieties with very few models

Table 1: **List of results** (continued)	
10.3	A characterization of finitely generated congruence meet-semi-distributive varieties with few models
13.4	**A characterization of locally finite varieties omitting type 1 with very few models**
14.6	A characterization of finitely generated, congruence modular, nilpotent varieties with few models
17.2	**A characterization of finitely generated varieties omitting type 1 with few models**
18.1	A characterization of finitely generated varieties of groups with few and very few models
18.2	A characterization of finitely generated varieties of commutative rings with unit that have few and very few models
19.1	A characterization of finitely generated varieties of semigroups with very few models
7.3	G-spectra for categorically equivalent varieties

Table 2: **Very Few Models**

	Variety or generating algebra(s)	$f_{\mathcal{V}}(k)$	$\mathrm{G}_{\mathcal{V}}(k)$
1.1	sets	k	k
3.9	G-sets		see 3.9
1.1	vector spaces over a finite field $\mathbf{F}$	$\|F\|^k$	$1+k$
13.5	modules over a finite ring of finite representation type		see 13.5
1.1	p-element group, p a prime	p^k	$1+k$
6.7	a locally finite variety $\mathcal{V}$ of Abelian groups	$f_{\mathcal{V}}(1)^k$	see 6.7

Table 3: **Few Models**

	Variety or generating algebra(s)	$f_{\mathcal{V}}(k)$	$\mathrm{G}_{\mathcal{V}}(k)$
1.1	Boolean algebras	2^{2^k}	$1+2^k$
3.10	one unary operation	$1+2k$	between $\Pi(k)$ and $(k+1)^2\Pi(k)$
5.6	6-element multiunary algebra	linear	$\geqslant 2^k$
5.13	pure discriminator	$\prod i^{S(k,i)}$	$\prod(1+S(k,i))$
6.5	ring $\mathbf{Z}_{p^2}$, for p a prime	$(p^{k+2})^{p^k}$	$\leqslant p^{(k+1)(k+2)^2}$
6.12	r–nilpotent, locally finite groups	$\leqslant 2^{ck^r}$	$\leqslant 2^{2c^2k^{2r}}$
6.18	modular ortholattices	see 6.18	see 6.18

Table 4: **Many Models**

	Variety or generating algebra(s)	$f_{\mathcal{V}}(k)$	$G_{\mathcal{V}}(k)$
4.1	semilattices	$2^k - 1$	$2^{\binom{k}{\lfloor k/2 \rfloor}(1+o(1))}$
4.2	(bounded) distributive lattices	$2^{\binom{k}{\lfloor k/2 \rfloor}(1+o(1))}$	$2^{2^k(1+o(1))}$
5.7	a finitely generated Abelian variety	$2^{k+1} + k - 1$	2–fold exp.
8.2	any non–Boolean variety of pseudo-complemented (semi-) lattices	$\geqslant$ 2–fold exp.	$\geqslant$ 2–fold exp.
10.4	any non–Boolean variety of Heyting algebras	$\geqslant$ 2–fold exp.	$\geqslant$ 2–fold exp.
10.4	any non–trivial variety of implication algebras	$\geqslant$ 2–fold exp.	$\geqslant$ 2–fold exp.
5.14	monadic algebras	$2^{2^k 2^{(2^k-1)}}$	$\sum_{i \leqslant 2^k}(1 + \binom{2^k}{i})$
5.3	4–element algebra of finite type	2–fold exp.	3–fold exp.
5.4	fin. gen. strongly solvable variety	2–fold exp.	3–fold exp.
9.4	fin. gen. meet semi-distributive variety	2–fold exp.	3–fold exp.
5.15	loc. fin. discriminator var. of finite type	$(m+1)$–fold exp.	m–fold exp.
5.10	loc. fin. locally strongly solvable var. of fin. type	m–fold exp.	$(m+1)$–fold exp.
5.10	locally finite, m binary ops.		$\geqslant m$–fold exp.
5.8	loc. fin. variety (of infinite type) exceeding any prescribed function		
5.9	loc. fin. variety of groupoids exceeding any prescribed function		
5.12	loc. fin. discriminator variety exceeding any prescribed function		
5.16	loc. fin. discriminator variety with an arbitrarily large gap between free– and G– spectra		

Table 5: **Relation between free and G-spectra in finitely generated varieties**

size of free–spectrum	size of G-spectrum	example
polynomial	polynomial	sets
polynomial	exponential	Ex. 5.6
exponential	polynomial	vector spaces over a finite field
exponential	exponential	finite nilpotent non–Abelian group
exponential	doubly exponential	semilattices
doubly exponential	exponential	Boolean algebras
doubly exponential	doubly exponential	distributive lattices
doubly exponential	triply exponential	Ex. 5.3 and 5.4

Bibliography

[1] V. B. Alexseev, The number of families of subsets closed with respect to intersections, (in Russian), *Diskrete Math.*, **1**(1989), 129-136.

[2] G. E. Andrews, *The Theory of Partitions*, Encyclopedia of Mathematics and Its Applications, Vol. 2, Addison-Wesley, Reading, MA, 1976.

[3] H. Bass, Finite monadic algebras, *Proc. Amer. Math. Soc.*, **9**(1958), 258–268.

[4] J. Berman, Free spectra gaps and tame congruence types, *International Journal of Algebra and Computation*, **5**(1995), 651–672.

[5] J. Berman and W. J. Blok, Free spectra of nilpotent varieties, *Algebra Universalis*, **24**(1987), 279–282.

[6] J. Berman and P. M. Idziak, Counting finite algebras in the Post varieties, *International Journal of Algebra and Computation*, **10**(2000), 323–337.

[7] J. Berman and R. N. McKenzie, Clones satisfying the term condition, *Discrete Math.*, **52**(1984), 7–29.

[8] J. Berman and S. Seif, An approach to tame congruence theory via subtraces, *Algebra Universalis*, **30**(1993), 479–520.

[9] M. Bilski, Generative complexity in semigroups varieties, *Journal of Pure and Applied Algebra*, **165**(2001), 137–149.

[10] G. Burosch, J. Demetrovics, G. O. H. Katona, D. J. Kleitman, and A. A. Sapozhenko, On the number of databases and closure operations, *Theoretical Computer Science*, **78**(1991), 377-381.

[11] G. Burosch, J. Demetrovics, G. O. H. Katona, D. J. Kleitman, and A. A. Sapozhenko, On the number of closure operations, in *Combinatorics, Paul Erdos is Eighty*, Vol. 1, 91-95, Bolyai Soc. Math. Stud., Janos Bolyai Math. Soc., Budapest, 1993.

[12] S. Burris, Spectrally determined first-order limit laws, in: *Logic and Random Structures*, eds. R. Boppana and J. Lynch, DIMACS Series in Discrete Mathematics and Theoretical Computer Science, vol. 33, pp. 33–52, Amer. Math. Soc., 1997.

[13] S. Burris, *Number theoretic density and logical limit laws.* Mathematical Surveys and Monographs, 86. American Mathematical Society, Providence, RI, 2001.

[14] S. Burris and K. Compton, Fine spectra and limit laws. I. First order laws, *Canad. J. Math.*, **49**(1997), 468–498.

[15] S. Burris, K. Compton, A. Odlyzko, and B. Richmond, Fine spectra and limit laws. II. First-order 0-1 laws, *Canad. J. Math.*, **49**(1997), 641–652.

[16] S. Burris and P. Idziak, A directly representable variety has a discrete first-order law, *International J. of Algebra and Computation*, **6**(1996), 269–276.

[17] S. Burris and H. P. Sankappanavar, *A Course in Universal Algebra*, Springer-Verlag, New York, 1981.

[18] D. Clark and P. Krauss, Plain para primal algebras, *Algebra Universalis*, **11**(1980), 365–388.

[19] K. Compton, personal communication.

[20] B. Davey and I. Rival, Exponents of lattice-ordered algebras, *Algebra Universalis*, **14**(1982), 87–98.

[21] R. Fagin, Generalized first-order spectra and polynomial-time recognizable sets, in: R.Karp, ed., *Complexity and Computation, SIAM-AMS Proceedings*, **7**(1974), 43–73.

[22] R. Freese and R. McKenzie, *Commutator Theory for Congruence Modular Varieties*, London Math. Soc. Lecture Notes, No. 125, Cambridge U. Press, Cambridge, 1987.

[23] O. Frink, Pseudo-complements in semi-lattices, *Duke Math. J.*, **29** (1962), 505 – 514.

[24] G. Grätzer, Composition of functions, in *Proceedings of the Conference on Universal Algebra, (Kingston, 1969)*, Queen's University, Kingston, Ont., 1970, pp. 1–106.

[25] G. Grätzer and A. Kisielewicz, A survey of some open problems on p_n-sequences and free spectra of algebras and varieties, in *Universal Algebra and Quasigroup Theory*, A. Romanowska and J.D.H. Smith (eds.), Heldermann Verlag, Berlin, 1992, pp. 57–88.

[26] H. P. Gumm, *Geometrical methods in congruence modular varieties*, Memoirs Amer. Math. Soc., **289**(1983).

[27] J. Hagemann and C. Herrmann, A concrete ideal multiplication for algebraic systems and its relation to congruence distributivity, *Archive der Mathematik*, **32**(1979), 234–245.

[28] B. Hart and M. Valeriote, A structure theorem for strongly abelian varieties with few models, *Journal of Symbolic Logic*, **56**(1991), 832–852.

[29] B. Hart, S. Starchenko and M. Valeriote, Vaught's conjecture for varieties, *Trans. Amer. Math. Soc*, **342**(1994), 173–196.

[30] M. Haviar, P. Konôpka, and C.B. Wegener, Finitely generated free modular ortholattices, II, *International Journal of Theoretical Physics*, **36**(1997), 2661–2679.

[31] G. Higman, The orders of relatively free groups, Proc. International Conf. Theory of Groups, Austral. Nat. Univ. Canberra, 1965, pp. 153–165.

[32] D. Hobby and R. McKenzie, *The Structure of Finite Algebras*, Contemporary Mathematics vol. 76, Amer. Math. Soc., Providence, RI, 1988.

[33] T. K. Hu, Stone duality for primal algebra theory, *Math. Z.*, **110**(1969), 180–198.

[34] P. Idziak, Varieties with decidable finite algebras II: Permutability, *Algebra Universalis*, **26**(1989), 247–256.

[35] P. Idziak, A characterization of finitely decidable congruence modular varieties, *Trans. of the Amer. Math. Soc.*, **349**(1997), 903–934.

[36] P. Idziak and R. McKenzie, Varieties with very few models, *Fundamenta Mathematicae*, **170**(2001), 53–68.

[37] P. Idziak, R. McKenzie and M. Valeriote, The structure of locally finite varieties with polynomially many models, manuscript 2003, 74pp.

[38] P. Idziak and J. Tyszkiewicz, Monadic second order probabilities in algebra. Directly representable varieties and groups, in: *Logic and Random Structures*, eds. R. Boppana and J. Lynch, DIMACS Series in Discrete Mathematics and Theoretical Computer Science, vol. 33, pp. 79–107, Amer. Math. Soc., 1997.

[39] N. Jacobson, *Basic Algebra II*, W. H. Freeman and Co., San Francisco, 1980

[40] J. Jeong, Type 2 subdirectly irreducible algebras in finitely decidable varieties, *J. of Algebra*, **174** (1995), 772–793.

[41] G. Kalmbach, *Orthomodular Lattices*, London Math. Soc. Monographs, 18, Academic Press, London, 1983.

[42] K. A. Kearnes, An order-theoretic property of the commutator, *International Journal of Algebra and Computation*, **3**(1993), 491-533.

[43] K. A. Kearnes, Congruence modular varieties with small free spectra, *Algebra Universalis*, **42**(1999), 165–181.

[44] K. A. Kearnes and E. W. Kiss, Modularity prevents tails, *Proc. Amer. Math. Soc.*, **127**(1999), 11–19.

[45] K. A. Kearnes and E. W. Kiss, Finite algebras of finite complexity, *Discrete Math.*, **207**(1999), 89–135.

[46] K. A. Kearnes, E. W. Kiss and M. Valeriote, A geometric consequence of residual smallness, *Ann. Pure Appl. Logic*, **99**(1999), 137–169.

[47] K. A. Kearnes and R. D. Willard, Finiteness properties of locally finite abelian varieties, *International Journal of Algebra and Computation*, **9**(1999), 157–168.

[48] L. Lovász, *Combinatorial Problems and Exercises, 2nd ed.*, North-Holland, Amsterdam, 1993.

[49] J. F. Lynch, Probabilities of first-order sentences about unary functions, *Trans. Amer. Math. Soc.*, **287**(1985), 543–568.

[50] R. McKenzie, Finite forbidden lattices. *Universal Algebra and Lattice Theory* (Puebla 1982), 176–205, Lecture Notes in Mathematics, 1004, Springer-Verlag.

[51] R. McKenzie, Nilpotent and solvable radicals in locally finite congruence modular varieties, *Algebra Universalis*, **24**(1987), 251–266.

[52] R. McKenzie, Narrowness implies uniformity, *Algebra Universalis*, **12** (1982), 67–85.

[53] R. McKenzie, An algebraic version of categorical equivalence for varieties and more general algebraic categories, in *Logic and Algebra*, ed. A. Ursini and P. Aglianò, Marcel Dekker Inc., 1996, pp. 211–243.

[54] R. McKenzie, Locally finite varieties with large free spectra, *Algebra Universalis*, **47**(2002), 303–318.

[55] R. McKenzie, G. McNulty, W. Taylor, *Algebras, Lattices, Varieties,* Wadsworth/Brooks Cole, Monterrey, CA, 1987.

[56] R. McKenzie and M. Valeriote, *The Structure of Decidable Locally Finite Varieties,* Birkhäuser, Boston, 1989.

[57] L. Moser and M. Wyman, An asymptotic formula for the Bell numbers, *Transactions of the Royal Society of Canada*, **39** ser. 3 (1955), 49–54.

[58] P. Neumann, Some indecomposable varieties of groups, *Quart. J. Math. Oxford*, **14**(1963), 46–50.

[59] R. W. Quackenbush, Structure theory for equational classes generated by quasi-primal algebras, *Trans. Amer. Math. Soc.*, **187**(1974), 127–145.

[60] R. W. Quackenbush, Algebras with minimal spectrum, *Algebra Universalis*, **10**(1980), 117–129.

[61] R. W. Quackenbush, Enumeration in classes of ordered structures, in: *Ordered Sets (Banff, Alta., 1981)*, pp. 523–554, Reidel, Dordrecht-Boston, MA, 1982.

[62] R. W. Quackenbush, Pseudovarieties of finite algebras isomorphic to bounded distributive lattices, *Discrete Math.*, **28**(1979), 189–192.

[63] W. R. Scott, *Group Theory*, Prentice Hall, Englewood Cliffs, NJ, 1964.

[64] S. Seif, A note on free algebras of discriminator varieties, *Algebra Universalis*, **27**(1990), 150-151.

[65] J. D. H. Smith, *Mal'cev Varieties*, Lecture Notes in Mathematics, 554, Springer-Verlag, Berlin, 1976.

[66] W. Taylor, The fine spectrum of a variety, *Algebra Universalis*, **5**(1975), 263–303.

[67] W. Taylor, Equational Logic, in: G. Grätzer, *Universal Algebra*, 2nd ed., pp. 378–400, Springer-Verlag, NY, 1979.

[68] R. S. Wilson, Representation of finite rings, *Pacific Journal of Mathematics*, **53**(1974), 643–679.

Editorial Information

To be published in the *Memoirs*, a paper must be correct, new, nontrivial, and significant. Further, it must be well written and of interest to a substantial number of mathematicians. Piecemeal results, such as an inconclusive step toward an unproved major theorem or a minor variation on a known result, are in general not acceptable for publication. Papers appearing in *Memoirs* are generally longer than those appearing in *Transactions*, which shares the same editorial committee.

As of January 31, 2005, the backlog for this journal was approximately 5 volumes. This estimate is the result of dividing the number of manuscripts for this journal in the Providence office that have not yet gone to the printer on the above date by the average number of monographs per volume over the previous twelve months, reduced by the number of volumes published in four months (the time necessary for preparing a volume for the printer). (There are 6 volumes per year, each containing at least 4 numbers.)

A Consent to Publish and Copyright Agreement is required before a paper will be published in the *Memoirs*. After a paper is accepted for publication, the Providence office will send a Consent to Publish and Copyright Agreement to all authors of the paper. By submitting a paper to the *Memoirs*, authors certify that the results have not been submitted to nor are they under consideration for publication by another journal, conference proceedings, or similar publication.

Information for Authors

Memoirs are printed from camera copy fully prepared by the author. This means that the finished book will look exactly like the copy submitted.

The paper must contain a *descriptive title* and an *abstract* that summarizes the article in language suitable for workers in the general field (algebra, analysis, etc.). The *descriptive title* should be short, but informative; useless or vague phrases such as "some remarks about" or "concerning" should be avoided. The *abstract* should be at least one complete sentence, and at most 300 words. Included with the footnotes to the paper should be the 2000 *Mathematics Subject Classification* representing the primary and secondary subjects of the article. The classifications are accessible from `www.ams.org/msc/`. The list of classifications is also available in print starting with the 1999 annual index of *Mathematical Reviews*. The Mathematics Subject Classification footnote may be followed by a list of *key words and phrases* describing the subject matter of the article and taken from it. Journal abbreviations used in bibliographies are listed in the latest *Mathematical Reviews* annual index. The series abbreviations are also accessible from `www.ams.org/publications/`. To help in preparing and verifying references, the AMS offers MR Lookup, a Reference Tool for Linking, at `www.ams.org/mrlookup/`. When the manuscript is submitted, authors should supply the editor with electronic addresses if available. These will be printed after the postal address at the end of the article.

Electronically prepared manuscripts. The AMS encourages electronically prepared manuscripts, with a strong preference for $\mathcal{A}\!\mathcal{M}\!\mathcal{S}$-LaTeX. To this end, the Society has prepared $\mathcal{A}\!\mathcal{M}\!\mathcal{S}$-LaTeX author packages for each AMS publication. Author packages include instructions for preparing electronic manuscripts, the *AMS Author Handbook*, samples, and a style file that generates the particular design specifications of that publication series. Though $\mathcal{A}\!\mathcal{M}\!\mathcal{S}$-LaTeX is the highly preferred format of TeX, author packages are also available in $\mathcal{A}\!\mathcal{M}\!\mathcal{S}$-TeX.

Authors may retrieve an author package from e-MATH starting from `www.ams.org/tex/` or via FTP to `ftp.ams.org` (login as `anonymous`, enter username as password, and type `cd pub/author-info`). The *AMS Author Handbook* and the *Instruction Manual* are available in PDF format following the author packages link from `www.ams.org/tex/`. The author package can be obtained free of charge by sending email to `pub@ams.org` (Internet) or from the Publication Division, American Mathematical Society, 201 Charles St., Providence, RI 02904, USA. When requesting an author package, please specify $\mathcal{AMS}$-LaTeX or $\mathcal{AMS}$-TeX, Macintosh or IBM (3.5) format, and the publication in which your paper will appear. Please be sure to include your complete mailing address.

Sending electronic files. After acceptance, the source file(s) should be sent to the Providence office (this includes any TeX source file, any graphics files, and the DVI or PostScript file).

Before sending the source file, be sure you have proofread your paper carefully. The files you send must be the EXACT files used to generate the proof copy that was accepted for publication. For all publications, authors are required to send a printed copy of their paper, which exactly matches the copy approved for publication, along with any graphics that will appear in the paper.

TeX files may be submitted by email, FTP, or on diskette. The DVI file(s) and PostScript files should be submitted only by FTP or on diskette unless they are encoded properly to submit through email. (DVI files are binary and PostScript files tend to be very large.)

Electronically prepared manuscripts can be sent via email to `pub-submit@ams.org` (Internet). The subject line of the message should include the publication code to identify it as a Memoir. TeX source files, DVI files, and PostScript files can be transferred over the Internet by FTP to the Internet node `e-math.ams.org` (130.44.1.100).

Electronic graphics. Comprehensive instructions on preparing graphics are available at `www.ams.org/jourhtml/graphics.html`. A few of the major requirements are given here.

Submit files for graphics as EPS (Encapsulated PostScript) files. This includes graphics originated via a graphics application as well as scanned photographs or other computer-generated images. If this is not possible, TIFF files are acceptable as long as they can be opened in Adobe Photoshop or Illustrator. No matter what method was used to produce the graphic, it is necessary to provide a paper copy to the AMS.

Authors using graphics packages for the creation of electronic art should also avoid the use of any lines thinner than 0.5 points in width. Many graphics packages allow the user to specify a "hairline" for a very thin line. Hairlines often look acceptable when proofed on a typical laser printer. However, when produced on a high-resolution laser imagesetter, hairlines become nearly invisible and will be lost entirely in the final printing process.

Screens should be set to values between 15% and 85%. Screens which fall outside of this range are too light or too dark to print correctly. Variations of screens within a graphic should be no less than 10%.

Inquiries. Any inquiries concerning a paper that has been accepted for publication should be sent directly to the Electronic Prepress Department, American Mathematical Society, 201 Charles St., Providence, RI 02904, USA.

Editors

This journal is designed particularly for long research papers, normally at least 80 pages in length, and groups of cognate papers in pure and applied mathematics. Papers intended for publication in the *Memoirs* should be addressed to one of the following editors. In principle the Memoirs welcomes electronic submissions, and some of the editors, those whose names appear below with an asterisk (*), have indicated that they prefer them. However, editors reserve the right to request hard copies after papers have been submitted electronically. Authors are advised to make preliminary email inquiries to editors about whether they are likely to be able to handle submissions in a particular electronic form.

***Algebra** to ALEXANDER KLESHCHEV, Department of Mathematics, University of Oregon, Eugene, OR 97403-1222; email: ams@noether.uoregon.edu

Algebraic geometry to DAN ABRAMOVICH, Department of Mathematics, Brown University, Box 1917, Providence, RI 02912; email: amsedit@math.brown.edu

***Algebraic number theory** to V. KUMAR MURTY, Department of Mathematics, University of Toronto, 100 St. George Street, Toronto, ON M5S 1A1, Canada; email: murty@math.toronto.edu

***Algebraic topology** to ALEJANDRO ADEM, Department of Mathematics, University of British Columbia, Room 121, 1984 Mathematics Road, Vancouver, British Columbia, Canada V6T 1Z2; email: adem@math.ubc.ca

Combinatorics and Lie theory to SERGEY FOMIN, Department of Mathematics, University of Michigan, Ann Arbor, Michigan 48109-1109; email: fomin@umich.edu

Complex analysis and harmonic analysis to ALEXANDER NAGEL, Department of Mathematics, University of Wisconsin, 480 Lincoln Drive, Madison, WI 53706-1313; email: nagel@math.wisc.edu

***Differential geometry and global analysis** to LISA C. JEFFREY, Department of Mathematics, University of Toronto, 100 St. George St., Toronto, ON Canada M5S 3G3; email: jeffrey@math.toronto.edu

Dynamical systems and ergodic theory to ROBERT F. WILLIAMS, Department of Mathematics, University of Texas, Austin, Texas 78712-1082; email: bob@math.utexas.edu

***Functional analysis and operator algebras** to MARIUS DADARLAT, Department of Mathematics, Purdue University, 150 N. University St., West Lafayette, IN 47907-2067; email: mdd@math.purdue.edu

***Geometric analysis** to TOBIAS COLDING, Courant Institute, New York University, 251 Mercer St., New York, NY 10012; email: traneditor@cims.nyu.edu

***Geometric analysis** to MLADEN BESTVINA, Department of Mathematics, University of Utah, 155 South 1400 East, JWB 233, Salt Lake City, Utah 84112-0090; email: bestvina@math.utah.edu

Harmonic analysis, representation theory, and Lie theory to ROBERT J. STANTON, Department of Mathematics, The Ohio State University, 231 West 18th Avenue, Columbus, OH 43210-1174; email: stanton@math.ohio-state.edu

***Logic** to STEFFEN LEMPP, Department of Mathematics, University of Wisconsin, 480 Lincoln Drive, Madison, Wisconsin 53706-1388; email: lempp@math.wisc.edu

Number theory to HAROLD G. DIAMOND, Department of Mathematics, University of Illinois, 1409 W. Green St., Urbana, IL 61801-2917; email: diamond@math.uiuc.edu

***Ordinary differential equations, and applied mathematics** to PETER W. BATES, Department of Mathematics, Michigan State University, East Lansing, MI 48824-1027; email: bates@math.msu.edu

***Partial differential equations** to PATRICIA E. BAUMAN, Department of Mathematics, Purdue University, West Lafayette, IN 47907-1395; email: bauman@math.purdue.edu

***Probability and statistics** to KRZYSZTOF BURDZY, Department of Mathematics, University of Washington, Box 354350, Seattle, Washington 98195-4350; email: burdzy@math.washington.edu

***Real analysis and partial differential equations** to DANIEL TATARU, Department of Mathematics, University of California, Berkeley, Berkeley, CA 94720; email: tataru@math.berkeley.edu

All other communications to the editors should be addressed to the Managing Editor, ROBERT GURALNICK, Department of Mathematics, University of Southern California, Los Angeles, CA 90089-1113; email: guralnic@math.usc.edu.

Titles in This Series

828 **Joel Berman and Paweł M. Idziak,** Generative complexity in algebra, 2005

827 **Trevor A. Welsh,** Fermionic expressions for minimal model Virasoro characters, 2005

826 **Guy Métivier and Kevin Zumbrun,** Large viscous boundary layers for noncharacteristic nonlinear hyperbolic problems, 2005

825 **Yaozhong Hu,** Integral transformations and anticipative calculus for fractional Brownian motions, 2005

824 **Luen-Chau Li and Serge Parmentier,** On dynamical Poisson groupoids I, 2005

823 **Claus Mokler,** An analogue of a reductive algebraic monoid whose unit group is a Kac-Moody group, 2005

822 **Stefano Pigola, Marco Rigoli, and Alberto G. Setti,** Maximum principles on Riemannian manifolds and applications, 2005

821 **Nicole Bopp and Hubert Rubenthaler,** Local zeta functions attached to the minimal spherical series for a class of symmetric spaces, 2005

820 **Vadim A. Kaimanovich and Mikhail Lyubich,** Conformal and harmonic measures on laminations associated with rational maps, 2005

819 **F. Andreatta and E. Z. Goren,** Hilbert modular forms: Mod p and p-adic aspects, 2005

818 **Tom De Medts,** An algebraic structure for Moufang quadrangles, 2005

817 **Javier Fernández de Bobadilla,** Moduli spaces of polynomials in two variables, 2005

816 **Francis Clarke,** Necessary conditions in dynamic optimization, 2005

815 **Martin Bendersky and Donald M. Davis,** V_1-periodic homotopy groups of $SO(n)$, 2004

814 **Johannes Huebschmann,** Kähler spaces, nilpotent orbits, and singular reduction, 2004

813 **Jeff Groah and Blake Temple,** Shock-wave solutions of the Einstein equations with perfect fluid sources: Existence and consistency by a locally inertial Glimm scheme, 2004

812 **Richard D. Canary and Darryl McCullough,** Homotopy equivalences of 3-manifolds and deformation theory of Kleinian groups, 2004

811 **Ottmar Loos and Erhard Neher,** Locally finite root systems, 2004

810 **W. N. Everitt and L. Markus,** Infinite dimensional complex symplectic spaces, 2004

809 **J. T. Cox, D. A. Dawson, and A. Greven,** Mutually catalytic super branching random walks: Large finite systems and renormalization analysis, 2004

808 **Hagen Meltzer,** Exceptional vector bundles, tilting sheaves and tilting complexes for weighted projective lines, 2004

807 **Carlos A. Cabrelli, Christopher Heil, and Ursula M. Molter,** Self-similarity and multiwavelets in higher dimensions, 2004

806 **Spiros A. Argyros and Andreas Tolias,** Methods in the theory of hereditarily indecomposable Banach spaces, 2004

805 **Philip L. Bowers and Kenneth Stephenson,** Uniformizing dessins and Belyĭ maps via circle packing, 2004

804 **A. Yu Ol'shanskii and M. V. Sapir,** The conjugacy problem and Higman embeddings, 2004

803 **Michael Field and Matthew Nicol,** Ergodic theory of equivariant diffeomorphisms: Markov partitions and stable ergodicity, 2004

802 **Martin W. Liebeck and Gary M. Seitz,** The maximal subgroups of positive dimension in exceptional algebraic groups, 2004

801 **Fabio Ancona and Andrea Marson,** Well-posedness for general 2×2 systems of conservation law, 2004

800 **V. Poénaru and C. Tanas,** Equivariant, almost-arborescent representation of open simply-connected 3-manifolds; A finiteness result, 2004

799 **Barry Mazur and Karl Rubin,** Kolyvagin systems, 2004

798 **Benoît Mselati,** Classification and probabilistic representation of the positive solutions of a semilinear elliptic equation, 2004

797 **Ola Bratteli, Palle E. T. Jorgensen, and Vasyl' Ostrovs'kyĭ,** Representation theory and numerical AF-invariants, 2004

796 **Marc A. Rieffel,** Gromov-Hausdorff distance for quantum metric spaces/Matrix algebras converge to the sphere for quantum Gromov-Hausdorff distance, 2004

795 **Adam Nyman,** Points on quantum projectivizations, 2004

794 **Kevin K. Ferland and L. Gaunce Lewis, Jr.,** The $RO(G)$-graded equivariant ordinary homology of G-cell complexes with even-dimensional cells for $G = \mathbb{Z}/p$, 2004

793 **Jindřich Zapletal,** Descriptive set theory and definable forcing, 2004

792 **Inmaculada Baldomá and Ernest Fontich,** Exponentially small splitting of invariant manifolds of parabolic points, 2004

791 **Eva A. Gallardo-Gutiérrez and Alfonso Montes-Rodríguez,** The role of the spectrum in the cyclic behavior of composition operators, 2004

790 **Thierry Lévy,** Yang-Mills measure on compact surfaces, 2003

789 **Helge Glöckner,** Positive definite functions on infinite-dimensional convex cones, 2003

788 **Robert Denk, Matthias Hieber, and Jan Prüss,** $\mathcal{R}$-boundedness, Fourier multipliers and problems of elliptic and parabolic type, 2003

787 **Michael Cwikel, Per G. Nilsson, and Gideon Schechtman,** Interpolation of weighted Banach lattices/A characterization of relatively decomposable Banach lattices, 2003

786 **Arnd Scheel,** Radially symmetric patterns of reaction-diffusion systems, 2003

785 **R. R. Bruner and J. P. C. Greenlees,** The connective K-theory of finite groups, 2003

784 **Desmond Sheiham,** Invariants of boundary link cobordism, 2003

783 **Ethan Akin, Mike Hurley, and Judy A. Kennedy,** Dynamics of topologically generic homeomorphisms, 2003

782 **Masaaki Furusawa and Joseph A. Shalika,** On central critical values of the degree four L-functions for GSp(4): The Fundamental Lemma, 2003

781 **Marcin Bownik,** Anisotropic Hardy spaces and wavelets, 2003

780 **S. Marmi and D. Sauzin,** Quasianalytic monogenic solutions of a cohomological equation, 2003

779 **Hansjörg Geiges,** h-principles and flexibility in geometry, 2003

778 **David B. Massey,** Numerical control over complex analytic singularities, 2003

777 **Robert Lauter,** Pseudodifferential analysis on conformally compact spaces, 2003

776 **U. Haagerup, H. P. Rosenthal, and F. A. Sukochev,** Banach embedding properties of non-commutative L^p-spaces, 2003

775 **P. Lochak, J.-P. Marco, and D. Sauzin,** On the splitting of invariant manifolds in multidimensional near-integrable Hamiltonian systems, 2003

774 **Kai A. Behrend,** Derived ℓ-adic categories for algebraic stacks, 2003

773 **Robert M. Guralnick, Peter Müller, and Jan Saxl,** The rational function analogue of a question of Schur and exceptionality of permutation representations, 2003

772 **Katrina Barron,** The moduli space of $N = 1$ superspheres with tubes and the sewing operation, 2003

771 **Shigenori Matsumoto,** Affine flows on 3-manifolds, 2003

770 **W. N. Everitt and L. Markus,** Elliptic partial differential operators and symplectic algebra, 2003

For a complete list of titles in this series, visit the
AMS Bookstore at **www.ams.org/bookstore/**.